Springer Proceedings in Physics 56

Springer Proceedings in Physics

Managing Editor: H. K. V. Lotsch

Volumes 1–43 are listed on the back inside cover

G. L. Harris M. G. Spencer C. Y.-W. Yang (Eds.)

Amorphous and Crystalline Silicon Carbide III

and Other Group IV – IV Materials

Proceedings of the 3rd International Conference,
Howard University,
Washington, D. C., April 11 – 13, 1990

With 265 Figures

Springer-Verlag
Berlin Heidelberg New York
London Paris Tokyo
Hong Kong Barcelona
Budapest

Professor Gary L. Harris, Ph.D.
Michael G. Spencer, Ph.D.
Materials Science Research Center of Excellence, Howard University,
School of Engineering, Washington, DC 20059, USA

Professor Cary Y.-W. Yang, Ph.D.
Microelectronics Laboratory, Santa Clara University,
Santa Clara, CA 95053, USA

Library of Congress Cataloging-in-Publication Data. Amorphous and crystalline silicon carbide III and other group IV-IV materials: proceedings of the 3rd international conference, Howard University, Washington, D.C., April 11-13, 1990 / G.L. Harris, M.G. Spencer, C.Y. Yang (eds.). p. cm. – (Springer proceedings in physics; 56) Includes bibliographical references and index.
ISBN-13: 978-3-642-84404-1 e-ISBN-13: 978-3-642-84402-7
DOI: 10.1007/ 978-3-642-84402-7
1.Silicon carbide–Congresses. I. Harris, Gary Lynn, 1953 –. II. Spencer, M.G. (Michael G.) III. Yang, C.Y.-W. (Cary Y.-W.), 1948 –. IV. Series: Springer proceedings in physics; v. 56. TP245.S5A69 1992 621.381'52–dc20 91-44285

Softcover reprint of the hardcover 1st edition 1992

Typesetting: Camera ready by authors
54/3140 - 543210 – Printed on acid-free paper

Preface

This volume contains written versions of the papers presented at the Third International Conference on Amorphous and Crystalline Silicon Carbide and Other Group IV-IV Materials (ICACSC 90), which was held at Howard University, April 11–13, 1990 in Washington, DC. The ICACSC continued to provide an international forum for discussion and exchange of ideas regarding the current state of research aimed at developing silicon carbide devices and circuits and related materials. ICACSC attracted over one hundred participants from seven countries. A special session was held in honor of the eight Soviet scientists who attended the conference.

The substantial increase in the number of papers compared with the previous year is an indication of the growing interest in this field. The conference also included a poster session for the first time. This volume contains 54 refereed contributions grouped into four parts.

Several exciting new results are reported for the first time here: SiC-based solid-solution growth and technology, the formation of SiGe heterostructures by ion implantation, 6H-SiC substrates grown by the sublimation method, explanation of the appearance of negative differential resistance in a N^+PN-SiC-6H transistor by the Wannier-Starck effect, the formation of amorphous SiC/Si heterojunctions by the polymer route, and the prospects of developing SiC bipolar transistors and thyristors.

Finally, on behalf of all conference participants we thank our sponsors – the National Science Foundation, 9004235, the Office of Naval Research, EMCORE, IEEE-Electron Devices Society, the Howard University School of Engineering, and the Graduate School of Arts and Sciences – for their generous financial support. This work relates to Department of Navy Grant N00014-90-J-1653 issued by the Office of Naval Research. The United States Government has a royalty-free license throughout the world in all copyrightable material contained herein. Additional support from Howard University Materials Science Research Center of Excellence and Santa Clara University Microelectronics Laboratory is gratefully acknowledged. The time and effort contributed by members of the conference committees are greatly appreciated.

Washington, DC
November 1990

G.L. Harris
M.G. Spencer
C.Y.-W. Yang

Contents

Part I Growth of Single Crystal SiC

Part II Growth and Characterization of Amorphous SiC

Part IV Techniques for Thin Film Growth

Part I

Growth of Single Crystal SiC

SiC-Based Solid Solutions: Technology and Properties

V.A. Dmitriev

A.F. Ioffe Physical Technical Institute, Academy of Sciences, 194021 St. Petersburg, Russia

Abstract. This paper is devoted to SiC-AlN, and SiC-GaN-AlN solid solutions. The paper composition includes: 1) a general discussion of the prospects for SiC-AlN and SiC-GaN solid solutions as electronic materials, 2) a detailed discussion of the experimental data available on the properties of SiC-AlN solutions, and finally, 3) an outline of the current techniques for growing monocrystalline layers of SiC-AlN solutions that are technologically suitable for the production of solid-state devices.

1. Introduction

Current progress in solid-state electronics largely depends on advances being made in R&D of semiconductor solid solutions. To develop a solid solution one essentially designs a material with definite physical properties, by tailoring fundamental semiconductor characteristics such as the bandgap. The compositional flexibility of solid solutions not only offers the opportunity to control the bulk properties of semiconductors, but it also allows one to develop devices based on two-dimensional effects. The most common applications for semiconductor solid solutions include (i) devices based on materials with designated bandgaps (e.g., diodes emitting light of a definite wavelength), (ii) heterojunction devices (semiconductor lasers, photodetectors with wide-gap windows, bipolar devices with wide-gap emitters etc.), (iii) devices utilizing superlattices and graded-gap structures.

The above concerns mainly solid solutions of III-V compounds and to some extent of II-VI compounds. Solid solutions containing silicon carbide were not studied until the end of the 1970s. We believe this is due to technological problems encountered in the growth of silicon carbide.

The first publication on SiC-based solid solutions appeared in Nature in 1978 [1]. Since then many technological advances have, however, been made in this area. It is the purpose of this review to summarize the state-of-the-art in the field and to evaluate the prospects for future development.

We shall consider only solid solutions having high (over 1%) solubility: neither solid solutions of impurities present in SiC, such as SiC<Al>, and SiC<N> nor the Si_xC_{1-x} system will be discussed. Solid solutions reported so far in the literature are SiC-AlN, SiC-GaN, Si-C-O-N-Al, SiC-BeO, and SiC-BN. Our discussion will be centered around the SiC-AlN system, perhaps the most studied system to date.

The composition of the paper is as follows:

- starting with a general discussion of the prospects for SiC-AlN and SiC-GaN solid solutions as semiconductor electronic materials,
- we proceed to the available experimental data on the properties of SiC-AlN solutions,
- and finally outline the current techniques for growing monocrystalline layers of SiC-AlN solutions that are technologically suitable for production of solid state devices.

It should be noted that a substantial part of the experimental data available on date on SiC-based solid solutions has been obtained using ceramic technologies (hot-pressing, for instance). Clearly, these methods are unsuitable for semiconductor devices.

2. The Anticipated Properties of SiC-AlN and SiC-GaN Solid Solutions

The general requirements for components of a solid solution are closely matched atomic dimensions, crystallographic type, and similarity in conditions of synthesis. Considering that (i) the covalent radii of Al,Ga and Si, on the one hand, and those of N and C, on the other, are commensurate, (ii) AlN and GaN both crystallize to a wurtzite structure, just as SiC-2H does, and (iii) the conditions for synthesis of these nitrides are rather similar to those required by SiC, it is reasonable to suppose that SiC, AlN and GaN can form solid solutions.

The width of the immiscibility regions of SiC-AlN and SiC-GaN systems have been calculated by Sukhanek et al [2]. Their results (Fig.1) indicate that at temperatures of 1400-1800°C, the temperatures typical for SiC technology, continuous solid solutions might form in the SiC-AlN system, and limited solid solutions might exist in the case of the SiC-GaN system.

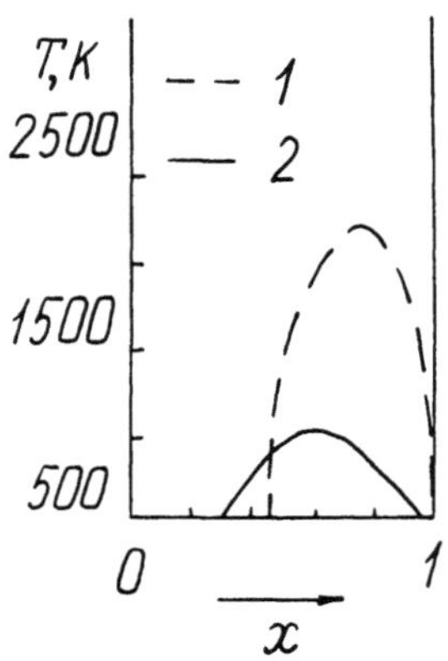

Fig.1. Boundaries of the immiscibility regions of $(SiC)_{1-x}$-$(GaN)_x$ (1) and $(SiC)_{1-x}$-$(AlN)_x$ (2) systems [2].

Now we shall consider the properties of these solid solutions as predicted from the most general considerations. Whether or not these predictions are valid will become clear in Section 3 where the appropriate experimental data are discussed.

Let us consider the crystallographic aspect of the problem. As already noted, AlN and GaN crystallize in a wurtzite-type lattice, and therefore, their crystallographic parameters could well be compared with those of SiC-2H. Indeed, the data of Table 1 indicate that AlN, and SiC have closely matching unit cell parameters, a and c. Figure 2 shows how the lattice parameters could change in response to changes of composition in hypothetical SiC-AlN and SiC-GaN solid solutions with 2H structure, provided that Vegard's law is fulfilled.

Bandgap variations with solid solutions is the next general consideration. For III-V solutions it has been found that the

Table 1. Lattice constants of SiC-based solid solution components at T = 300 K.

Material	Lattice constant, Å	
	a	c
AlN	3.114	4.986
GaN	3.180	5.166
SiC-2H	3.076	5.048
SiC-4H	3.080	10.081
SiC-6H	3.081	15.117

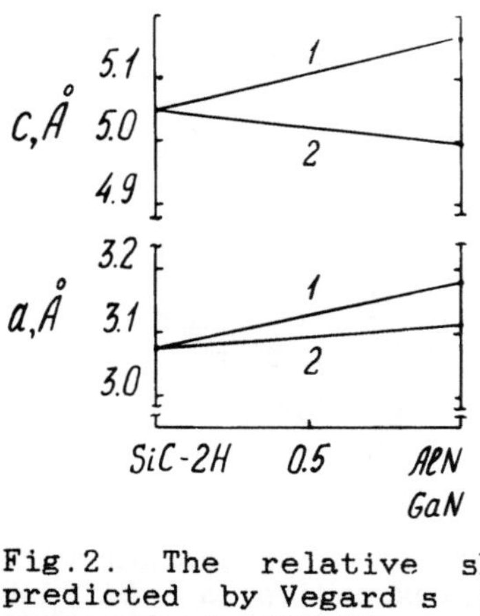

Fig.2. The relative shift of a and c lattice parameters predicted by Vegard s law for hypothetical solid solutions composed of SiC-2H and GaN (1) and AlN (2) as the second component.

dependence of Eg on composition is describable by a linear interpolation between the values for elemental components and the additional term which is quadratic in x (the content of one of the components) and vanishes at x=0 and x=1. In III-V compounds the quadratic term can be neglected if the lattice mismatch between the solid solution is less than 0.1 Å [3]. Therefore, to a first approximation we assume the linear dependence of Eg on composition for solid solutions formed between SiC-2H on one side, and AlN and GaN, on the other (Fig.3a). It is seen that for certain compositions a transition from indirect to direct band structure can be expected.

The possibility of controlling the band structure seems to be most easily realized in a five-component system of Si-C-Al-Ga-N. Using the linear approximation mentioned above it can be shown that the range of bandgap values in solid solutions having direct band structure is considerably wider in this case (Fig.3b). It is thus probable that SiC-AlN and SiC-GaN solid solutions will allow the capabilities of semiconductor electronics to be substantially enhanced by exploiting the range of bandgap values from ≈ 3 to ≈ 6 eV.

An important question arises as to whether or not device structures based on the solid solutions under consideration are feasible. In view of the problems still faced in the growth of SiC, to say nothing of AlN or GaN, it seems that a practical way of growing solid solution device structures could be by hetero-epitaxy. In the hetero-epitaxial growth of solid solutions, SiC crystals could be used as substrates, both because of the availabity of good quality SiC crystals and because SiC has a high chemical inertness. Lattice mismatch between (SiC-2H)-AlN solid solutions and the SiC-6H substrate

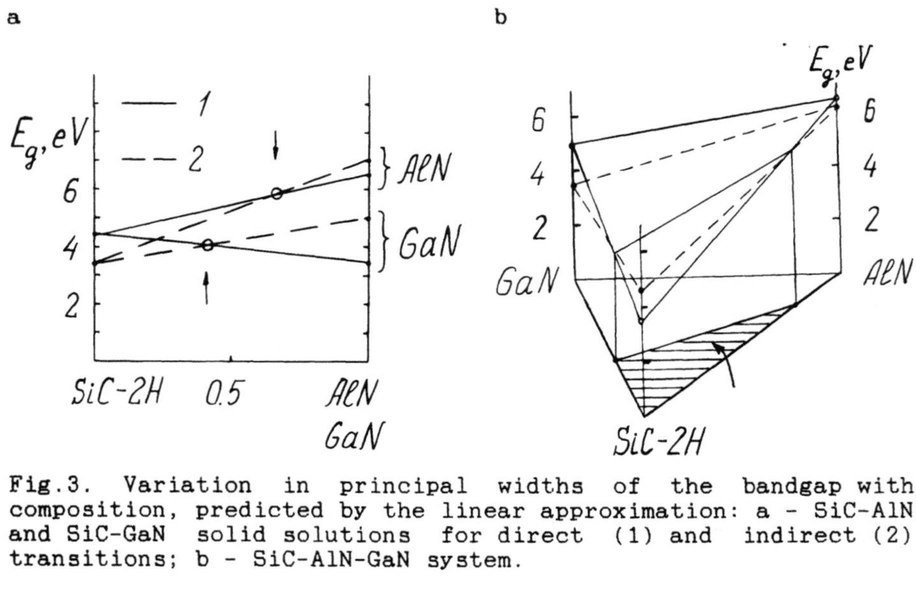

Fig.3. Variation in principal widths of the bandgap with composition, predicted by the linear approximation: a - SiC-AlN and SiC-GaN solid solutions for direct (1) and indirect (2) transitions; b - SiC-AlN-GaN system.

does not exceed 1.2% (at room temperature) (Fig.4). This value is greater than the value for the ideal GaAs-AlAs heteropair (0.12%) but less than that for the GaAs-Si pair (4%) which has been used in fabrication of a number of devices. Thus, SiC-6H substrates (using crystals with {0001} orientation of the basal plane) appear to be quite acceptable for growth of solid solutions. The second choice would be substrates made of monocrystalline AlN layers grown on sapphire. In this case it is possible to grow five-component Si-C-Ga-Al-N solid solutions isoperiodic with AlN (Fig.5).

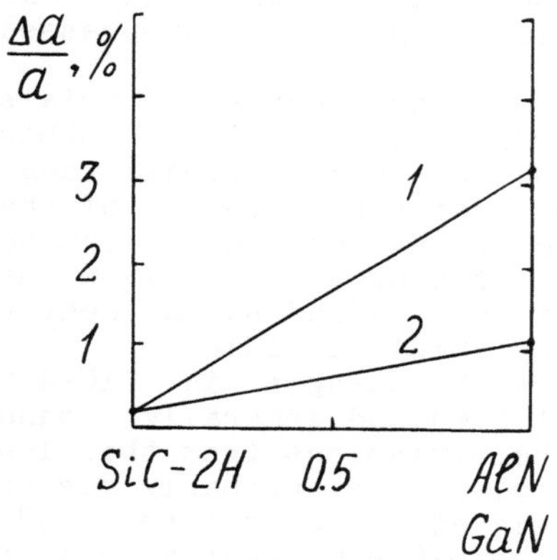

Fig.4. Lattice mismatch for (SiC-2H)-GaN (1) and (SiC-2H)-AlN (2) layers on SiC-6H substrate on the assumption that Vegard's law is fulfilled.

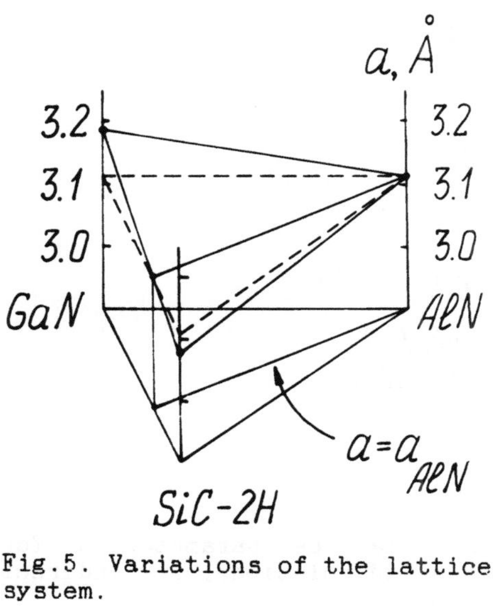

Fig.5. Variations of the lattice parameter a for SiC-AlN-GaN system.

Thus, on the basis of the above criteria for the formation of solid solutions, it is possible to predict and design SiC-AlN and SiC-GaN solid solutions, which would greatly enhance the capabilites of semiconductor materials (Fig.6).

3. Experimental Data on the Properties of SiC-AlN Solid Solutions

There have been several studies of the phase relationships in the SiC-AlN system [1,4-8]. It seems that the most reliable data has been obtained by Zangvil and Ruh [8]. These investigators determined a tentative SiC-AlN phase diagram (Fig.7) by using X-ray diffraction analysis and analytical scanning transmission electron microscopy on hot-pressed samples. Diffusion couples were studied.

This tentative diagram is specific to the source materials and the method of hot-pressing. However, this as well as other known diagrams provide information on the compositional range where solid solutions crystallize preferably to one of the existing polytype structures: 6H, 4H or 2H. The last polytype appears to have the broadest range of composition. From the data reported in [8], the solid solution studied has a nearly stoichiometric composition, defined by $(SiC)_{1-x}(AlN)_x$.

Changes of lattice parameters with changes in SiC-AlN composition have also been studied [1,5,6] and generally found to conform to Vegard's law (Fig.8).The deviations from this law observed in thermally treated samples are probably due to decomposition of solid solutions at elevated temperatures [6]. In general, the phase stability in the SiC-AlN system remains an open question.

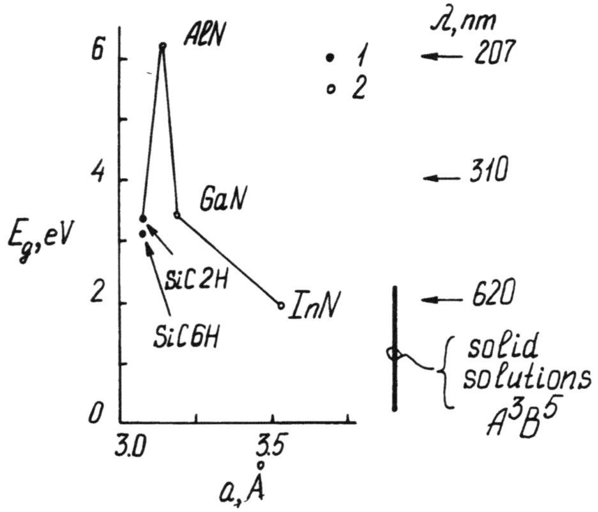

Fig.6. Bandgap width values and lattice parameter a for predicted components of SiC-based solid solutions; 1 - indirect band materials, 2 - direct band materials.

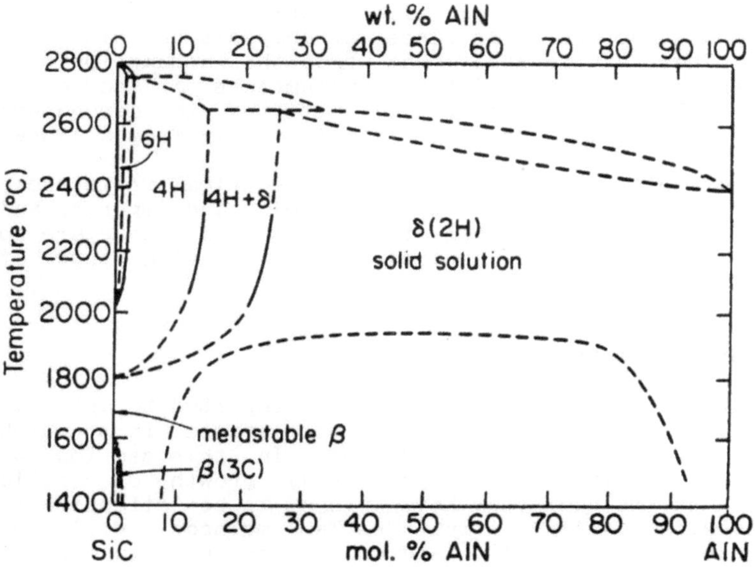

Fig.7.Tentative SiC-AlN phase diagram [8].

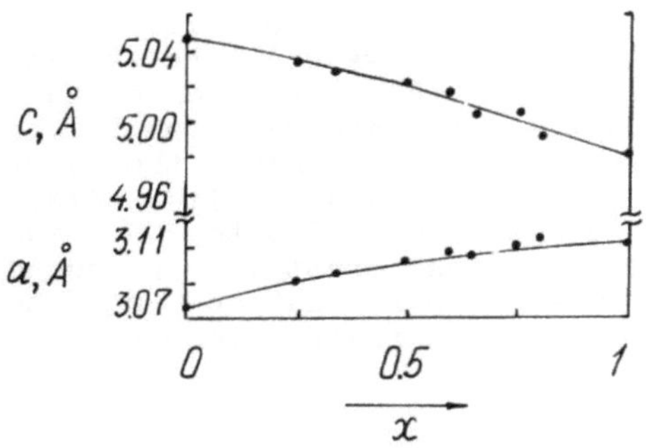

Fig.8. Lattice parameters of the SiC-AlN solid solution versus composition [5].

Nearly linear compositional dependences have been obtained for microhardness [6], Young's modulus [4], and linear expansion coefficient [4].

Data is scarce on the electrophysical properties of solid solutions. Both n- and p-type conductivity have been demonstrated for solid solutions [9, 10] - a significant fact, considering that attempts to obtain p-type AlN have not yet been successful.

Optical properties of SiC-AlN solid solutions have been studied on epitaxial layers grown on SiC and Al_2O_3 substrates by sublimation [11, 12]. Transmission spectra taken in the region of the fundamental absorption edge at 300 K and cathodoluminescence spectra recorded at 80 K were investigated.

According to Nurmagomedov et al. [12] transition from indirect to direct band structure takes place for compositions with 70% AlN (Fig.9). Thus, it is seen that while the assumption of a linear dependence of the lattice parameters a and c on composition of SiC-AlN is fairly well justified, the dependence of Eg on composition deviates significantly from a linear relationship.

Aspects of solid solutions such as electrical resistance, carrier mobility, effects of doping level, and impurity and defect concentrations are yet to be studied. There are no reports on the ways of controlling the changing type of conductivity in these solid solutions.

4. SiC-AlN Growth Techniques

As noted above, the bulk of the work reported to date has been with solid solutions obtained in the ceramic form, which is unsuitable for semiconductor devices. In this section we shall present results on the epitaxial growth of solid solutions by the methods that have proved to be effective in producing device-quality structures. Two methods will be reviewed, sublimation [9,13] and container-free liquid-phase epitaxy [14].

The first successful attempt at growing SiC-AlN layers in the semiconductor form was made by Rutz [13]. The method consisted in subliming SiC and AlN in a flowing hydrogen/nitrogen mixture (Fig. 10). Growth was carried out on a variety of substrates, including monocryctalline W(111), SiC, and AlN (previously grown of a sapphire wafer). The growth temperature was in the range 1900°-2020°C. The temperature difference between the source and the substrate was from 10 to 100°C. Typical growth rates were around 1 μm/min; the layer thickness was 1 - 10 μm. The method yielded $(SiC)_{1-x}(AlN)_x$ crystals that had x in the range from 0.5 to 0.8 and were of the 2H polytype. Most of the crystals produced were highly conducting, but further purification of the source materials and use of high growth temperatures

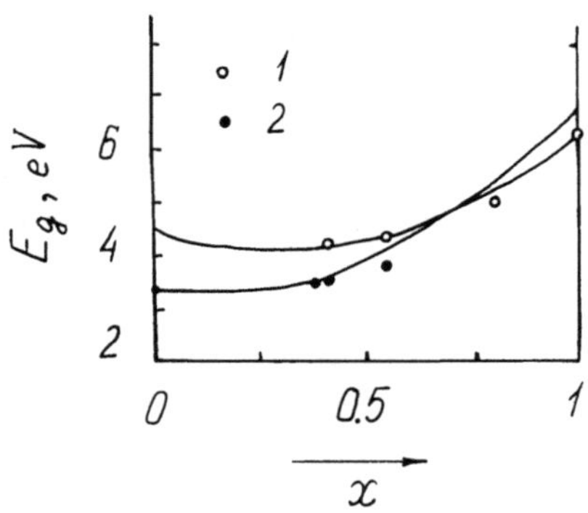

Fig.9. Widths of direct (1) and indirect (2) gaps as functions of compositions in SiC-AlN solid solutions [12].

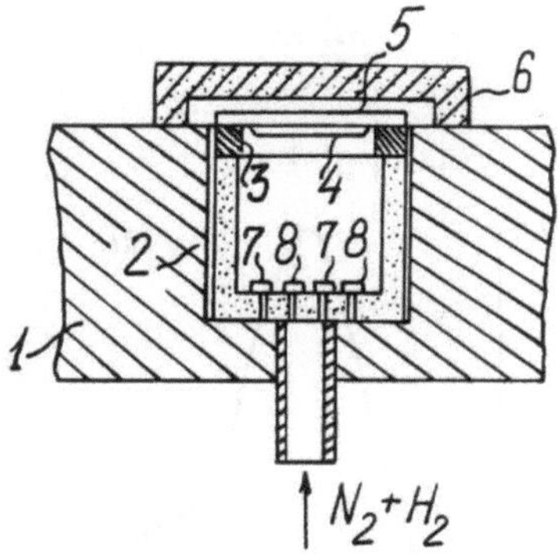

Fig.10. Schematic cross-sectional view of an apparatus suitable for forming SiC-AlN crystals [13]: 1 - tungsten crucible, 2 - carbon crucible, 3 - substrate support, 4 - solid solution, 5 - substrate, 6 - lid, 7 - SiC source, 8 - AlN source.

allowed the production of n-type material with resistivity as low as 10^4 Ohm cm.

Nurmagomedov and Safaraliev [9] have grown epitaxial SiC-AlN layers by sublimation at 1900-2100°C on monocrystalline SiC (6H or 15R) and sapphire substrates in the atmosphere of argon, nitrogen, and argon/nitrogen mixture. Polycrystalline SiC-AlN raw materials of various compositions were used as sources. The rates of growth varied from 25 to 75 μm/h, depending on the AlN content in the source. This finding was later shown to be consistent with the results which the authors obtained in the thermodynamic analysis of the SiC-AlN system [15].

It was thus demonstrated that by varying the composition of the source it is possible to obtain layers with the desired value of the x parameter over the range from 0.1 to 1.0 (Fig.11). At $x > 0.2$, the layers had the 2H structure and

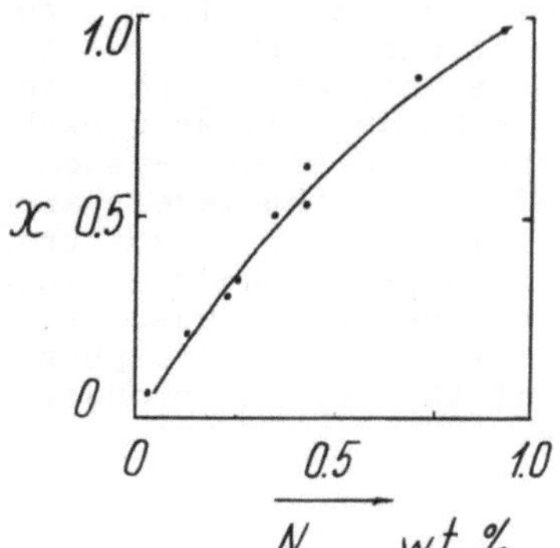

Fig.11. Composition of epitaxially grown $(SiC)_{1-x}-(AlN)_x$ solid solution layers versus AlN content in the SiC-AlN polycrystalline vapor source [11].

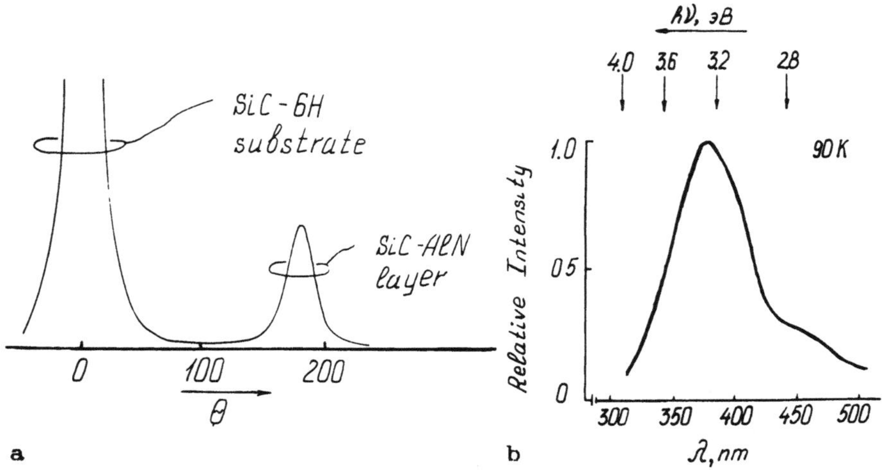

Fig.12. X-ray diffraction (a) and cathodoluminescence (b) spectra from SiC-AlN epitaxial layers grown by LPE on SiC-6H substrates.

exhibited conductivity of both types. The physical characteristics of the layers have been studied in [9,12].

Liquid-phase epitaxial growth of SiC-AlN solid solutions is the subject of an investigation recently undertaken by the Leningrad group. The project is based on the use of the container-free liquid-phase epitaxy method [14] which is now a well-established technique for fabricating SiC pn junctions designed for a broad spectrum of applications.

Layers of SiC-AlN solid solution (1-50 μm in thickness) have been grown at 1400-1600°C on SiC-6H substrates having the (0001)Si orientation, with the growth rate ranging from 0.5 to 5.0 μm/min.

The preliminary results can be summarized as follows. We have obtained $(SiC)_{1-x}$-$(AlN)_x$ solid solutions with $x = 0.1$, 0.2, 0.6, 0.7. Auger-spectroscopic analyses indicated that in addition to SiC-AlN solid solutions the technique allows growth of five-component solid solutions in the Si-C-Al-N-O system. X-ray diffraction studies demonstrated that the layers are monocrystalline (Fig.12a). Cathodoluminescence spectra (Fig.12b) revealed the presence of broad emission bands lying in the UV, violet, and blue regions of the spectrum. Thus, the feasibility of fabricating SiC-AlN layers from the liquid phase has been demonstrated. Characterization of the layers produced by this technique is presently under way.

5. Summary

1. The published data suggest that SiC- based solid solutions, primarily SiC-AlN, SiC-GaN and SiC-AlN-GaN, certainly hold great promise as materials for semiconductor electronic devices.

2. SiC-AlN solid solutions have been demonstrated over the entire range of compositions, with smooth changes in lattice constant and bandgap width.
3. Thermal probe studies have established that SiC-AlN solid solutions can have both n- and p-type conductivity.
4. SiC-AlN solid solutions have been found to crystallize predominantly in the wurtzite structure (2H) though both 4H and 6H structures could be obtained as well.
5. The value around 0.7 has been identified for the $(SiC)_{1-x}$-$(AlN)_x$ solution as the value of the x parameter associated with the transition from indirect to direct band structure.
6. The methods of sublimation and liquid-phase epitaxy have been shown to be effective in producing monocrystalline layers of SiC- AlN solid solutions on SiC substrates.
7. There is a lack of data on electrophysical and optical properties of SiC-based solid solutions as well as device fabrication processes (metallization, etching etc.). The question of whether or not it is possible to change the type of conductivity in these solid solutions in a controllable manner still remains to be addressed. The predictions of the performance that can be expected from SiC-AlN solid solution device structures must therefore await the results of the appropriate investigations. One prediction, however, can be made already : it is that the devices themselves, primarily UV light emitting diodes and, possibly, lasers will be produced in the near future.

Acknowledgments

Valuable discussions with Ya.V.Morozenko, G.K.Safaraliev and S.A.Nurmagomedov are gratefully acknowledged. The author thanks A.E.Cherenkov, L.B.Elfimov, I.P.Nikitina and I.Yu.Linkov for technological experiments and measurements and would also like to thank I.D.Kovalenko for her help with the preparation of this review.

References

1. Cutler I.B., Viller P.D., Rafaniello W., Park H.K., Thompson D.P., Jack K.H.: Nature, v.275, N 5679, pp.434-435,(1978)
2. Sukhanek G., Tairov Yu.M. and Tzvetkov V.F.: Sov. Tech. Phys. Lett., v.9, N 9, pp.317-318 (1983)
3. Glisson T.H., Hauser J.R., Littlejohn M.A., Williams C.K.: J.Electron. Mat.,v.7, N1, pp.1-15, (1978)
4. Rafaniello W., Cho K., Virkar A.V.: J. of Materials Science, v.16, N12, pp.3479-3488 (1981)
5. Ruh R., Zangvil A.: J. of the Am. Ceram. Soc., v.65, N5, pp.260-265 (1982)
6. Rafaniello W., Plichta M.R., Virkar A.V.: J. of the Am. Ceram. Soc., v.66, N4, pp.272-276 (1983)
7. Safaraliev G.K., Sukhanek G., Tairov Yu.M. and Tzvetkov V.F.: Izv.AN USSR. Inorganic Mat.,Ser., v.22, N11, pp.1883-1941 (1986) [in Russian]
8. Zangvil A., Ruh R.: J. of the Am. Ceram. Soc., v.71, N10, pp.884-890 (1988)

9. Nurmagomedov S.A. and Safaraliev G.K.: 6th All-Union. Conf. Crystal Growth, Abstracts, Erevan, 1985, v.3, pp.116-117 [in Russian]
10. Safaraliev G.K., in: Wide-gap Semiconductors (Makhachkala, 1988), pp.34-44 [in Russian]
11. Nurmagomedov S.A., Pikhtin A.N., Rasbegaev V.N., Safaraliev C.K., Tairov Yu.M. and Tzvetkov V.F.: Sov. Tech. Phys. Lett., v.12, N 9,pp.431, (1986)
12 .Nurmagomedov S.A., Pikhtin A.N., Razbegaev V.N., Safaraliev G.K., Tairov Yu.M. and Tzvetkov V.F.: Sov. Phys. Semicond., v.23, N1, pp. 162-164 (1989) [in Russian]
13. Rutz R.F. United States Patent, N 4382837, May 10, 1983.
14. Dmitriev V.A., Ivanov P.A., Korkin I.V., Morozenko Ya.V., Sidorova T.A., Strel'chuk A.M., Chelnokov V.E.: Sov.Tech. Phys.Lett., v.11, N2, pp.98-99 (1985)
15. Nurmagomedov S.A., Safaraliev G.K., Sorokin N.D., Tairov Yu.M. and Tzvetkov V.F.: Izv. An USSR. Inorganic Mat. Ser., v.22, N10, pp.1672-1674 (1986) [in Russian]

Growth and Morphology of 6H-SiC Prepared by the Sublimation Method

S. Nishino, Y. Kojima, and J. Saraie

Department of Electronics and Information Science, Faculty of Engineering and Design, Kyoto Institute of Technology,
Matsugasaki, Sakyo-ku, Kyoto 606, Japan

6H-SiC was grown on 6H-SiC by using the modified sublimation method. Crystal growth was carried out at various pressures ranging from 200 Torr to 760 Torr in an Ar gas. The growth rate was strongly influenced by the pressure. Substrates of 6H-SiC with various orientations such as (0001)Si plane,(000$\bar{1}$)C plane, a-planes (planes perpendicular to a-axis) and highly c-oriented polycrystalline 3C-SiC were used. The temperatures of SiC source and substrate were 2400 ^{o}C and 2100 ^{o}C, respectively. Single crystalline ingots of 4H-,6H- and 15R-SiC were obtained. Polytypes of the grown layers were examined by RHEED, X-ray diffraction and photoluminescence methods.

1.Introduction

Silicon carbide (SiC) is an interesting material for electronic and optical device applications because of its high temperature stability and large energy gap. A high frequency performance is expected in 6H-SiC MOS devices based on its high saturation velocity (2x10^{7} cm/sec) and the high breakdown electric field (5x10^{6} V/cm).[1] Recently, bulk-crystal growth of 6H-SiC using 6H-SiC seed crystals by a modified sublimation method has been widely studied, and several successful results have been reported.[2-4] A large size of SiC wafer would aid the development of SiC technology. Epitaxial growth of 3C-SiC on Si has been studied widely and has encountered the anti-phase domain problem [5,6] This problem was resolved by using a substrate with off-orientation,[7] however, it is difficult to reduce the number of stacking faults which are caused by the large lattice mismatch between SiC and Si.[7] To reduce stacking faults and other types of defects, homoepitaxial growth is needed using 6H-SiC substrates. Single crystals of 6H-SiC ingot are necessary for the substrates grown by CVD (chemical vapor deposition) and LPE (liquid phase epitaxial). In CVD and LPE, 6H-SiC substrates with off-axis orientations are used for obtaining good device performance.[8,9] So, it is worthwhile investigating the surface morphology on the various orientations of the substrates.

In this paper, we report the crystal growth, surface morphology , crystallinity and optical properties of single crystalline 6H-SiC. Identification of SiC polytypes by X-ray diffraction is also described.

2. Experiment

Single crystal growth of SiC was carried out using the modified sublimation method in a graphite crucible filled with SiC powder by changing growth temperature and pressure in an Ar atmosphere. The growth temperature and pressure were 1800 ^{o}C-2400 ^{o}C and 200-760 Torr, respectively. Source material used was SiC powder, green carborundum which was made for abrasive applications. When a p-type ingot was needed, Al_4C_3 powder was added together with the host SiC powder. Substrates were 6H-SiC made by the Acheson method for industrial use. The substrates of $(11\bar{2}0)$ and (1100) planes were prepared by cutting an SiC ingot made by the modified sublimation method. Polycrystalline 3C-SiC prepared by CVD was also used as the substrate. The crucible, with a diameter of 40 mm and a height of 60 mm, was surrounded by a carbon felt used as a radiation shield. The substrate of 6H-SiC was attached to a graphite cap of the crucible by carbonizing sugar. The distance between the substrate and source was about 20 mm. The crucible was set in the reaction chamber, which was made of clear fused quartz with a water cooled jacket. The crucible was heated externally by RF induction with a frequency of 300 kHz. The temperature of the substrate (T_s:top of the crucible) and source material (T_b: bottom of the crucible) were measured by an optical pyrometer. The entire system was evacuated by an oil rotary pump and backfilled with Ar gas. After evacuating again at a pressure of 0.1 Torr, the crucible was heated to 1700 ^{o}C for 30 min to purify the source material. The crucible was then heated to 2400 ^{o}C and crystal growth commenced.

To characterize the grown layer, single-crystal wafers of approximately 600 μm thick were prepared using the following procedure:1) The as-grown SiC ingot was removed from the cap of the crucible and was then cut into wafers 600-1000 μm thick by a diamond saw. 2) The wafers were lapped and polished for optical measurements. 3) The polytypes and crystallographic plane of grown layers were examined by X-ray diffraction, RHEED (reflection high energy electron diffraction) and PL (photoluminescence methods).

3. Result and Discussion

The growth rate was strongly dependent on the pressure of the reaction chamber. The growth rate changed exponentially when the pressure decreased from 700 Torr to 200 Torr. The growth rate at 200 Torr was about 3.5 mm/h. In over 50 runs we obtained a yield of about 80 percent for the desired 6H-SiC.

The morphology of the grown layer was observed by optical microscope (Nomarski interference contrast). Spiral patterns were observed on the (0001)Si plane of 6H-SiC as shown in Fig.1(a), however, flat and wavy patterns were observed on the $(000\bar{1})$C plane as shown in Fig.1(b). Spiralling and facetting with six-fold symmetry were observed as shown in Fig.1(c). These spiral patterns are related to screw dislocations. The difference of surface morphology was related to the polarity of the basal plane. Sometimes, small dots or protrusions were observed as shown in Fig.1(d). These droplets are Si, which has been confirmed by electron probe microanalysis. This protrusion is caused by a fluctuation of supersaturation

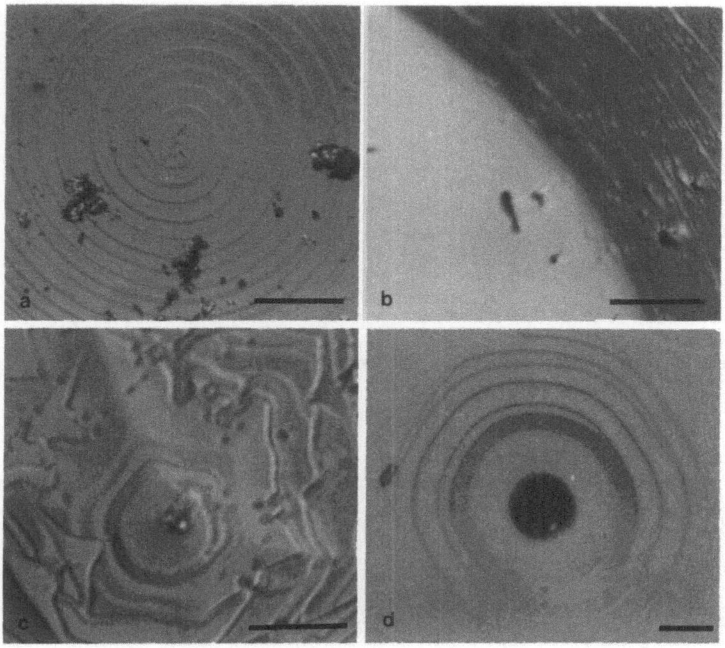

Fig.1. Surface morphology of the grown layer:(a) on (0001)Si plane, (b) on $(000\bar{1})$C plane, (c) on $(000\bar{1})$C plane and (d) on (0001)Si plane with Si droplet.

during the crystal growth. The growth takes place by the VLS (vapor-liquid-solid) mechanism[10].

Four substrates were attached side by side to the cap of the crucible. The growth was carried out under standard conditions, 200 Torr, T_s=2100 °C, T_b=2400 °C for one hour. Surface morphologies of the grown layers are shown in Fig.2. The grown layer on $(11\bar{2}0)$ was a roof-shape and the middle of the grown layer was elongated like a roof, which was in the $\langle 11\bar{2}0 \rangle$ direction as shown in Fig.2(a). The grown layer on (1100) plane was almost flat and showed parallel lines whose direction is $\langle 11\bar{2}0 \rangle$ as shown in Fig.2(b). The growth rate dependence of the planes was larger in the order of $(11\bar{2}0)$,(1100),$(000\bar{1})$C and (0001)Si planes. This order is reasonable considering the shape of the basal plane of a hexagon.

Identification of the various polytypes grown is important. Polytypes were examined by RHEED, X-ray diffraction and PL measurements. To obtain a spot pattern using the RHEED technique, the surface must be roughened by grinding with SiC powder and etching in a molten KOH. By choosing a good incident beam direction which shows double diffraction spots, it is easy to identify the stacking order of the c-direction for 4H-,6H- and 15R-SiC. For example, there are five spots between (0006) and (00012) along the $\langle 0001 \rangle$ direction for 6H-SiC. Three spots were observed between (0004)

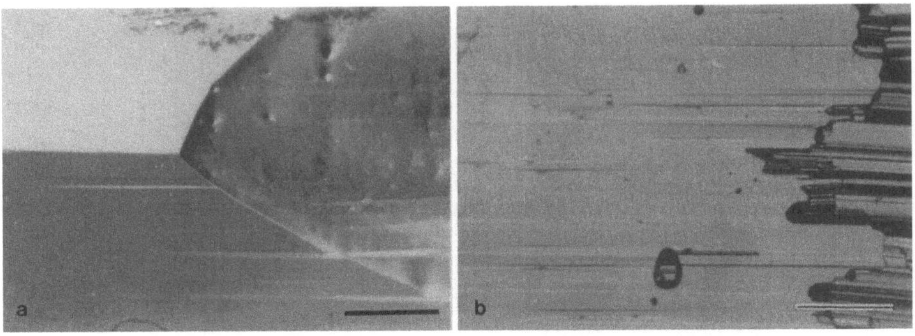

Fig.2. Surface morphology of the grown layer on a-planes: (a) $(11\bar{2}0)$ plane and (b)(1100) plane.

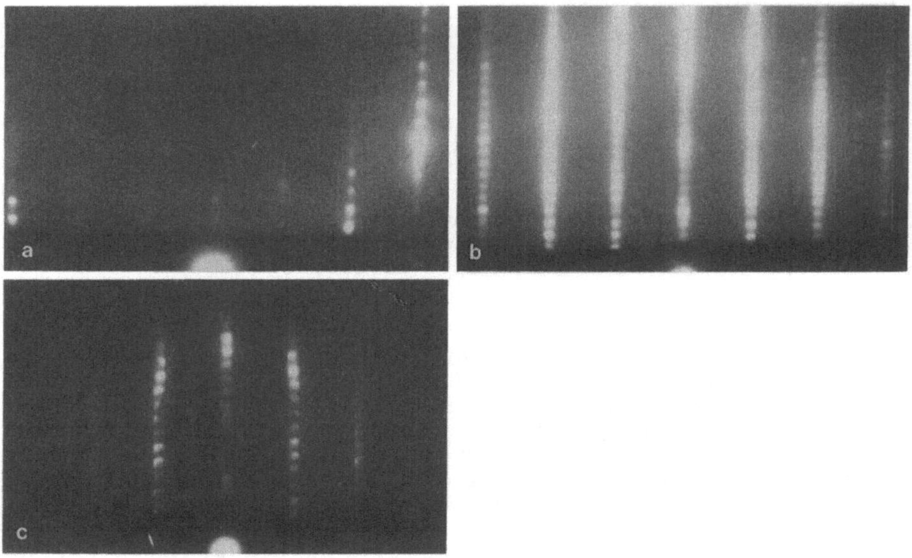

Fig.3. RHEED patterns of SiC single crystals : (a) 4H-SiC, (b) 6H-SiC and (c) 15R-SiC ; These patterns contained double diffraction spots. $<11\bar{2}0>$ azimuth.

and (0008) for 4H-SiC. Four spots were observed between (00015) and (00030) for 15R-SiC. RHEED patterns of these polytypes are shown in Fig.3.

Using an X-ray diffraction method, observation of small periodic peaks of 4H- and 6H-SiC was first reported by Kanaya et al. [11] However, they did not explain the origin of the small peaks. We precisely examined this phenomena and found directional dependence of the small peaks. In the normal θ-2θ diffraction method using an X-ray apparatus (RAD-C, Rigaku) operated with Cu target at 30 kV, 20 mA, when the incident beam direction was parallel to <1100>, small periodic peaks appeared but when the incident beam was parallel to <11$\bar{2}$0>, small peaks were not observed. This directional dependence is similar to the case of RHEED measurement. In this observation, main peaks like (0006) and (00012) for 6H-SiC always appeared , but the appearance of small peaks changed periodically when the direction of incident beam was changed. Namely ,the small peaks had a directional dependence on the incident X-ray beam. So we speculate that these small periodic peaks cause a double diffraction effect.[12] The intensity of the small peaks was very weak, where the intensity ratio of the main peak to small peak was about 10,000. Double diffraction is frequently observed in RHEED measurement because the wavelength is short in RHEED measurement. In X-ray measurement, the wavelength is not short, however, SiC has a large lattice distance along the c-axis, 10.0 Å for 4H-SiC, 15.1 Å for 6H-SiC and 128.1 Å for 15R-SiC, and as a result a double diffraction condition could be easily established. As an example, the diffraction pattern is shown in Fig.4. When the double diffraction patterns are different in the various SiC polytypes, this method is applicable to identify the various SiC polytypes.

We found an interesting result using polycrystalline 3C-SiC, where single crystals of 4H-SiC were grown on the highly c-oriented 3C-SiC. RHEED patterns of the substrate of polycrystalline 3C-SiC and grown layer are shown in Fig.5. The crystallinity of the grown layer was almost single crystal , but a small misoriented pattern was superimposed. In the rocking curve measurement of X-ray diffraction, the half-width of the substrate was about 3° but it was 0.01^o for the grown layer. We speculate that 4H-SiC grew on the 6H-SiC layer according to the following procedure. The surface of the polycrystalline 3C-SiC was changed to 6H-SiC through solid-state transformation at the beginning stage of the crystal growth and then 4H-SiC grew on the 6H-SiC layer.[13]

For PL measurement, a filtered high pressure Hg-lamp with a major line of 365 nm was used. The PL spectra were measured at 10 K. In the PL spectrum of 4H-SiC, the main peak was broad at 510 nm, which is related to the impurity level. In the PL spectrum of 15R-SiC, two peaks were observed, one was at around 490 nm and the other at around 660 nm. The band at the shorter wavelength relates to D-A pair emission and the broad band at the longer wavelength relates to nitrogen impurity or a defect center. The PL spectrum of 6H-SiC showed mainly D-A pair emission.

Optical properties of the grown layer were studied at room temperature using transmittance and reflectivity spectra in the visible range (wavelength of 300 nm to 900 nm). Absorption coefficients of the SiC polytypes were calculated from the transmission, reflectivity and thickness of the crystal. Optically obtained band gaps were 3.2 eV for 4H-SiC, 2.96 eV for 6H-SiC and 2.90 eV for 15R-SiC, respectively.

Electrical properties of the grown layer were measured by using the van der Pauw method. Typical ranges of carrier concentration, mobility and resistivity were 10^{17} cm^{-3}, 50-80 $cm^2/V.s$, 0.03-0.3 ohm.cm, respectively.

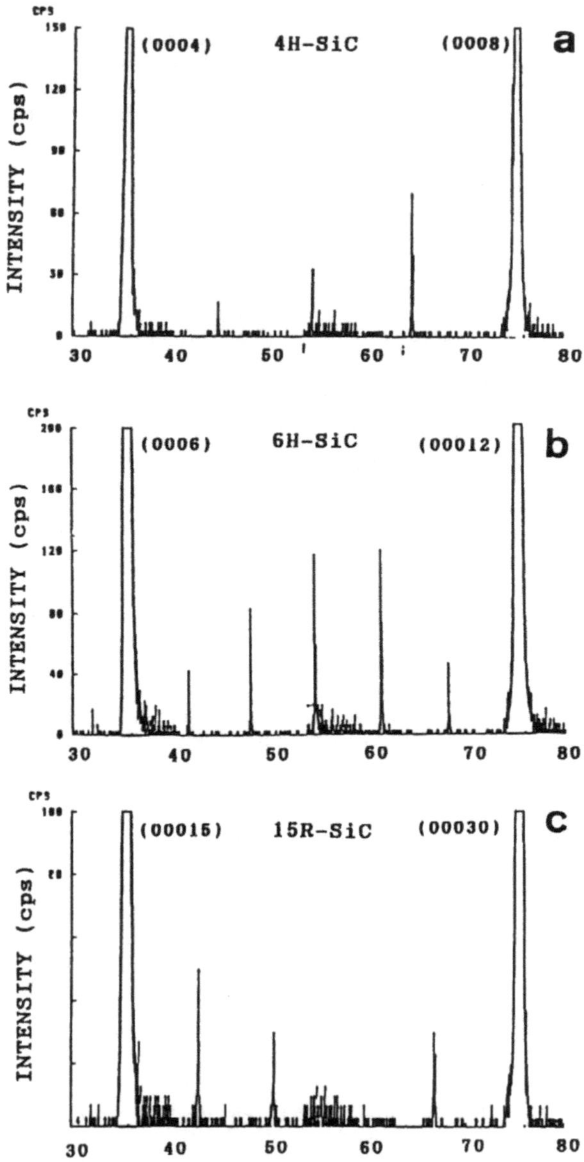

Fig.4.
X-ray diffraction patterns of SiC single crystals: (a) 4H-SiC, (b) 6H-SiC and (c) 15R-SiC.
When the incident direction of X-ray beam was parallel to <1100> in real space, small periodic peaks were observed between main peaks: 3 peaks in 4H-SiC, 5 peaks in 6H-SiC and 4 peaks in 15R-SiC. These peaks relate with double diffraction effect. In these figures(a,b,c), vertical axis is the intensity of reflected X-ray and lateral axis is two times of Bragg angle in degree. X-ray was operated at 30 kV, 20 mA with Cu-target.

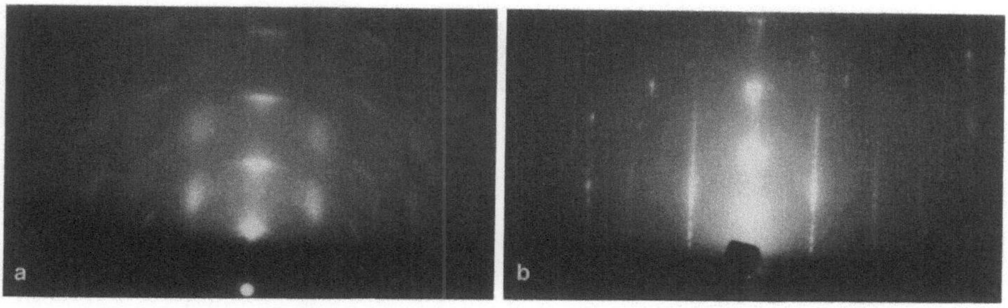

Fig.5. RHEED Patterns: (a) poly-3C-SiC substrate, and (b) 4H-SiC grown layer.

4.Conclusion

6H-SiC was grown by the modified sublimation method. 4H- and 15R-SiC were also grown using the same technique. Growth on $(11\bar{2}0)$ and (1100) planes was also carried out. SiC polytypes were identified by X-ray diffraction and the directional dependence of small peaks was explained by using double diffraction effects. Optical and electrical properties were characterized.

References

1) W.v.Muench and E.Pattenpaul:J.Appl.Phys,**48**,4823,(1987).
2) Y.M.Tairov and V.F.Tsvetkov: J.Crystal Growth, **52**, 146(1981).
3) G.Ziegler, P.Lanig, D.Theis and C.Weyrich: IEEE Electron Device**ED-30**,277(1983).
4) K.Koga,T.Nakata and T.Niina:Ext.Abst.17th Conf. SSDM,Tokyo,246(1985).
5) S.Nishino, J.A.Powell and H.A.Will: Appl.Phys.Lett.**42**, 460(1983).
6) K.Shibahara, S.Nishino and H.Matsunami: J.Crystal Growth, **78**,538,(1986).
7) K.Shibahara, S.Nishino and H.Matsunami: Appl.Phys.Lett. **50**,1888,(1987).
8) K.Shibahara, N.Kuroda, S.Nishino and H.Matsunami: Jpn.J.Appl.Phys.**26**,L1815,(1987).
9) Y.Matsushita, T.Nakata, T.Uetani, T.Yamaguchi and T.Niina:Jpn.J.Appl.Phys.**29**,L343,(1990).
10) Y.M.Tairov and V.F.Tsvetkov and I.F.Khlebnikov: J.Crystal Growth,**20**,155(1973).
11) M.Kanaya, Y.Fujiwara, A.Moritani and S.Shinoyama: Extended Abst. 50th Fall Meeting of Japan Society of Applied Physics, Fukuoka, 28a-W-10,236(1989).
12) A.H.Gomes de Mesquita: Acta Cryst.**23**,610(1967).
13) W.S.Yoo, S.Nishino and H.Matsunami: Springer Proceedings in Physics[Springer-Verlag], vol.**43**,35(1989).

Growth of 6H-SiC Single Crystals by the Modified Sublimation Method

Hyeong Joon Kim and Dong Wook Shin

Department of Inorganic Materials Engineering, Seoul National University, Seoul 151-742, Korea

6H-SiC single crystals have been grown in the graphite vacuum furnace by the sublimation method. During the growth of 6H-SiC single crystals, the temperature gradient, the growth temperature and the pressure ranges employed in this experiment were 44°C/cm, 1800-1990°C and 50-1000 mTorr, respectively. Single crystals grown by the Acheson method were used as seed crystals. The grown single crystals were epitaxially grown on the seed crystals with a growth rate of about 1 mm/h. Polarizing microscopy and back-reflection X-ray Laue diffraction were used to examine the structural relationship of the grown crystal with the seed crystal. XRD analysis showed that 6H-SiC was grown at a temperature above 1840°C and 3C-SiC was grown at a lower temperature or under high supersaturation of vapor. The carrier type, concentration and mobility were measured to be hole (p-type), 7.6×10^{14} cm^{-3} and 19 $cm^2V^{-1}sec^{-1}$, respectively, by the van der Pauw method.

1. Introduction

Since silicon carbide (SiC) has high hardness, thermal and chemical stability, oxidation resistance and high temperature strength, it has been mainly applied to abrasive materials or high temperature structural elements. In addition, the possibility of its application to semiconductor materials has been examined since the early 1960s because of its high thermal conductivity 3.9 W/cm°C and high breakdown field 2.5×10^7cm/V. But its application as a semiconductor has been limited because the grown crystals contain many crystallographic defects which deteriorate the electrical properties of SiC crystals. Recently, high-quality 3C-SiC thin films have been grown on Si substrates by chemical vapor deposition [1-3] and intense effort to develop high power and high temperature devices using 3C-SiC is being made around the world. But problems with SiC crystals still remain although many advances have been made. Attempts to reduce crystal defects have been tried with the homoepitaxial growth of 3C-SiC thin films on 6H-SiC substrates. On the other hand, 6H-SiC is a promising semiconductor material for a blue light-emitting diode since it has a wider band gap than 3C-SiC. In addition, it can be used as a substrate for the homoepitaxial growth of 3C-SiC [4]. The need for 6H-SiC bulk crystals is steadily increasing.

The most widely employed method of growing bulk and relatively pure 6H-SiC single crystals is the sublimation method, owing to its simplicity, efficiency, flexibility and controllability [5,6,7]. This method allows the preparation of reasonably large, moderately high purity and 6H-SiC dominant crystals. In this method, graphite crucible is inductively or resistively heated with the appropriate temperature gradient. The sublimed vapor (mainly Si, Si_2C and SiC_2) from the vapor source loaded in the hotter portion of crucible migrates onto the seed crystal located in the somewhat cooler portion of crucible. Consequently, crystal growth takes place on the seed by the condensation of supersaturated vapor. In this experiment, we grew 6H-SiC single crystals by the modified sublimation method and analyzed them crystallographically, chemically and electrically.

2. Experiment

SiC crystal has been grown in the graphite vacuum furnace, which is resistively heated. Temperature gradients of 23°C/cm and 44°C/cm were maintained in the upper part and the lower part, respectively, of the furnace. The configuration of the growth crucible was similar to that used by Ziegler and Lanig [7]. The furnace was evacuated with a mechanical vacuum pump and the growth pressure was controlled by the leak of Ar gas. Prior to growth, high pressure was kept during heating to the growth temperature in order to prevent unintentional sublimation of the source powder and seed crystal at a temperature lower than the growth temperature.

The vapor source used in this experiment was an abrasive SiC power, which was a mixture of 3C- and 6H-SiC, of 98% purity (major impurities:Al, Fe) and of which the average particle size was 2.5mm. The adoption of seed crystals in the sublimation method was an important advance in the growth of 6H-SiC, since it enhanced the controllability of the polytype, reproducability and efficiency of the yield. The seed crystal was 6H-SiC crystal grown by the Acheson process, and its well-developed {0001} face was used as a crystal growth site.

The grown crystals were structurally analyzed by an X-ray powder diffractometer, the back-reflection X-ray Laue method and a polarized microscope, chemically by secondary ion mass spectrometry (SIMS) and electrically by the four-point probe method and van der Pauw method.

3. Results and Discussion

A 6H-SiC crystal was epitaxially grown on seed crystal at a growth temperature above 1840°C. Below 1940°C, small 3C-SiC crystallites were grown between the grown crystal and a surface of graphite crucible. The growth temperature and the phase of the grown crystal and the surrounding crystallites are listed in Table 1. As you can see in Table 1, the phase of the grown crystal differs from that of the surrounding crystallites, although they were grown at the same temperature. This result cannot be explained exactly at this point. However, the polytype of the seed crystal is likely to affect that of the grown crystal. The color of the grown 6H-SiC crystals was tint blue green or colorless transparent and that of the 3C-SiC crystallites was light yellow. A tint blue green color of the 6H-SiC is attributed to the introduction of impurities, especially Al, from the vapor source.

Table 1. The phase of single crystal grown on the seed and crystallites grown on the graphite at various temperatures

Growth temperature(C)	Phase of crystal grown on the seed	Phase of crystalline grown on the graphite
1990	6H	6H
1940	6H	3C
1890	6H	3C
1840	6H	3C
1800	6H+3C	3C

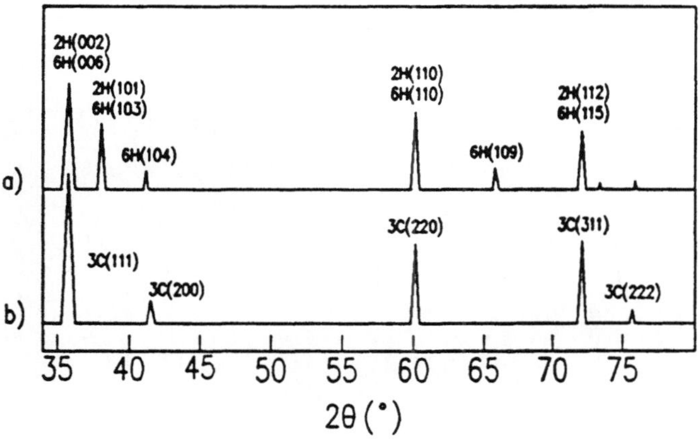

Fig.1 X-ray diffraction pattern of a) blue crystal grown on the seed and b) yellowish crystallites grown on the graphite

XRD patterns of blue-green colored 6H-SiC crystals and yellowish colored 3C-SiC crystals are shown in Fig. 1. The pattern of 3C-SiC (Fig. 1b) is a typical pattern of 3C-SiC with regard to the peak angle and intensity, while that of 6H-SiC (Fig. 1a) doesn't coincide with any patterns of polytypes. Therefore, the grown 6H-SiC crystal is not composed of only 6H type. But the XRD pattern of the grown 6H-SiC doesn't show a characteristic peak of 4H and 15R polytypes in 2θ = 34.6-34.8° . Therefore, we exclude the existence of 4H and 15R polytypes in the grown 6H-SiC crystals. However, it is very difficult to distinguish the 2H from the 6H polytype in the XRD pattern, and thus it is hard to say that 6H-SiC crystal doesn't contain the 2H polytype. Anyhow, it is strange that the 2H phase, which generally is stable at temperatures lower than 1400°C [7,8], is grown at a temperature higher than 1800°C.

In order to distinguish 6H-SiC from 3C-SiC, the grown crystals were examined by polarized microscope. Extinction under polarized microscope occurs when the incident plane polarized beam is normal to the unique optical axis of the crystal under a closed Nicol prism. Since 3C-SiC doesn't have a unique optical axis of the crystal, it does not show extinction. But α-SiC including 6H-SiC has a unique optical axis and shows extinction. Fig. 2(a) shows that extinction occurs in the 6H polytype region but not in the triangle-shaped 3C polytype region of the SiC crystal grown at the relatively low growth temperature of 1800°C. 3C-SiC appears on a surface and in an interface of the 6H-SiC crystal with a particular shape. But in the case of 45° rotation extinction disappears, as shown in Fig. 2(b). Therefore, the unique optical axis of the grown crystals is parallel to the growth direction of the grown crystal, marked by an arrow in Fig. 2(a). The back-reflection Laue pattern also gives information about the polytype of the grown crystal. Fig. 3 shows the back-reflection Laue pattern of the grown crystal with an incident X-ray beam parallel to the growth direction. The resulting diffraction pattern of the grown crystal shows 6-fold symmetry and is the same as that of a seed crystal. This result also indicates that the grown crystal is 6H-SiC and epitaxially grown on a seed crystal.

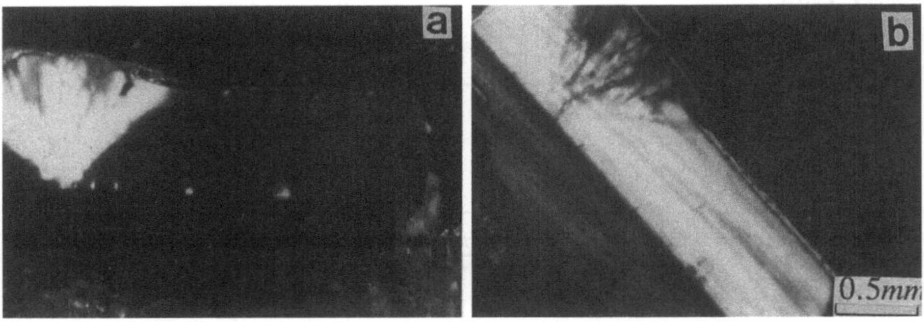

Fig.2 Polarized light microscope photograph showing extinction. a) Extinction does not occur in the a layer. b) Extinction occurs in the a layer (grown at 1800°C)

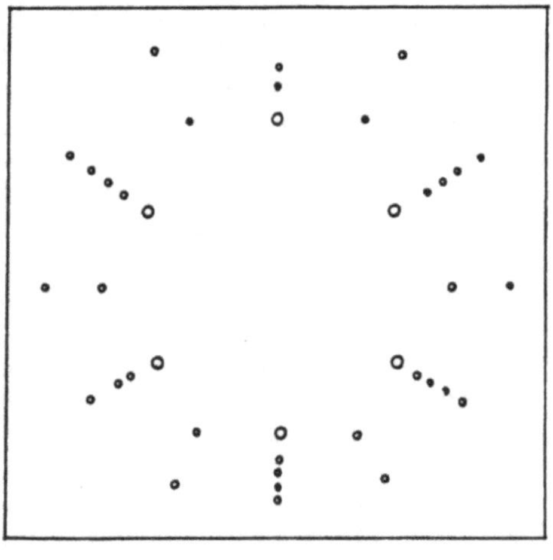

Fig.3 Back reflection Laue pattern of single crystal, where c-axis is parallel to the direction of incident X-ray beam

Fig. 4 shows a polarizing micrograph of a heterostructure consisting of 6H-SiC and 3C-SiC layers. The upper thin layer of 3C-SiC was produced immediately after the furnace power was turned off, since the growth temperature was lowered and the source vapor was highly supersaturated due to the sudden decreasing of the growth temperature. This result indicates that the 3C-SiC polytype is stable below 1800°C and under high supersaturation conditions. This is consistent with the result of the previous report [9].

The temperature dependence of the crystal growth rate, as shown in Fig. 5, is less than that in the sandwich method reported by Tairov et al. [9]. This discrepancy originates from the non-steady-state behavior of the sublimation of SiC. The sublimation from the SiC surface remains C-layer, which restricts sublimation of Si or Si_2C vapors, on the vapor source. Consequently, the vapor pressure or supersaturation produced from the vapor source decreases drastically during the growth process. The rate of decrease of the supersaturation becomes fast as the sublimation temperature is elevated.

The pressure dependence of the crystal growth rate is shown in Fig. 6. The growth rate decreases with increasing Ar pressure. Since the mean free path of vapor molecules decreases with increasing pressure, the rate

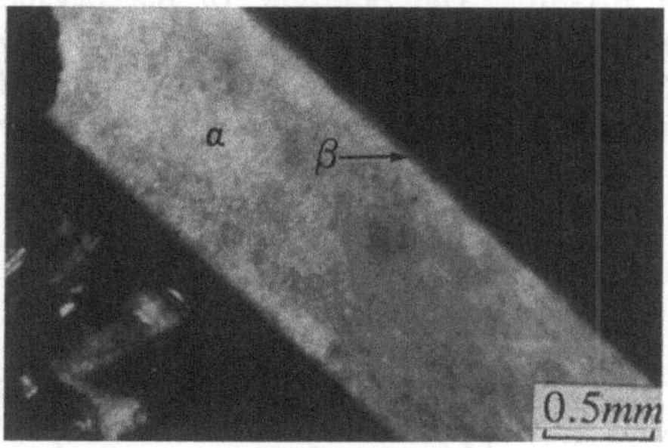

Fig.4 Polarised light microscope photograph of 6H epitaxial layer grown at 1990°C

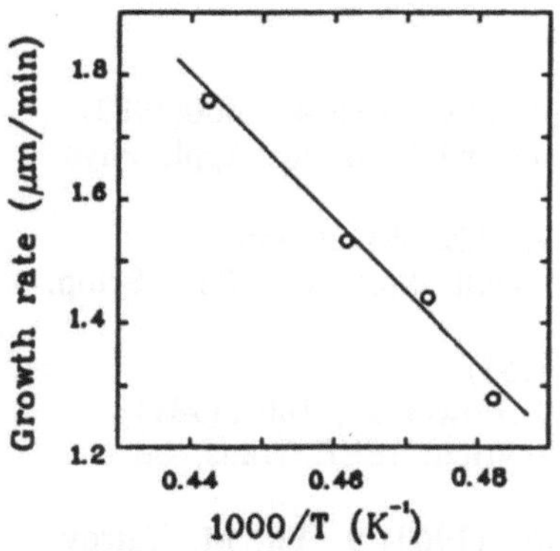

Fig.5 The temperature dependence of epitaxial layer growth rate (pressure is 100mTorr)

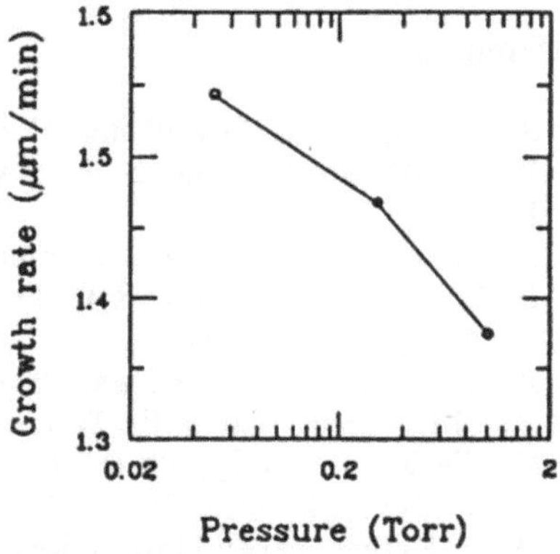

Fig.6 The pressure dependence of epitaxial layer growth rate (growth temperature is 1840°C)

determining step of the crystal growth is the gas-state diffusion of the vapor molecules in the high pressure growth condition. However, if the mean free path of the vapor molecules is large enough in the low pressure condition, the rate determining step is the chemical reaction on the seed surface. The pressure dependence of the growth rate in this experiment implies that the rate determining step in the pressure range from 10^{-1} to 10^{-2} mTorr is the gas state diffusion of vapor molecules.

The SIMS spectra, which are not shown here, illustrate that the grown crystal is relatively pure except for Al impurity. The resistivity of the grown crystal is about 6 Ω-cm. The grown crystal is p-type and has a hole concentration of $7.6x10^{14}$ cm^{-3} and hole mobility of 19 $cm^2V^{-1}sec^{-1}$.

4. Conclusions

6H-SiC single crystals were epitaxially grown by the modified sublimation method. The experimental conditions of crystal growth are relatively low temperatures (between 1800 - 1990 °C), the growth pressure of 50 - 1000 mTorr and the temperature gradient of 44°C. The rate of crystal growth was 1.1 mm/h - 3.5 mm/h, depending on the deposition condition. The growth direction is normal to the basal plane of 6H-SiC. The polytype of the crystal grown in the temperature range 1840 - 1990°C was 6H-SiC, while below 1800°C the crystal was a mixture of 6H- and 3C-SiC. The crystal showed p-type conduction with a hole concentration of $7.6x10^{14}cm^{-3}$ and hole mobility of 19 $cm^2V^{-1}sec^{-1}$.

Acknowledgement

The authors wish to thank Mr. S. Misawa in ETL of Japan for his help with the Hall measurement. This work is supported by 1986 and 1987 Grants-in-Aid from Interuniversity Semiconductor Research Center in Korea.

References

1. S. Nishino, J.A. Powell and H.A. Will: Appl. Phys. Lett.42, 460(1983)
2. K. Sasaki, E. Sakuma, S. Misawa, S. Yoshida and S. Gonda:Appl. Phys. Lett. 45, 72 (1984)
3. P. Liaw and R.F. Davis: J. Electrochem. Soc. 132, 642 (1985)
4. H.S. Kong, J.T. Glass, R.F. Davis and S.R. Nutt: Mat. Res. Soc. Symp. Proc. 77, 405 (1987)
5. A. Lely: Ber. Deut. Keram. Ges. 32, 229 (1955)
6. Yu. M. Tairov and V.F. Tsvetkov: J. Cryst. Growth 52, 146 (1981)
7. G. Ziegler, P. Lanig, Ph.D. Thesis and C. Weyrich: IEEE Trans. on Elec. Dev. ED-30, 277 (1983)
8. W.F. Knippenberg: Phillips Res. Rept. 18, 161 (1963) 9. Yu. M. Tairov, V.F. Tsvetkov, S.K. Lilov and G.K. Safalaliev:J. Cryst. Growth 36, 147 (1976)

A System for Growth of Bulk Beta-SiC by Sublimation

V. Shields, K. Fekade, P. Johnson, M. Spencer, M. Aluko, and G. Harris

Materials Science Research Center of Excellence, Department of Electrical Engineering, School of Engineering, Howard University, Washington, DC 20059, USA

Abstract. Silicon Carbide (SiC) has received considerable attention as a material for fabricating high power microwave transistors. Beta SiC is the polytype of choice due to its high, low field mobility. Single crystal beta SiC substrates with a low defect density are required in order to produce high quality epitaxial material. In this paper, a system for vacuum sublimation growth of bulk beta SiC is described. This system includes a two zone resistively heated graphite furnace and a ten inch diffusion pump for evacuating the furnace. The growth system has a useable diameter of one inch with possible growth temperatures up to 2400 ^{o}C. The thermal gradient within the growth region is 43 ^{o}C per inch along the axial direction with a maximum horizontal gradient of 15 ^{o}C. Both thermal gradients are independent of argon pressure.

Introduction

Among the many SiC polytypes, the beta form is distinguished due to its potential as a substrate for gallium nitride and diamond. These desirable characteristics make beta SiC an important semiconductor material. The objective of this research is bulk growth of the single crystal beta SiC polytype. To accomplish this task, a two temperature zone, resistively heated graphite furnace has been designed. The furnace utilizes a modified Lely growth method whereby sublimation occurs onto the surface of a seed crystal. This paper describes the furnace design and thermal characterization. Present efforts have involved the determination of the growth conditions. The preliminary growth rate achieved was 0.5 mm per hour of polycrystalline material.

System Description

The crystal growth system utilizes a water cooled vacuum chamber containing a two zone furnace assembly. This system was designed in conjunction with Mellen Company, Inc. The high temperature furnace is capable of operating at temperatures up to 2400 ^{o}C in a 380 Torr argon atmosphere. An illustration of the furnace is provided in Figure 1. The interior of the furnace assembly is composed of an all graphite construction consisting of a thermal shield, heater elements, crucible and seed holder. The diameter of the thermal shield is 61 cm with a length of 76 cm. Two pyrometer ports are provided for determining and controlling the zone temperatures. These two ports are located at the seed crystal and source levels along the crucible axis.

There are two heater elements within the furnace. The upper element is designated zone 1 while the lower element is referred to as zone 2. The crucible is inserted into the heater elements and has an internal diameter of 2.5 cm. The seed crystal is located above the source at the point of the desired temperature. A temperature controller maintains the thermal set points during the growth cycle by use of two pyrometers. The internal argon pressure within the chamber is maintained by way of a throttle valve feedback loop.

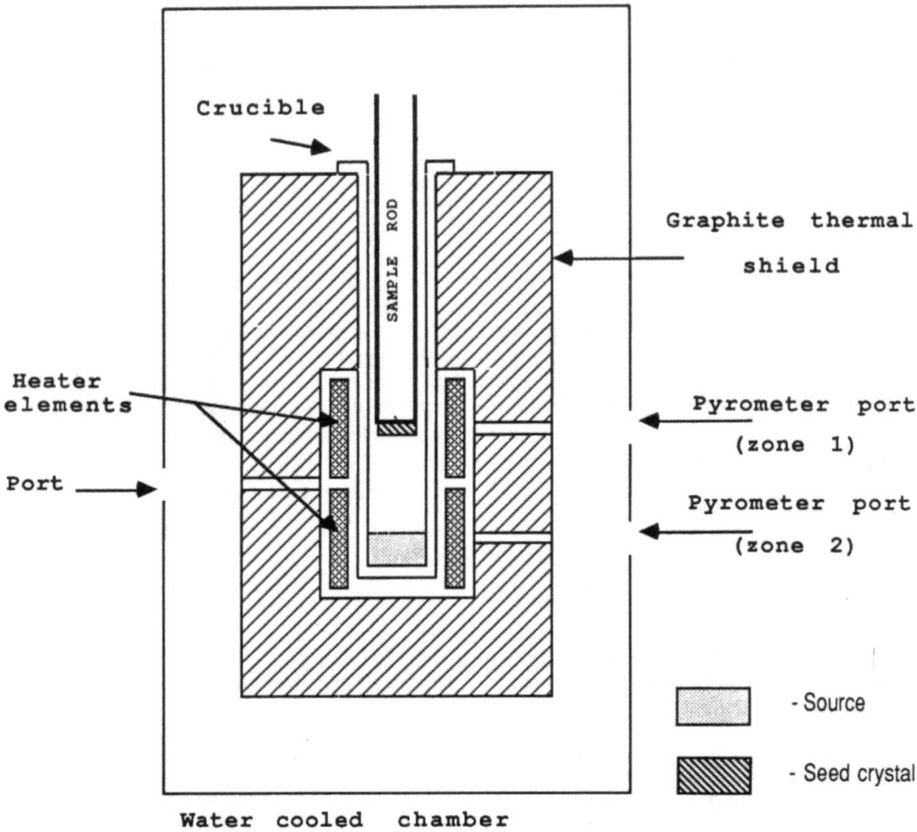

Figure 1. The internal diagram of the reactor. The source material is located in the bottom of the crucible while the seed crystal is attached to the sample rod.

Experimental Procedures and Results

Prior to conducting the initial growth experiments, an accurate determination of the internal crucible temperatures and the thermal gradients was necessary. Internal temperature measurements were made for correlation with the external pyrometer readings. To facilitate these measurements, W/Re thermocouples with molybdenum sheathing were inserted into the interior of the crucible under actual growth conditions. Thermal profiles of the interior of the crucible were prepared along both the axial (vertical) and the radial (horizontal) directions within the crucible. Measurements were made at approximately 0.1 Torr and 380 Torr to ascertain the effects of pressure upon the thermal gradients. The zone 1 temperature control point was varied over a temperature range of 1400 ^{o}C to 1800 ^{o}C. The corresponding zone 2 temperature control point was varied over a range of 1600 ^{o}C to 2000 ^{o}C. The pyrometer readings were compared with the thermocouple measurements. The external gradient between zones 1 and 2 was maintained at 200 ^{o}C for all the thermal profiles. The maximum temperature was limited to 2000 ^{o}C due to the formation of a eutectic between the thermocouple molybdenum sheath and the graphite in the crucible at 2206 ^{o}C. Temperature measurements were made along the perimeter of the interior of the crucible at four locations

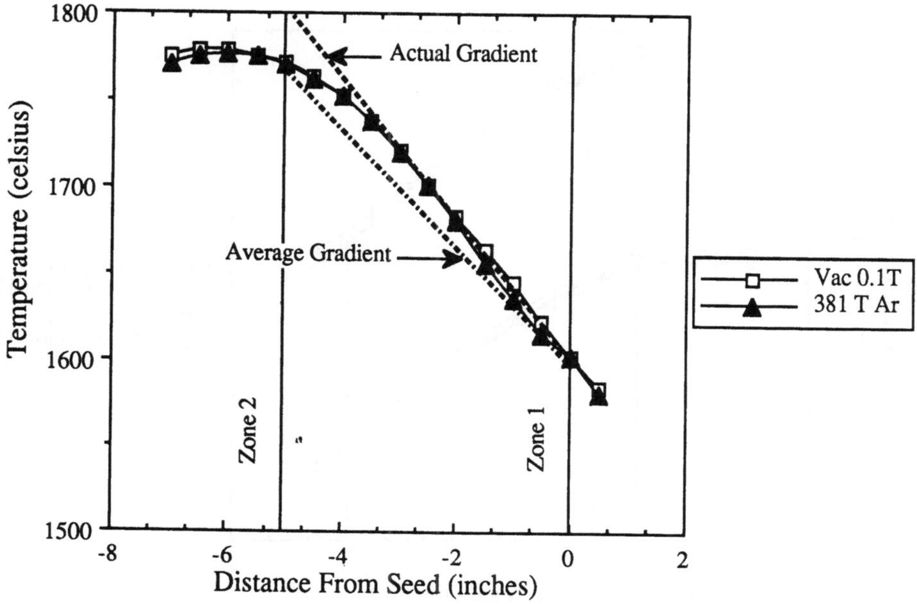

Figure 2. Typical axial temperature gradient within the crucible.

spaced 90 degrees apart and in the center. The horizontal gradient was determined as the difference between the average perimeter and the center measurement.

Figure 2 shows the typical axial thermal variation within the crucible. The internal temperatures corresponding to both zones 1 and 2 pyrometer readings are noted. Corresponding measurements at 0.1 Torr and 380 Torr are shown. The thermal gradient as set by the furnace control system displays no significant difference due to pressure. The thermal set points for this figure were 1800 ^{o}C and 2000 ^{o}C for zones 1 and 2 respectively. Due to inherent thermal limitations within the furnace, the actual temperature difference obtained between zones 1 and 2 is less than the 200 ^{o}C value set by the controller. The maximum temperature difference obtained was between 175 ^{o}C and 190 ^{o}C. The temperature difference between the pyrometer and the thermocouple readings at both zones varied from 3 ^{o}C to a maximum of 27 ^{o}C.

The axial thermal gradient at the maximum and minimum temperature range at which measurements were conducted is displayed in Figure 3. No significant variation in the temperature gradient was observed over the measurement range. Also shown in this figure is the gradient between the inner perimeter of the crucible and its center. The maximum difference between the average perimeter and center temperature measurements was 15 ^{o}C.

Conclusions

Several conclusions can be derived from these results. The effective or actual gradient in the growth region of the furnace was determined to be 43 ^{o}C per inch compared to an average value of 34 ^{o}C. In addition, the temperature feedback control provides a gradient which is independent of argon pressure for a constant difference between temperature zones. Finally, the maximum horizontal gradient in the growth region of the crucible was determined to be 15 ^{o}C. Due to the stability of these thermal gradients over a wide range of temperatures, these results provide confidence in a reliable extrapolation into the expected growth range.

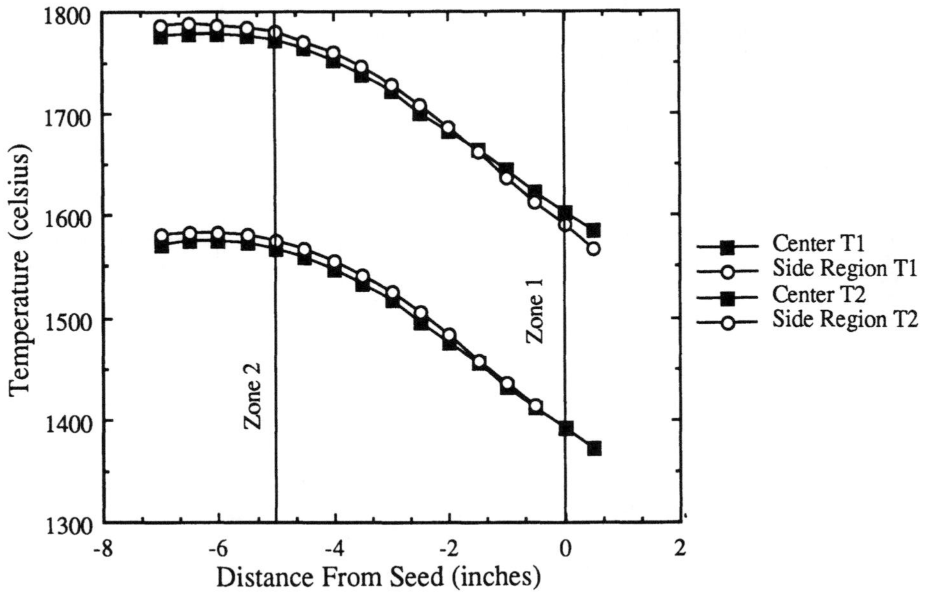

Figure 3. Variation in axial and horizontal thermal gradients with temperature. Measurements were made at 0.1 Torr vacuum pressure

Acknowledgements

This work was supported by the National Science Foundation, NASA and the Jet Propulsion Laboratory.

Sublimation Vapor Transport Growth of Silicon Carbide

D.L. Barrett[1], *R.G. Seidensticker*[1], *W. Gaida*[1], *R.H. Hopkins*[1], *and W.J. Choyke*[2]

[1]Westinghouse Science and Technology Center, 1310 Beulah Rd., Pittsburgh, PA 15235, USA

[2]Department of Physics and Astronomy, University of Pittsburgh, Pittsburgh, PA 15260, USA

Abstract. Recent advances in the growth of SiC boules yielding wafers up to 25 mm in diameter have led to renewed efforts to capitalize on the unique thermal and electrical properties of SiC semiconducting and optoelectronic devices. We report the design of a sublimation vapor transport growth system for SiC boules, and the growth of 33 mm diameter boules with large 25 mm single crystal regions.

1. Background

In 1955, Lely [1] developed a laboratory sublimation process for the growth of α-SiC platelets at random nucleation sites inside a heated cavity of grain SiC. This technique was subsequently used [2,3] to prepare platelets for device fabrication using grown junction and diffused junction technology. Lely platelets such as the grown junction platelets shown in Fig. 1, were small in size, irregular in shape, and proved to be very difficult to fabricate into reproducible devices. Different crystal polytypes also were often found in layers in the Lely-grown platelets. Since these polytypes have different electron mobilities [4], the slightly different characteristics of each platelet presumably led to nonuniform device results. It should be pointed out that even with these difficulties, devices were prepared which routinely operated in 500°C ambients.

In 1978, Tairov and Tsvetkov [5] described a technique for the growth of SiC boules from which wafers could be fabricated for device preparation. The authors described the growth of an 8 mm diameter by 8 mm long boule of SiC on a seed crystal located within a quasi-closed cavity of SiC grain. The growth occurred when vapor from the subliming source was transported under near vacuum conditions to the seed located at a lower temperature. In a later publication [6] these authors reported the growth of SiC boules up to 14 mm diameter and 18 mm in length, and the use of wafers from the boules to fabricate boron-diffused or aluminum-implanted luminescent diodes exhibiting yellow, green, and blue colors.

In 1983, **Ziegler** [7] described a 20 mm diameter, 24 mm long 6H SiC crystal grown by a method similar to that produced by Tairov, but with a modified cavity geometry. In 1987, Carter [8] described the seeded growth of 15 mm diameter, 8 mm thick 6H polytype SiC single crystals. The vacuum-sublimation method has also been used by Nakata [9] to grow 6H polytype crystals 33 mm in diameter and 14 mm long. The crystal growth rate varied from 1 to 4 mm/hr when near vacuum (0.001 to 5 torr) conditions were used. While these growth rates are slow, they compare reasonably with the 6-9 mm/hr commonly used in commercial liquid-encapsulated Czochralski growth of GaAs.

The wide interest in SiC boule growth and wafer preparation is evidenced by papers presented at this conference by Nishino, Kim, and Spencer as well

Springer Proceedings in Physics, Volume 56 Amorphous and Crystalline Silicon Carbide III
Eds.: G.L. Harris, M.G. Spencer, and C.Y. Yang

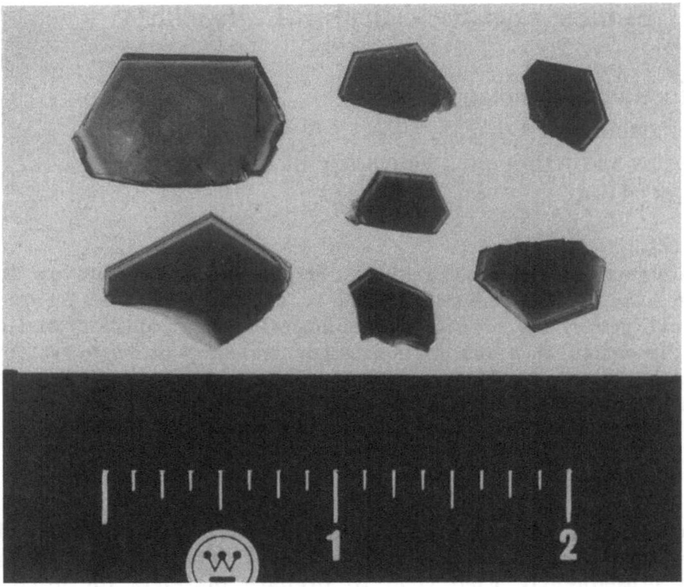

Figure 1. Grown junction platelets prepared by the Lely technique.

as the present work. Advances in device fabrication were recently reviewed by Edmond [10] at a symposium on Single Crystal SiC, where a series of SiC-based discrete diodes, MESFET, MOSFET, and high-frequency FET devices and their characteristics were described.

2. Crystal Growth System Design

The sublimation vapor transport system we designed for high-purity SiC boule growth is shown in Fig. 2. In this design, a quartz-wall induction-heated system was adopted using high-purity graphite and carbon parts within the hot zone. A degassable rigid carbon foam was used as thermal insulation to provide a clean environment while avoiding coupling effects with the 10 KHz induction field. The system is vacuum-tight and capable of high-temperature bakeout to remove nitrogen, a pernicious source of shallow donors in SiC.

The growth system consists of concentric 170 mm od and 150 mm od quartz tubes sealed with vacuum-tight integrity into stainless steel end flanges using vacuum-pumped, double o-ring seals on ground surfaces of the quartz tubes. Cooling water is circulated into the bottom end flange, between the concentric quartz tubes and out the top flange. A liquid nitrogen-trapped, 4-inch silicone oil diffusion pump will maintain a vacuum of 2×10^{-7} torr at room temperature after bakeout, and mid-10^{-6} torr at 2000°C. Access into the tube is easily achieved through an 8-inch top flange for easy loading and unloading. The top and lower plates provide ports for windows, probes, and mechanical devices for moving the seed and source.

The induction coil is a 9-turn, 8-inch high, and 7-1/2-inch id coil rigidly held by insulating members to an insulating coil holder attached to an xyz milling head. This feature permits the coil to be moved in any direction for

Figure 2. Sublimation vapor transport system showing quartz tube furnace, induction coil, vacuum system, and control panel.

adjustment of thermal gradients within the hot zone. Furnace power is provided by a highly stabilized 50 kW motor generator set operating at 10 KHz frequency.

3. Crystal Growth

The basic arrangement of the growth cavity is shown schematically in Fig. 3, where the cavity is defined by a porous carbon member similar to that shown by **Ziegler** [7]. The SiC source feedstock is located in the hot region which is typically 2100-2400°C, depending on the process and polytype desired, while the seed is positioned several cm away in a colder region of the cavity. Growth occurs on the seed when the system is maintained at reduced pressures of less than 20 torr with a temperature gradient of 20-35°C/cm between source and seed. The graphite crucible also functions as the susceptor.

The axial thermal gradient in this system was evaluated using W:W-Rh thermocouple probe measurements in the annular space between the crucible/susceptor and the rigid carbon foam insulation. The gradient was determined for several temperatures lower than the growth temperature as a function of the induction coil voltage. These gradients were then extrapolated to the growth voltage as shown in Fig. 4. Under these conditions, a linearized gradient of 32°C/cm was obtained for our boule growths.

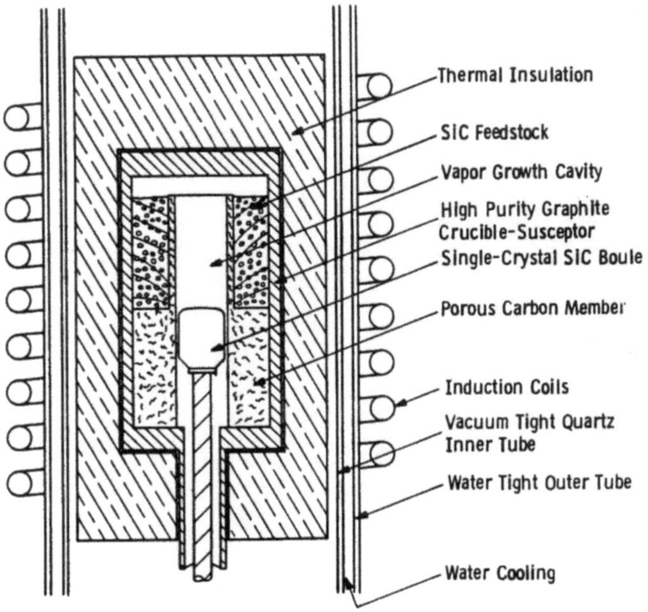

Figure 3. Schematic of sublimation vapor transport growth cavity.

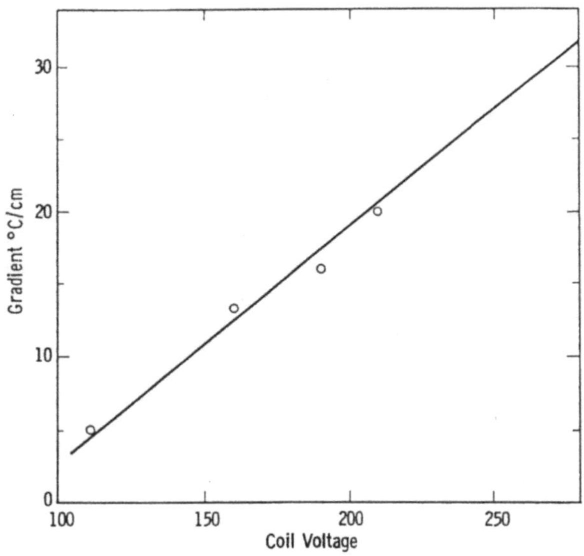

Figure 4. **Linearized temperature gradient between the source and seed region, extrapolated to 270 V to obtain 32°C/cm gradient at growth conditions.**

Figure 5 shows a 33 mm diameter boule grown using this sublimation vapor transport technique. Figure 6 shows a wafer sliced from this boule. A 5 x 8 mm Lely-grown platelet was used as the seed, resulting in a 20 x 25 mm single-crystal region shown in the wafer sliced from near the top surface of the boule. The single-crystal region increased over 10 times in area as the boule grew 18 mm in length. The multiple polytype region at the outer perimeter of the wafer nucleated on the carbon rod used as the seed holder. The single-crystal area was identified by Laue transmission patterns, shown in Fig. 7 to be pure 6H polytype: Laue patterns repeated at locations across the single-crystal surface area show no evidence of other polytypes.

Figure 5. SiC boule grown from a 5 x 8 mm Lely seed.

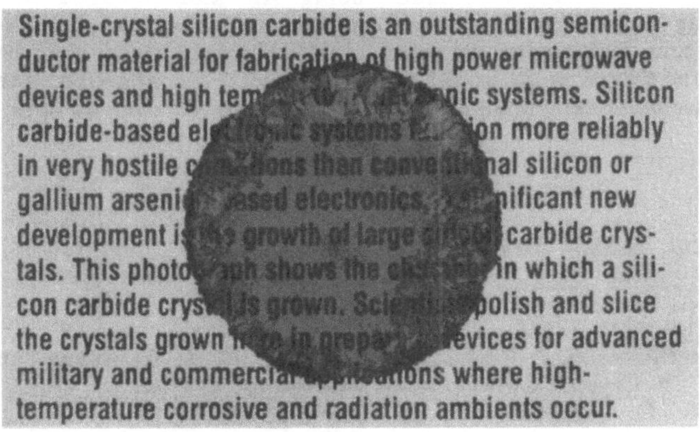

Figure 6. A 33 mm diameter wafer with a 25 x 20 mm single-crystal area.

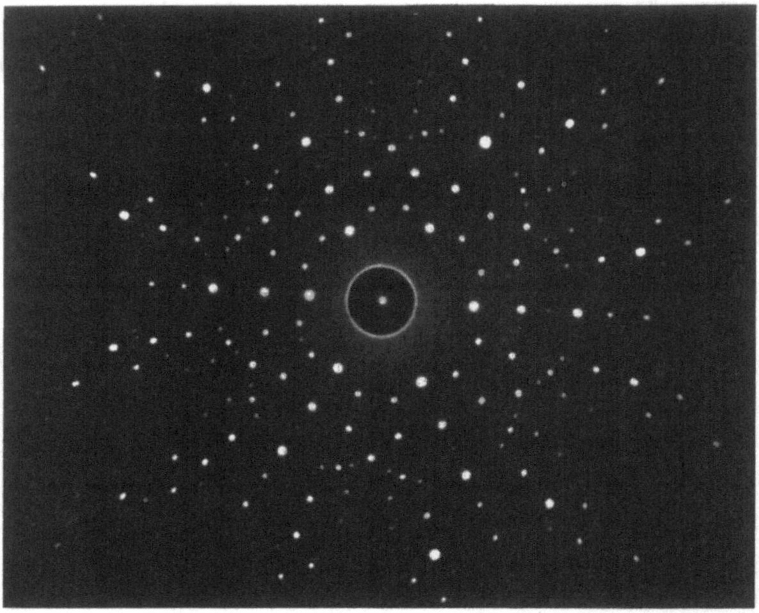

Figure 7. 6H polytype Laue transmission pattern obtained from single-crystal region of SiC boule.

4. Conclusions

A system for sublimation vapor transport growth of SiC and preliminary growth results have been described. The use of large-diameter seeds in place of the small Lely platelet should result in 25 mm diameter single-crystal regions which can be processed into "silicon-like" wafers for device processes. While the system described was designed to be capable of high-purity growth, the preliminary growths described here were made using Al- and N-doped SiC charge material resulting in wafers with mixed conduction. The use of high-purity charge material, and high-purity degassed graphite components, will be a specific objective of our future work in SiC crystal growth. In addition, the system is designed to accommodate scale-up of boule diameter to over 50 mm.

5. References

1. J. A. Lely: Ber. Deut. Keram. Ges., 32, 229 (1955).
2. D. R. Hamilton: In Silicon Carbide (Pergamon Press, New York (1960), p. 43.; L. J. Kroko: J. Electrochem. Soc. 113, 801 (1966); D. L. Barrett: J. Electrochem. Soc. 113, 1215 (1966).
3. W. F. Knippenberg: Philips Res. Rept. 18, 161 (1963); C. J. Kapteyns, and W. F. Knippenberg: J. Crystal Growth 7, 20 (1970).
4. D. L. Barrett and R. B. Campbell: J. Appl. Phys. 38, 53 (1967).
5. Yu. M. Tairov and V. F. Tsvetkov: J. Crystal Growth 43, 209 (1978).
6. Yu. M. Tairov and V. F. Tsvetkov: J. Crystal Growth 52, 146 (1981).

7. G. Ziegler, P. Lanig, D. Theis, and C. Weyrich: Trans. Electron. Dev. ED-30, 277 (1983).
8. C. H. Carter, Jr., L. Tang, and R. F. Davis: Presented at the Fourth National Review Meeting on the Growth and Characterization of SiC, Raleigh, NC (1987); also see J. A. Powell and L. G. Matus: Amorphous and Crystalline Silicon Carbide, Springer Proceedings in Physics 34, 2 (1989).
9. T. Nakata, K. Koga, Y. Matsushita, Y. Ueda, and T. Niina, Amorphous and Crystalline Silicon Carbide and Related Materials II, Springer Proceedings in Physics 43, 26 (1989).
10. L. A. Tang, J. A. Edmond, J. W. Palmour, and C. H. Carter, "Discrete Devices in 6H-Silicon Carbide," Abs. 484, J. Electrochem. Soc. 137, 407C (1989).

Defects and Polytypism of SiC Bulk Single Crystals

Yu.M. Tairov and V.F. Tsvetkov

V.I. Ulyanov (Lenin) Electrical Engineering Institute,
197022 St. Petersburg, P-22, Russia

Abstract. The investigation results of structural defects and crystallochemical properties of SiC polytypes single crystals in relation to their production conditions are presented. New methods of SiC single crystal growth under controllable conditions first developed for polytype substances permitted the crystallichemical properties of a polytype variety to be studied. The reference data on the lattice parameters (a, c, c/n) and their temperature dependences, the basic components content (C_{Si} and C_C) of the crystals grown in equilibrium conditions, the X-ray and practical densities, the chemical displacement of silicon ($Si_{K}\alpha$) analytical line, Debye characteristic temperatures, and the enthalpy and entropy of polytypes have been obtained. The relations established between the polytype structure and the crystal properties are of great importance for evaluating the role of the nearest environment in physical properties of the materials.

For comparison a study was made of dislocation type defects on the single crystals grown both by Lely method (by spontaneous nucleation) and by seed crystallization. The joint use of X-ray topography and transmission electron microscopy enabled the basic types of dislocations to be detected and identified. From the investigations performed it was concluded that Lely method in nature does not permit controllable growth of a specific single crystals polytype structure having the stable density and uniform distribution of dislocations throughout the crystal. The analysis of the dislocation structure of the bulk single crystals grown by seed crystallization made it possible to establish the effect of basic technological factors and to determine the regularities and mechanisms of dislocation generation.

The methods of EMP, ESR, chemical analysis, precision measurement of X-ray and practical densities were used for theoretical and experimental studies of point defects and deviations from stoichiometry and SiC polytype single crystals in relation to the conditions of their production. The joint investigations of crystals by the methods of optical absorption, ESR and photoconductivity indicated that the vacancy-nature defects exert a significant effect on electrical and optical properties of SiC.

Springer Proceedings in Physics, Volume 56 **Amorphous and Crystalline Silicon Carbide II**
Editors: G.L. Harris, M.G. Spencer, and C.Y. Yang

Part II

Growth and Characterization of Amorphous SiC

Growth Mechanism of Hydrogenated Amorphous SiC from a Glow-Discharge Plasma

A. Matsuda

Electrotechnical Laboratory, 1-1-4 Umezono, Tsukuba 305, Japan

Abstract. Growth mechanism of hydrogenated amorphous silicon-carbon alloys (a-SiC:H) prepared from a SiH_4/CH_4 glow-discharge plasma is discussed and compared with that of the hydrogenated amorphous silicon (a-Si:H) from a SiH_4 plasma using a radical-separation method. It is suggested that the carbon-related radical such as CH_3 contributes to the growth of a-SiC:H films through a quite different reaction behavior from SiH_3 on the film-growing surface. On the basis of the surface reaction model for the film growth, a guiding principle for preparing highly photosensitive a-SiC:H is proposed.

1. Introduction

Hydrogenated amorphous silicon (a-Si:H) is the first structure sensitive amorphous semiconducting material whose conduction type can be controlled to p or n type by the addition of a small amount of impurities such as boron or phosphorus into the amorphous network, respectively. Therefore, a variety of device applications are proposed by making good use of the nature of the amorphous phase, in which large area and uniform materials can be prepared easily. In this sense, solar cells, photoreceptors and thin-film transistor arrays are the typical application examples of a-Si:H. Another advantage of amorphous materials is that the properties of materials can be continuously controlled by changing their compositions. Since this advantage is very important in the application of the materials to photoelectronic devices, hydrogenated amorphous silicon based alloy materials such as hydrogenated amorphous silicon-carbon alloys (a-SiC:H) have attracted increasing attention to cover the wide range spectrum response in the devices.

However, in contrast to a-Si:H, a-SiC:H thin films prepared from a conventional glow-discharge decomposition of silane (SiH_4) and hydrocarbons such as methane (CH_4), ethylene (C_2H_4) and acetylene (C_2H_2) did not show properties favorable for photoelectronic device applications, i.e., these materials were neither highly photoconductive nor photosensitive [1].

In this paper, we discuss, firstly, the surface-reaction process in the film-formation process of a-Si:H from the SiH_4-glow-discharge plasma, where the final network structure is determined, and we propose a guiding principle for preparing a high density network structure showing a high photoconductivity in the deposition of a-Si:H. Secondly, the difference of the surface-reaction process between the growth of pure a-Si:H and that of a-SiC:H alloys is suggested. Finally, we demonstrate the drastic improvement of the photoelectric properties of a-SiC:H on the basis of the understanding of the characteristic reaction behavior of carbon related radicals (CH_3) on the film-growing surface.

Springer Proceedings in Physics, Volume 56 **Amorphous and Crystalline Silicon Carbide III**
Eds.: G.L. Harris, M.G. Spencer, and C.Y. Yang

2. Surface-Reaction Process of SiH_3 Radicals in the Formation of a-Si:H

In general, a network structure of a thin film is principally determined by the energy-relaxation process of adsorbed film precursors on the growing surface of the film [2]. A defective material with a low density network structure is likely to grow when the surface-diffusion length of the precursors is too short to find their energetically favorable sites. This is undesirable because materials with a low density network structure have lower free carrier mobility and lifetime, thereby yielding poor electronic properties. Therefore, in the film-formation process for producing highly photosensitive materials with a dense network structure, the precursors should be able to diffuse over great distances on the growing surface. A longer surface-diffusion distance (l) of radicals can be achieved either by its larger surface-diffusion coefficient (D_s) or a longer staying time (τ_s) on the surface according to the relations of $l = (2D_s\tau_s)^{1/2}$ and $D_s = va_o^2\exp(-E_s/kT)$, where v is the vibrational frequency of radicals, a_o the jump distance and E_s the activation energy for the site jumping [2]. The staying time τ_s is considered to be constant when the recombination time of the precursors on the surface is sufficiently shorter than the monolayer-deposition time. Thus D_s might be a key factor for determining the network structure of resulting film. It is clear from the relations listed above that a higher surface temperature and a smaller E_s result in a larger D_s. The magnitude of E_s has a strong correlation with the reactivity of adsorbed radicals on the growing surface.

For the growth of a-Si:H, the main film precursor has been considered to be SiH_3 [3] suggested from its longer reaction-lifetime in SiH_4-gas ambient. In order to investigate the reaction behavior of SiH_3 radicals and the surface H-covering condition during the

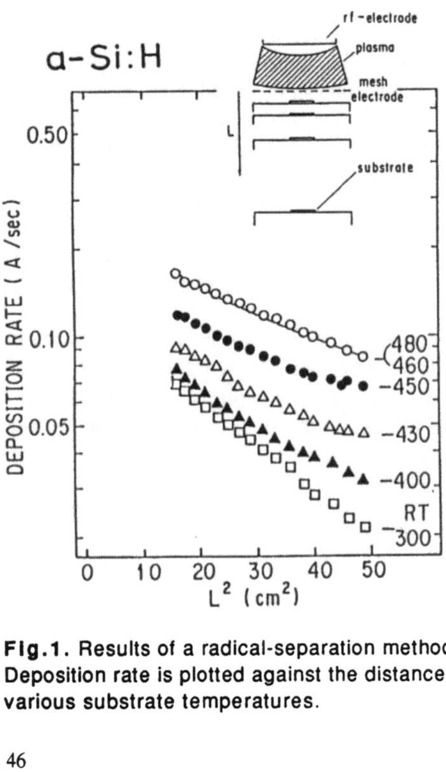

Fig.1. Results of a radical-separation method in the deposition of a-Si:H. Deposition rate is plotted against the distance from the mesh electrode for various substrate temperatures.

growth of the film, a-Si:H films were deposited from the SiH_4 plasma using a radical-separation technique [3]. Figure 1 shows the deposition rate of a-Si:H plotted against the square of the distance L between the mesh and substrate in a triode reactor (see the inset) for various substrate temperatures. When the distance L is large enough to eliminate the short lifetime radicals such as SiH_2, SiH and Si, only SiH_3 radicals must be the responsible species for the growth of the film. The deposition rate reveals a strong temperature dependence, in a whole scanned range of L^2, in the substrate-temperature range between 350°C and 460°C, while it shows no temperature dependence at the substrate temperature lower than 300°C as well as higher than 460°C. This drastic increase in the deposition rate above 350°C originates from the change in the sticking coefficient of SiH_3 radical via a decrease of the H-coverage factor [3]. The saturation at 460°C means that the number of H-deficient sites is large enough to trap all adsorbed SiH_3 radicals as the chemisorbed states, namely, the deposition rate of a-Si:H at this temperature can be regarded as the adsorption rate of SiH_3 radicals on the film-growing surface [4]. It should be noticed that the slope of the curve in the figure changes from steeper to flatter on elevating the substrate temperature. This implies that the SiH_3 radicals contribute to the growth of the film through a bimolecular reaction in the temperature range below 300°C (steeper slope) and a monomolecular reactivity starts to appear in the growth process of the film by SiH_3 radicals (slope becomes flatter) above 350°C [3]. From this temperature dependence of the reaction behavior of SiH_3 radicals on the film-growing surface, it is indicated that the SiH_3 radicals contribute to the growth of the film through a bimolecular reaction below 300°C, in which one SiH_3 radical abstracts H atoms from the Si-H bond on the surface and the other (next) SiH_3 radical finds the H-deficient site created by the previous SiH_3 and forms a Si-Si bond on the surface. On the other hand, above 350°C, H-elimination reaction takes place thermally and an H-deficient site starts to appear on the surface, which traps SiH_3 without any help of abstraction reaction of H by another SiH_3, which is the reason why the monomolecular reactivity starts to appear at higher temperatures above 350°C [3].

Figure 2 shows the photo and dark conductivity of deposited a-Si:H as a function of the substrate temperature during the growth of the film. Photoconductivity increases gradually with an increase in the substrate temperature from room temperature to 300°C, which can be thought of as the enhancement of the surface-diffusion coefficient of SiH_3 radicals with increasing temperature. The reduction of the photoconductivity above 350°C can be explained by the surface-diffusion length of the SiH_3 radicals on the surface being effectively reduced by the appearance of H-deficient sites on the surface which prevent the diffusion of SiH_3 radicals.

On the other hand, other radicals such as SiH_2 can easily stick to the surface through an insertion reaction even when the surface is covered with H. Thus, SiH_2 radicals are considered to have a smaller D_s on the growing surface, and eventually the columnar-like structure appears in the a-Si:H network when the SiH_2 fraction is dominant in a total flux density of the the possible precursors depending on the plasma condition [5].

Consequently, as a guiding principle for obtaining a highly photoconductive a-Si:H with a dense network structure, the following items can be proposed [6]; (1) less reactive radicals having a larger surface-diffusion coefficient like SiH_3 radicals should be selectively guided to the film-growing surface, and (2) the surface should be covered with H even at higher temperatures.

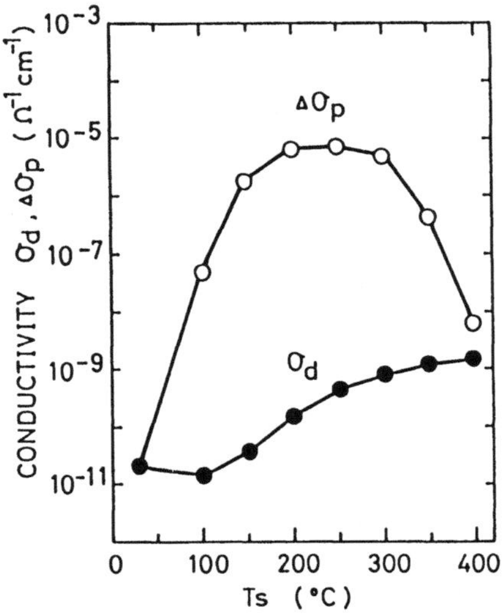

Fig.2. Photo and dark conductivities of a-Si:H as a function of the substrate temperature during the growth of the film.

3. Properties of a-SiC:H and the Surface-Reaction Process

Figure 3 shows the relationship between the photo and dark conductivity and their optical gap determined using a Tauc's plot for a-SiC:H films deposited by a conventional glow-discharge-decomposition method of SiH_4 plus a variety of hydrocarbons including CH_4. Photoconductivity is strongly reduced when the optical gap increases by increasing the carbon content in the film. This reduction in the photoconductivity is considered to be due to the increase in the recombination centers for photogenerated carriers originating from the Si-dangling bonds in the film which was confirmed by the electron-spin-resonance (ESR) measurements [1]. At the same time, the carbon polyhydride configuration in the bonding configuration of hydrogen is strongly enriched in the film, which was observed by infrared-absorption spectroscopy [1]. Since the films have low refractive indices, a-SiC:H deposited by a conventional method must have a less dense network structure.

A less dense network structure is likely to grow when the surface-diffusion length of the film precursors is too short to find their energetically favorable sites as was mentioned above. However, because of the strong binding energy of C with H, surface hydrogen on the a-SiC:H growing surface is more strongly bound than that on the a-Si:H growing surface. Furthermore, the film precursors must be SiH_3 and CH_3 considered from the strong reactivity of other radicals such as C, CH and CH_2 with CH_4. Nevertheless, the surface-diffusion coefficient of precursors seems to be smaller in the case of a-SiC:H growing surface as compared to the case of pure a-Si:H growing surface. Therefore, the apparent reduction of the surface diffusion coefficient of radicals must be caused by the characteristic reaction behavior of CH_3 radicals coexisting with SiH_3 on the film-growing surface.

Figure 4 shows the typical substrate-temperature dependence of C and Si partial deposition rate calculated by the proportional allotment of the total deposition rate using the composition ratio determined by Auger electron spectroscopy. As shown in the figure, the Si partial deposition rate is almost independent of substrate temperature, while the C partial deposition rate reveals a strong temperature dependence. This strong temperature de-

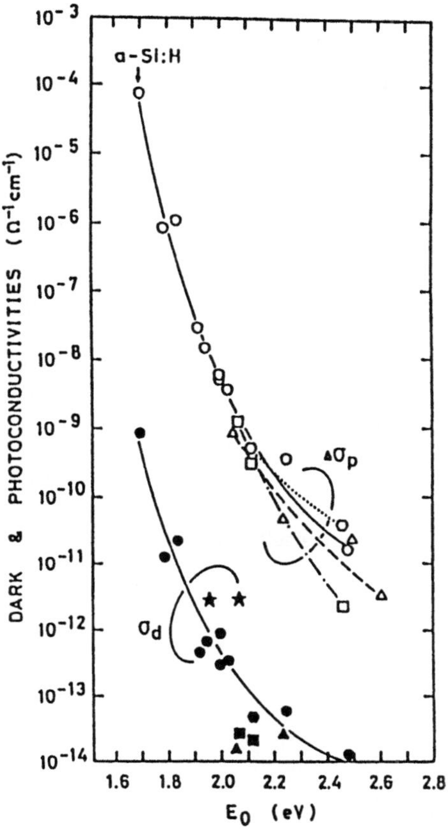

Fig.3. Photo (open symbols) and dark (solid symbols) conductivities of a-SiC:H films deposited from a conventional glow-discharge decomposition of SiH_4 and a variety of hydrocarbons (circles for CH_4, squares for C_2H_4, triangles for C_2H_2 and hexagons for CH_3SiH_3) plotted against their optical gaps (E_0).

pendence of C is commonly observed in the C deposition from a glow-discharge method as well as a photochemical vapor deposition method, which has been explained by the concept of "etching" of C atoms by the atomic hydrogens reaching the film-growing surface. However, the etching process during the growth of the C film can be ruled out by the following experimental facts. Namely, (1) a similar temperature dependence is observed even when the C film is deposited using a triode method in which the hydrogen atoms do not reach the film-growing surface, and (2) the deposition rate of C is not reduced even when the number of atomic hydrogens is increased intentionally.

In order to know the reaction behavior of CH_3 radicals on the a-SiC:H film-growing surface, the radical-separation method [3] was applied to the SiH_4+CH_4 glow-discharge plasma. Figure 5 shows the partial deposition rate of Si and C as a function of the distance L^2 from the mesh electrode for a variety of substrate temperatures. As shown in the figure, the slope of C-partial-deposition rate vs. L^2 is rather flatter for all substrate temperatures, while the slope of Si-partial-deposition rate is steeper in the temperature range below 200°C and it tends to be flatter above 300°C. According to the discussion mentioned above in the case of the deposition of a-Si:H, a flatter slope indicates the monomolecular reactivity of the radicals on the film-growing surface and a steeper slope means the bimolecular reactivity [6]. Therefore, the CH_3 radicals reveal the monomolecular reactivity on the surface in all temperature ranges, while the SiH_3 radicals show the bimo-

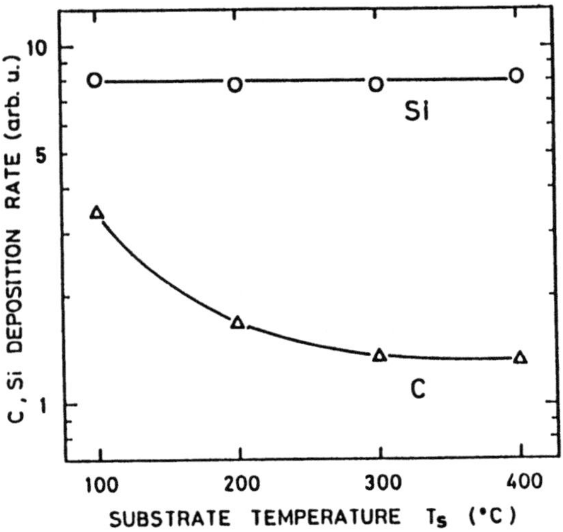

Fig.4. Partial deposition rate of Si and C as a function of the substrate temperature during the growth of a-SiC:H films.

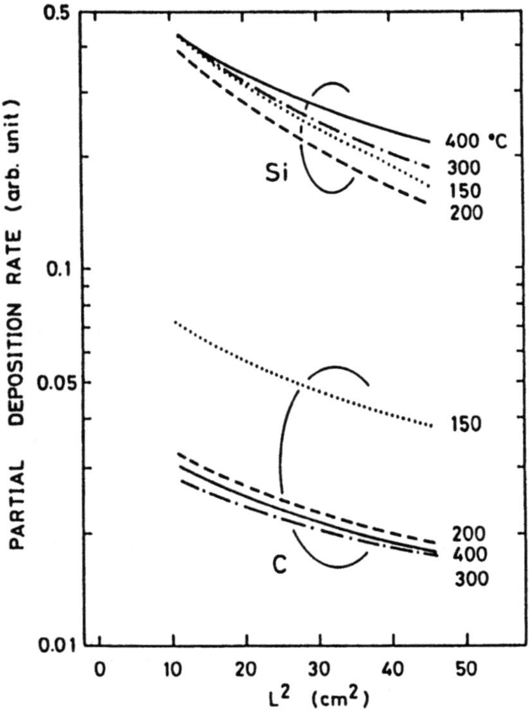

Fig.5. Results of the radical-separation method in the deposition of a-SiC:H. Partial deposition rate is plotted against the distance from the mesh electrode for various substrate temperatures.

lecular reactivity in the temperature range below 200°C and the change of its reactivity to the monomolecular type by elevating the temperature. It is indicated in the figure that the monomolecular reactivity starts to appear for the SiH_3 radicals from a lower temperature range in the presence of CH_3 radicals on the film-growing surface in comparison to the pure a-Si:H deposition condition as shown in Fig. 1.

The monomolecular reactivity of SiH_3 appears when the H-deficient sites are exposed on the film-growing surface as discussed in Fig. 1. Therefore, CH_3 radicals show indeed different reaction behavior on the surface from the SiH_3 radicals. Namely, CH_3 radicals do not show a bimolecular reactivity on the surface but do abstract the surface H, leading to a partially bald surface during the growth of the film. Thus formed H-deficient sites prevent the surface-diffusion of SiH_3 radicals on the surface, resulting in the less dense network structure of the deposited a-SiC:H.

4. Improvement of Photoelectric Properties of a-SiC:H

The reason for the less dense network structure of the conventionally prepared a-SiC:H is suggested from the results using the radical-separation method; that the CH_3 radicals abstract the H from the surface during the growth of the film and the appearance of H-deficient sites prevents the surface diffusion of SiH_3 radicals from finding their energetically favorable sites. In order to control the surface-H-coverage condition in the presence of CH_3 radicals, a sufficient amount of H atoms should be supplied from the gas phase (plasma). Namely, so-called "hydrogen-dilution method" might be effective to improve the surface-H-coverage condition. Figure 6 shows the photo and dark conductivity of a-SiC:H films plotted against their optical gaps for both the conventional method and the hydrogen-dilution method. Deposition conditions are listed in Table 1. As is clearly seen in the figure, the photoconductivity of a-SiC:H films prepared by the hydrogen-dilution method is drastically improved by five orders of magnitude in the film having the optical gap of 2.0eV as compared to those prepared by the conventional methods.

Table 1. Deposition conditions for the growth of a-SiC:H.

Method	Starting gas	Flow rate	Pressure	Power dens.	T_s
Conventional	CH_4/SiH_4 C_2H_4/SiH_4 CH_3SiH_3/SiH_4	5 SCCM	45-140 mTorr	0.03-2.1 W/cm^2	300°C
H_2 Dilution	$H_2/CH_4/SiH_4$ $H_2/C_2H_2/SiH_4$	6-201	200-800	0.13	300-350

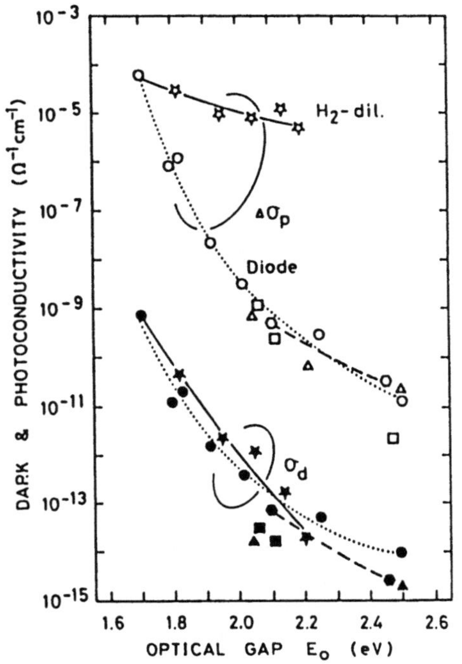

Fig.6. Photo and dark conductivities of a-SiC:H films prepared both by the conventional method (symbols are the same as in Fig.3) and the hydrogen-dilution method (stars).

5. Conclusion

It was suggested that the origin of the inferior photoelectric properties of a-SiC:H prepared by the conventional method lie in the fact that the surface covered hydrogens are abstracted by CH_3 radicals, leaving H-deficient sites on the surface. Thus formed H-deficient sites prevent the surface diffusion of SiH_3 radicals, resulting in the growth of the film showing a defective network structure. To compensate the H-deficiency on the film-growing surface, supply of H atoms from the plasma using the hydrogen-dilution method was effective. Finally, it can be pointed out that the selection of the carbon source for producing radicals showing no abstraction of surface H is an important means of further improving the photoelectric properties of a-SiC:H.

References

1. A.Matsuda, T.Yamaoka, S.Wolff, M.Koyama, Y.Imanishi, H.Kataoka, H.Matsuura and K.Tanaka: J. Appl. Phys., 60 (1986) 4025.
2. K.L.Chopra in "Thin Film Phenomena" (McGraw-Hill, 1969) p.137.
3. A.Matsuda and K.Tanaka: J. Appl. Phys., 60 (1986) 2351.
4. A.Matsuda, K.Nomoto, Y.Takeuchi, A.Suzuki, A.Yuuki and J.Perrin: Surface Science 227 (1990) 50.
5. C.C.Tsai, J.C.Knights, G.Chang and B.Wacker: J. Appl. Phys., 59 (1986) 2998.
6. A.Matsuda and K.Tanaka: J. Non-Cryst. Solids, 97&98 (1987) 1367.

Amorphous Silicon-Carbon Alloy Prepared by the Controlled Plasma Magnetron Method

M. Nishikuni, M. Kameda, K. Ninomiya, S. Okamato, T. Takahama, Y. Hishikawa, S. Tsuda, M. Ohnishi, S. Nakano, and Y. Kuwano

Functional Materials Research Center, SANYO Electric Co., Ltd., 1-18-13 Hashiridani, Hirakata, Osaka 573, Japan

Abstract. This paper describes our investigation into high efficiency a-Si solar cells using a-SiC p-layers and a unique a-SiC fabricated by the CPM method. To improve conversion efficicency, $B(CH_3)_3$ is used as a p-type doping gas and a module efficiency of 10.2% has been obtained for a 10cm×10cm integrated-type a-Si solar cell submodule. For further improvement in conversion efficiency, a-SiC has been investigated using the CPM method. The obtained films have many Si-C bonds and low Si-H bond density. P-type a-SiC was fabricated using the post-doping technique, and it was found that this a-SiC has a low absorption coefficient and a conductivity higher than $10^{-5}(\Omega\ cm)^{-1}$. These properties are excellent for a window layer of solar cells and this p-type a-SiC has been found to be effective for a-Si solar cells fabricated at a high temperature.

1. Introduction

a-SiC alloy is a basic material for p-layers of a-Si solar cells. We have, therefore, been investigating high quality a-SiC films. Consequently, we have achieved a high module efficiency. For further improvement in conversion efficiency, a more transparent window layer with high conductivity is required. To achieve this, a unique a-SiC has been investigated using the CPM method[1].

2. Present status of high efficiency a-Si solar cells

Figure 1 shows the illuminated I-V characteristics of an integrated type a-Si solar cell submodule (10cm×10cm) whose p-layer was doped with $B(CH_3)_3$. In addition, a high quality buffer layer prepared using the super chamber method was adopted at the p/i interface and highly textured TCO(Transparent Conductive Oxide) was used[2]. A total area efficiency of 10.2% has been obtained, representing the highest value in the world for $100cm^2$ a-Si solar cells.

3. Unique a-SiC using the CPM method to achieve further improvements in efficiency

3-1. Experimental

To achieve further improvements in conversion efficiency of a-Si solar cells, one of the key factor is the development of new high-quality window layers. Therefore, unique a-SiC films are prepared using the CPM method[1]. The CPM method is based on the capacitively-coupled RF plasma CVD method with parallel-type electrodes, and a electromagnet is placed under an electrode opposite the substrate. A magnetic field confines the plasma, resulting in a high electron density. The material gases are SiH_4 and CH_4 (diluted 10% in H_2) and the p-type doping gas is B_2H_6 (diluted 0.1% in H_2). The gas ratio of silane to methane is 0.1 and magnetic fields are 0 to about 100 Gauss (Magnetron current : Img=2A). The optical emission intensity of carbon-hydride in the plasma(Img=2A) increases to about 1 order of magnitude compared with that of 0A.

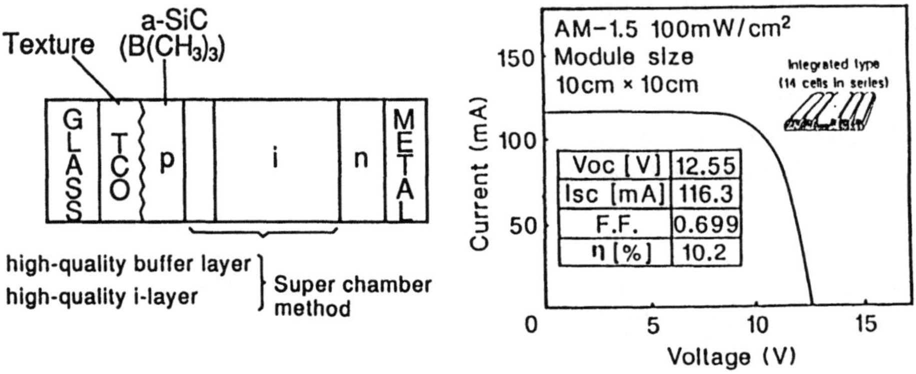

Fig.1 Illuminated I–V characteristics of an a–Si solar cell submodule

The substrates are glass (Corning 7059), c–Si(>1000 Ω cm) and SnO_2/glass. Both transmittance (T) and reflectance (R) of a–SiC on glass were measured, and the absorption coefficient (α) was calculated by the T/(1–R) method[3]. Hydrogen content and carbon content in the films were estimated from IR and XPS (X–ray Photoelectron Spectroscopy) measurements, respectively. Conductivity was determined using a co–planer Al electrode.

3–2 Non–doped films

Figure 2 shows the α of a–SiC films prepared at 400℃. It is found that the α decreases as the magnetic fields increase, and the α of the a–SiC film obtained with a magnetron current of 2A is about 1 order of magnitude lower than that of the conventional a–SiC. To investigate the origin of low α, the composition of the a–SiC was measured. Hydrogen content, estimated from IR measurements, decreases to a level 1/4 below that of conventional a–SiC. Carbon content, bonded in Si–C, is found to drastically increase as the magnetic field increased from the XPS spectra. Thus the low absorption coefficient of this a–SiC is thought to be caused by the large number of Si–C bonds. These results indicate that films which are nearly stoichiometric can be easily produced by using the CPM method.

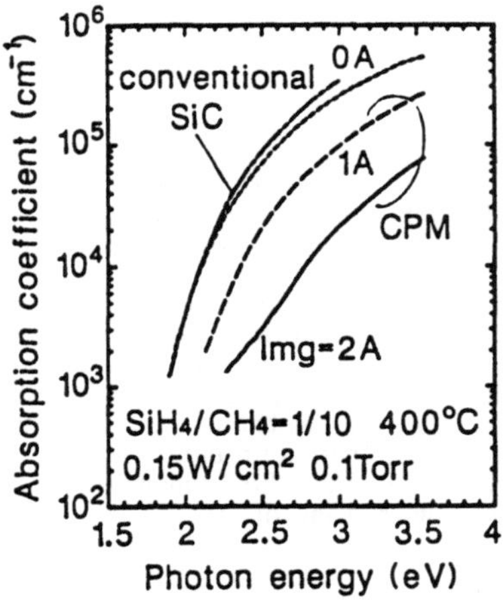

Fig.2 Absorption coefficient of a–SiC prepared using the CPM method

3-3 Boron doped films

Both an in-situ doping method, which is a conventional method for a-Si or a-SiC fabrication, and a post doping technique were used to prepare p-type a-SiC films. In a post-doping technique, non-doped a-SiC films were exposed to the B_2H_6 plasma at 400℃ to obtain p-type a-SiC films. Figure 3 shows the boron (B) depth profile in p-type a-SiC by SIMS. B concentration near the surface in post-doped films is high and effective thickness acting as a p-layer is thought to be about 200Å. This thickness is almost equal to that of the p-layer of a-Si solar cells. Furthermore, dark conductivity greater than $10^{-5}(\Omega\,cm)^{-1}$ is obtained for a post-doped films, which is enough for a p-layer (conventional:~$10^{-7}(\Omega\,cm)^{-1}$).

The α of the boron-doped a-SiC was measured. Although the α of p-type a-SiC fabricated by the CPM method increases with boron doping, it remains much lower compared with that of conventional p-type a-SiC, especially in the case of the post-doped films. These results indicate that the CPM method is effective for the preparation of p-type a-SiC with a high conductivity and a low absorption coefficient.

3-4. Application to solar cells

In order to use a-Si solar cells in power generating systems, not only high efficiency but also high reliability is important. Light-induced degradation in a-Si films is one of the most serious problems facing the reliability of a-Si solar cells. As reported elsewhere[4], high temperature deposition(≧300℃) has been found to be effective in reducing this problem. In addition, a high temperature deposited film has high α, and therefore, the conversion efficiency of a-Si solar cells can be improved. However, applying this method to the i-layer of a-Si solar cells is difficult, because the properties of the conventional p-SiC are easily changed during high temperature deposition of the i-layer.

These changes are thought to be caused by the large amount of hydrogen included in the films due to low temperature deposition(~200℃). The hydrogen content can be reduced by high temperature deposition, but this raises another problem, namely, impurity diffusion from the TCO.

A new-type of p-SiC, fabricated using the CPM method, however, has low hydrogen content and a large number of Si-C bonds. This indicates that these films show high stability against thermal annealing. Furthermore, impurity diffusion from the TCO is found to be negligible even at a high

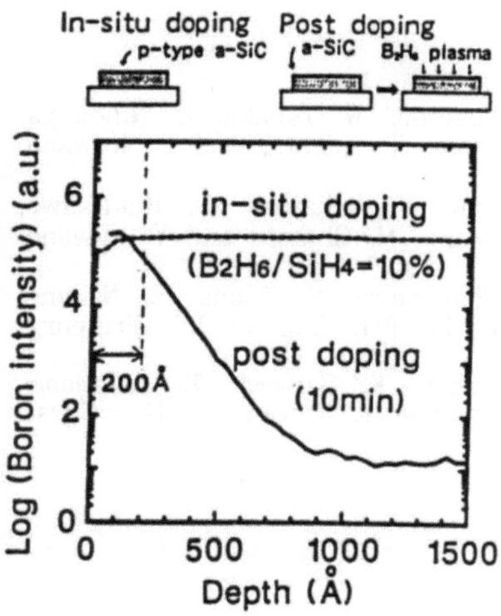

Fig.3 Boron profile in p-SiC by SIMS

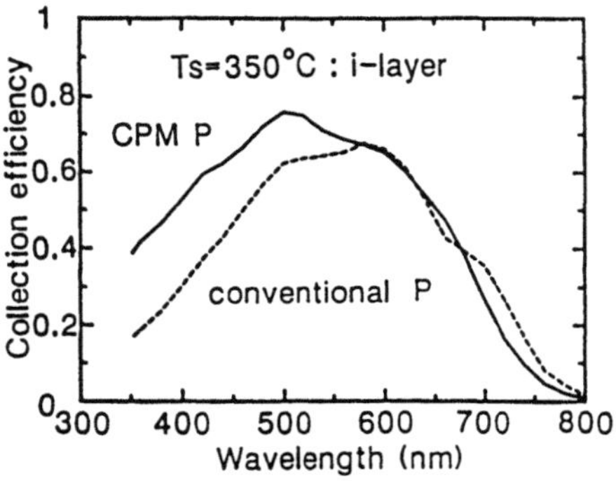

Fig.4 Collection efficiency of a-Si solar cells fabricated at 350℃

temperature(400℃) as suggested by the depth profile of AES. These p-SiC are, therefore, thought to be suitable for a-Si solar cells prepared at a high temperature.

Figure 4 shows the collection efficiency of a-Si solar cells, whose i-layers were fabricated at a high temperature(350℃). The collection efficiency in the short wavelength region was improved through using the a-SiC fabricated by the CPM method. This indicates that the a-SiC fabricated by the CPM method is stable for thermal treatment and highly reliable a-Si solar cells can be expected.

4. Conclusion

a-Si solar cell submodule with world's highest efficiency of 10.2% have been produced using high quality a-SiC doped with $B(CH_3)_3$.

Carbon-rich a-SiC with a high transparency and high conductivity can be obtained using the CPM method. This a-SiC film contains a large number of Si-C bonds compared with conventional a-SiC. Furthermore, this a-SiC is effective in improving the efficiency of high temperature deposited a-Si solar cells.

Acknowledgement

This work is supported by NEDO (New Energy and Industrial Technology Development Organization) as a part of the Sunshine Project under the Ministry of International Trade and Industry.

References

[1] M. Ohnishi, H. Nishiwaki, K. Uchihashi, K. Yoshida, M. Tanaka, K. Ninomiya, M. Nishikuni, N. Nakamura, S. Tsuda, S. Nakano, T. Yazaki, and Y. Kuwano, Jpn. J. Appl. Phys., 27(1), 40 (1988)

[2] S. Tsuda, T. Takahama, M. Isomura, H. Tarui, Y. Nakajima, Y. Hishikawa, N. Nakamura, T. Matuoka, H. Nishiwaki, S. Nakano, M. Ohnishi and Y. Kuwano, Jpn. J. Appl. Phys., 26(1), 33 (1987)

[3] Y. Hishikawa, S. Okamoto, K. Wakisaka, N. Nakamura, S. Tsuda, S. Nakano, M. Ohnishi, and Y. Kuwano, Proceedings of the 9th E.C. PVSEC, Freiburg, 37 (1989)

[4] M. Nishikuni, H. Haku, M. Isomura, S. Okamoto, K. Sayama, T. Takahama, T. Matsuoka, S. Tsuda, S. Nakano, M. Ohnishi, and Y. Kuwano, J. Non-Cryst. Solids, 114, 181 (1989)

Study of Substrate Doping and Amorphous SiC:H/Crystalline Si Interface with a MIS Structure

T. Harjono[1], *K.H. Lui*[1], *F.E. Pagaduan*[1], *H. Inokawa*[2], *M.M. Rahman*[1], *C.Y. Yang*[1], *and D. Sugiarto*[3]

[1]Microelectronics Laboratory, Santa Clara University, Santa Clara, CA 95053, USA
[2]On leave from NTT Labs., Atsugi-shi, Kanagawa 243-01, Japan
[3]Greyhawk Systems, Inc., Milpitas, CA 95035, USA

Abstract. The performance and reliability of silicon-based heterostructure devices depend critically on their interfacial characteristics. We report here the results from a study of substrate doping and interface electrical properties obtained with C–V measurements on an Al/undoped a–SiC:H/p–Si metal–insulator–semiconductor (MIS) structure. A density of interface traps D_{it} of 10^{11} eV^{-1} cm^{-2} at midgap is obtained. The substrate doping profile less than an extrinsic Debye length from the interface is computed from an iterative fit to the C–V data.

1. Introduction

Recent advances in the field of heterostructure devices involving silicon carbide and other group IV–IV materials with silicon have demonstrated considerable potential in their applications to high–speed, high–operating–temperature transistors [1]. For an amorphous film such as hydrogenated silicon carbon alloy (a–SiC:H), there are added benefits. Aside from utilizing relatively simple fabrication techniques and depositing at low temperatures, one can perform bandgap engineering easily with a–SiC:H. Deposition of device–quality amorphous thin films exhibiting acceptable electronic properties is achievable with existing process technology [2]. However, in applications to heterostructures with crystalline Si, the properties of the a–SiC:H/c–Si interface are of critical importance in determining device performance and reliability. In previous investigations [3] we have studied the morphology of the a–SiC:H/c–Si interface. It is equally important to evaluate other properties of the heterostructure such as the density of interface traps and substrate doping concentration near the interface. The density of interface traps is related to the surface recombination velocity and consequently affects the carrier transport in the heterostructure devices. In this paper we present results on the density of interface traps D_{it} and the substrate doping profile near the a–SiC:H/c–Si interface using an MIS capacitor as the test structure.

2. Fabrication of MIS Structure

MIS capacitors were fabricated by depositing a–SiC:H thin films on p–type Si (100) substrates having resistivity ranging from 1 to 3 ohm–cm in a capacitively–coupled RF PECVD system. Source gases were SiH_4 and CH_4. Other deposition conditions and film properties were: gas flow rate ratio $CH_4/(CH_4 + SiH_4) = 0.6$, pressure = 0.7 Torr, RF power = 0.115 W/cm^2, substrate temperature = 320°C, optical bandgap = $1.97eV$, film thickness $= 0.43\mu m$, dark conductivity = 5 10^{-14} S/cm, photoconductivity = 6×10^{-10} S/cm, measured with a He–Ne laser light source (λ = 630 nm) and at an intensity of 3 mW/cm^2. Circular Al dots, $0.3\mu m$ thick and 0.46 cm in diameter were

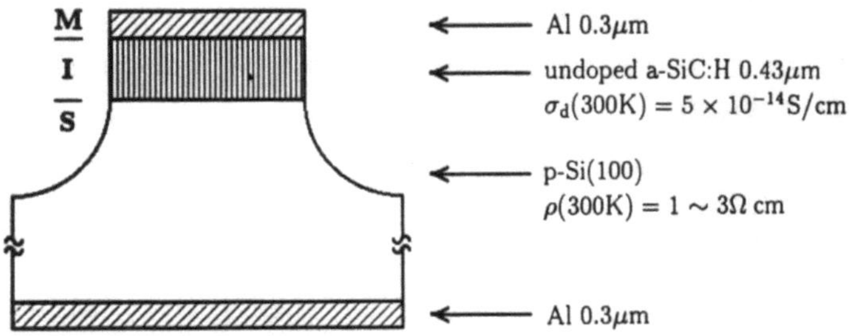

Figure 1. Schematic of the mesa–structure Al/undoped a–SiC:H/p–Si MIS capacitor.

deposited by vacuum evaporation method on to the a–SiC:H surface through a hard mask, followed by deposition of Al as the substrate contact. The specimens were then plasma–etched with CF_4 to yield the mesa–structured MIS capacitors. Figure 1 shows the cross section of the MIS capacitor fabricated. 100–kHz C–V measurements of the MIS structure were performed at room temperature using an LCR meter.

3. Doping Profile and D_{it} from Capacitance Measurements

In order to measure the effect of the plasma deposition near the substrate surface a doping profile up to the interface must be obtained. Conventional doping profiling technique [4] from C–V data based on the depletion approximation is not capable of determining accurately substrate doping concentration near the interface. Using an MIS structure, the conventional technique is acceptable for distances farther than a few extrinsic Debye lengths (λ_D) from the semiconductor surface. Toyabe et al. [5] have demonstrated that using an iterative technique based on measured and calculated C–V data, substrate doping concentration less than $2\lambda_D$ from the interface can be extracted.

Prior to C–V measurements, I–V characteristics of the MIS capacitors were measured to determine the leakage currents through the a–SiC:H film. Typical dark leakage current densities of about $60\mu A/cm^2$ and $10\mu A/cm^2$ were observed for applied biases of +10V and -10V, respectively. Figure 2 shows a typical C–V curve. From the capacitance in accumulation, the electrode area, and the thickness of the a–SiC:H film, dielectric constant of the a–SiC:H film was determined to be approximately 8. Using Toyabe's "C–V–fit" and conventional C–V profiling techniques, and assuming zero interface trap density, a substrate doping profile was obtained for distances as close as 200Å from the interface. This profile is shown in Figure 3. The apparent depletion of dopants near the interface is probably due to radiation–enhanced diffusion during the deposition process and/or dopant compensation due to defects in the substrate surface created by the plasma. This is consistent with the relatively high magnitude of leakage current through the undoped a–SiC:H film.

We have utilized the high–frequency capacitance method developed by Terman [6] to extract the density of interface traps of the a–SiC:H/c–Si system. Assuming a uniform doping concentration of $5 \times 10^{15}/cm^3$ throughout the substrate (Figure 3), an ideal MIS capacitance versus surface potential behavior was computed. Combining this

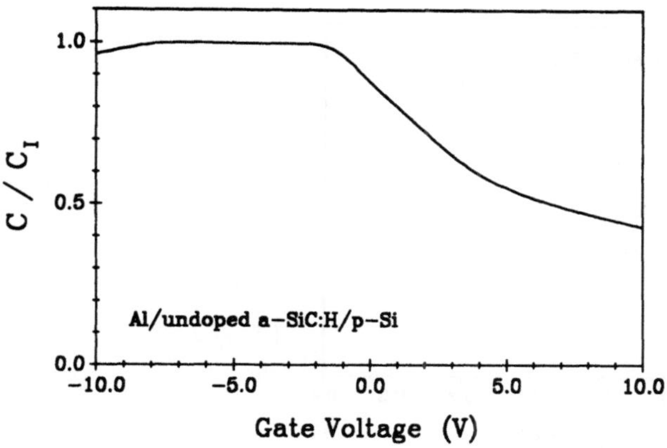

Figure 2. Capacitance–voltage characteristics of an Al/undoped a–SiC:H/ p–Si MIS capacitor. The capacitance is normalized to the insulator capacitance.

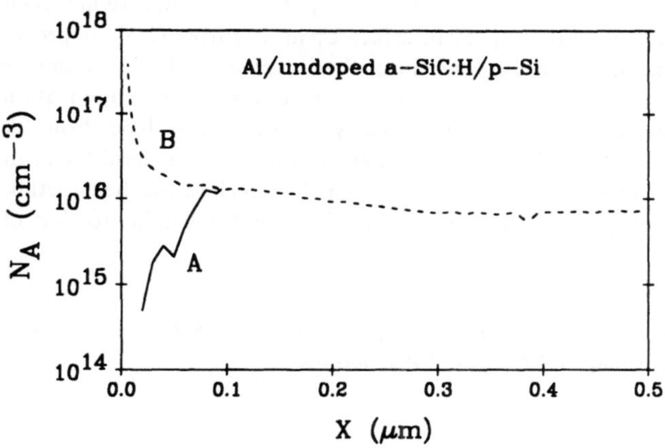

Figure 3. The substrate doping profiles of the MIS structure. Curve A shows the doping profile obtained for distances within twice the extrinsic Debye lengths using Toyabe's C–V–fit method. Curve B is obtained using the conventional technique.

with the measured C–V data shown in Figure 2, the HF capacitance method yields the density of interface traps versus electron energy curve as shown in Figure 4. The D_{it} curve in Figure 4 shows a value of approximately 10^{11} eV^{-1} cm^{-2} at midgap. These D_{it} curves reveal a trend similar to that obtained for the SiO_2/Si interface [7].

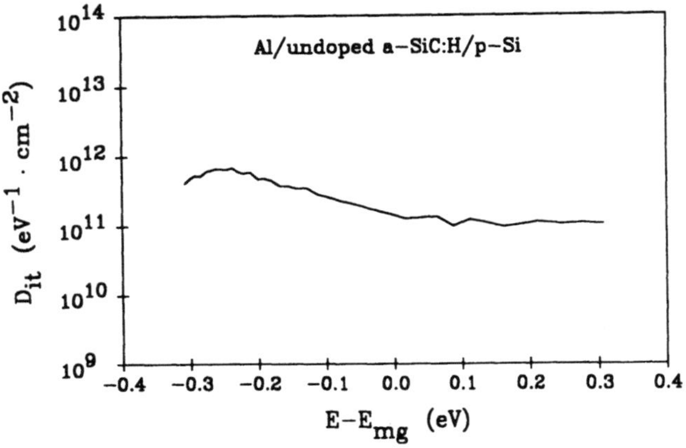

Figure 4. The density of interface traps for a-SiC:H/p-Si as a function of electron energy in the Si bandgap.

4. Summary and Conclusions

The substrate doping profile in an Al/ undoped a-SiC:H/p-Si MIS structure has been determined within 200 Åof the a-SiC:H/p-Si interface by an iterative C-V-fit procedure. Depletion of dopants has been observed near the interface. This depletion is attributed to radiation-induced out-diffusion of dopant species and/or compensation of dopants by plasma-induced defects during the early stages of film deposition. A midgap D_{it} of approximately 10^{11} eV^{-1} cm^{-2} has been obtained. A general trend in the density of interface traps behavior for the a-SiC:H/p-Si interface has been established using the high-frequency capacitance method. This trend is similar to that of the SiO_2/Si interface.

Acknowledgments. The authors thank Dr. C. C. Tsai of Xerox- PARC for assistance in plasma-etching of the samples and for useful discussions.

References

1. K. Sasaki, S. Furukawa, and M. M. Rahman, IEDM Tech. Dig. (1985) 294.
2. Amorphous Silicon Technology-1989, edited by A. Madan, M.J. Thompson, P. C. Taylor, Y. Hamakawa, and P.G. LeComber, (Mater. Res. Soc. Proc. 149, Pittsburgh, PA 1989).
3. M. M. Rahman, C. Y. Yang, D. Sugiarto, A. S. Byrne, M. Ju, K. Tran, K. H. Lui, T. Asano, and W. F. Stickle, J. Appl. Phys. 67 (1990) 7065.
4. S. M. Sze, Physics of Semiconductor Devices, 2nd ed. (John Wiley, New York, 1981), pp. 371-379.
5. T. Toyabe, H. Matsuo, S. Yamamoto, and H. Masuda, IEDM Tech. Dig. (1989) 699.
6. L. M. Terman, Solid-State Electron. 5 (1962) 285.
7. M. H. White and J. R. Cricchi, IEEE Trans. Electron Devices ED-19 (1972) 1280.

Effects of Operating Temperatures on Optical and Electrical Properties of Amorphous SiC Deposited by Photo-CVD

J.S. Park[1], *Y.S. Kim*[1], *C.H. Oh*[1], *G.Y. Kim*[2], *W.Y. Kim*[2], *and M.K. Han*[1]

[1]Department of Electrical Engineering, College of Engineering, Seoul National University, Kwanak-Ku Shinrim-Dong, Seoul 151-742, Korea

[2]Central Research Laboratory, Gold Star Co., LTD., Seoul, Korea

Abstract. We report the experimental results regarding the effects of operating temperatures as well as deposition conditions on optical and electrical properties of boron-doped p-type a-SiC films prepared by photo-CVD. The a-SiC films, of which optical bandgaps and photoconductivities exceed 2.0 eV and 10^{-5} S/cm, have been successfully deposited by optimizing the gas mixing ratios. Optical bandgaps and conductivities have been measured at various temperatures from room temperature to about $70^{o}C$. Temperature coefficient of optical bandgaps for p-type a-SiC films is found to be $8\sim9\times10^{-4}$ eV/K.

1. Introduction

Boron doped p-type amorphous silicon carbide (a-SiC) thin film has attracted considerable attention due to its application in various opto-electronic devices such as amorphous silicon (a-Si) solar cells, image sensors, and hetero-junction bipolar transistors.[1-3] A wide optical bandgap with a highly conductive material is essential to enhance the spectral response of the ultraviolet (UV) regime in a-Si devices. Since solar cells and image sensors operate at elevated temperature up to $70^{o}C$ due to illumination, it may be important to investigate the operating temperature effects on optical absorption coefficients and optical bandgaps of a-SiC films quantitatively. The optical and electrical properties of a-SiC films regarding deposition environments such as the choice of gases, gas mixing ratios and deposition temperatures have been intensively investigated. However, most of the previously reported experimental results have been concerned with room temperature and the optical and electrical properties of a-SiC thin films at practical device operating temperatures should be investigated.

The purpose of this work is to report the effects of various operating temperatures on the optical properties such as optical absorption coefficients and bandgaps of boron doped wide bandgap a-SiC films prepared by a photo-assisted chemical vapor deposition (photo-CVD) method without mercury sensitization. Conductivities and activation energies of various p-type a-SiC films are reported.

2. a-SiC Film Deposition

The a-SiC films used in this work were prepared by a low temperature (200 ~ 250°C) photo-CVD rather than plasma-enhanced CVD (PECVD). The advantages of photo-CVD over widely used PECVD may be the reduction of defect states due to the negligible ion damage so that a considerable improvement of conductivities may be achieved. In order to improve the film quality, the widely used mercury sensitization has not been employed. Of course, it should be noted that the deposition rate may be substantially decreased without it. The deposition conditions in this work used to prepare a-SiC films by photo-CVD are summarized in Table 1. The Al and Cr electrodes used to measure the electrical conductivities of a-SiC films were deposited by electron-beam evaporator.

Table 1. Deposition conditions of a-SiC films prepared by photo-CVD.

base and operating pressure[Torr]	$1\sim2\times10^{-6}$, $0.5\sim1$
substrate temperature[°C]	200~250
reactant gas flow rate[sccm]	Si_2H_6=5~10 C_2H_2=B_2H_6=5~30
light intensity[mW/cm^2]	5 at 185 nm, 15 at 254 nm

3. Experimental Results and Discussion

The important material properties of a-SiC films are conductivity and optical bandgap. The optical absorption coefficients, the optical bandgaps and the conductivities were measured systematically as a function of temperatures and mixing ratios of reactant gases such as C_2H_2, B_2H_6, and Si_2H_6, in order to investigate the effects of various operating temperatures as well as deposition conditions on the properties of a-SiC films. As shown in Fig. 1, the increase of C_2H_2 gas in the mixture causes a substantial increase of the optical bandgap, E_{opt}, from 1.72 to 2.2 eV and a large decrease of the photoconductivity, σ_{ph}, from 10^{-5} to 10^{-9} (S/cm).

Fig. 2 shows the changes in optical bandgaps and conductivities with varying volume ratios of $C_2H_2/Si_2H_6 = B_2H_6/Si_2H_6$ from 1 to 5. At lower gas mixing ratios below about 2, the optical bandgaps exhibited low values of 1.4 ~ 1.5 eV and the conductivities increased with increasing mixing ratios. When the mixing ratios exceeded 2, however, the optical bandgaps increased rapidly to about 2.0 ~ 2.1 eV, but the conductivities decreased. This result means that the dominant atoms determining the properties of a-SiC films may be boron rather than carbon at lower gas mixing ratios, while carbon rather than boron plays a dominant role to dictate the properties of a-SiC films prepared at higher ratios. It may be suggested that the volume ratio of mixing gases may be in the range of 4 ~ 5 in order to obtain the p-type a-SiC films with higher σ_{ph} and larger E_{opt}.

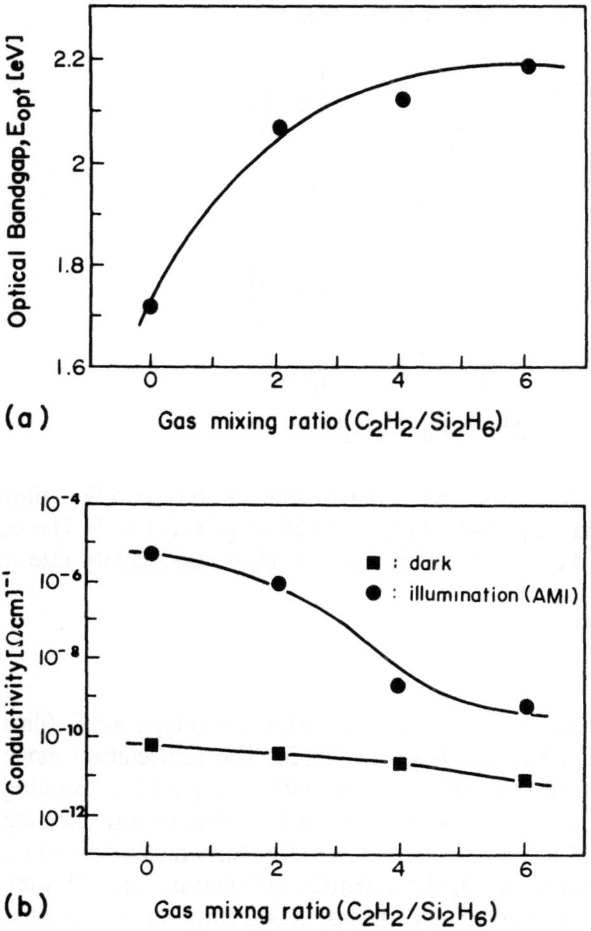

Fig. 1 Changes in optical bandgaps (a) and conductivities (b) of a-SiC films deposited by photo-CVD with varying volume ratios of C_2H_2/Si_2H_6 from 0 to 6. The deposition temperature was 250°C.

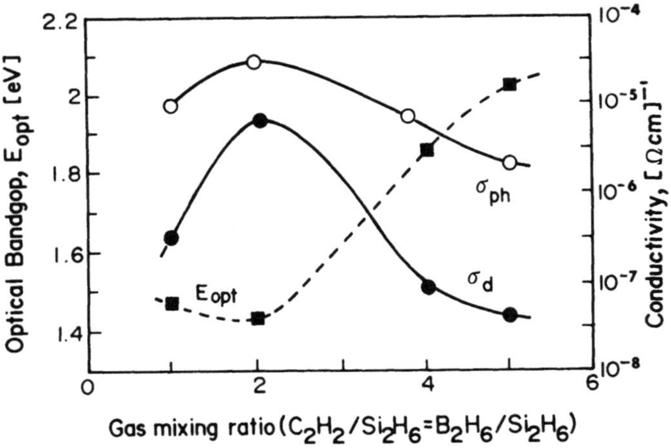

Fig. 2 Variations in optical bandgaps and conductivities of p-type a-SiC films deposited at 250^oC changing $C_2H_2/Si_2H_6 = B_2H_6/Si_2H_6$ from 1 to 5. The σ_d and σ_{ph} indicate the dark conductivitiy and the photoconductivity due to illumination at AM 1

The temperature dependences of dark conductivities for p-type a-SiC films deposited at 250^oC by photo-CVD are shown in Fig. 3. The temperature range has changed from room temperature to 70^oC, which may be the practical operating temperatures in a-Si solar cells. It is well known that the activation energy dictating the position of the Fermi level has been obtained from the Arrhenius model of $\sigma_d = \sigma_o \exp(-E_a/kT)$. As shown in Fig. 3, the linearities between σ_d and 1/T were well obtained and the activation energies of p-type a-SiC films were estimated from the slopes. As summarized in Table 2, p-type a-SiC films exhibited the lowest activation energy when the volume gas mixing ratio was about 2. These results may support those of Fig. 2, where the largest photoconductivities are obtained for a-SiC films prepared at $C_2H_2/Si_2H_6 = B_2H_6/Si_2H_6 = 2$.

We have characterized the effects of operating temperature on the optical absorption coefficients and the optical bandgaps of p-type a-SiC films. The optical transmittances were measured at various elevated temperatures by inserting the specially designed heating pad into the sample holder of the spectrophotometer.

Table 2. Activation energies of p-type a-SiC as a function of gas mixing ratios.

$(C_2H_2, B_2H_6) / Si_2H_6$	1/1	2/1	4/1
$E_a = E_F - E_V$ [eV]	0.62	0.57	0.86

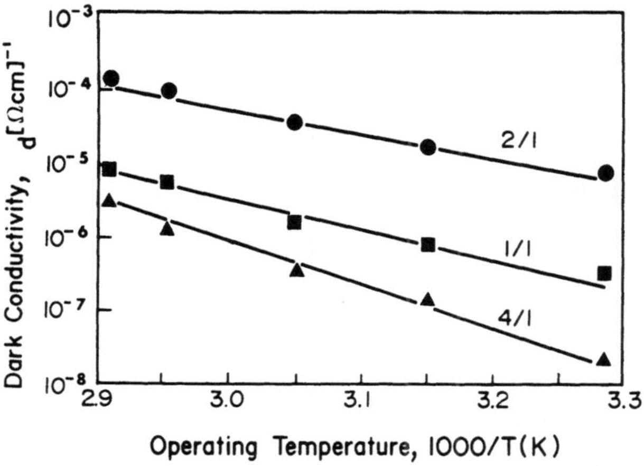

Fig. 3 Operating temperature dependence of dark conductivities for p-type a-SiC films deposited by photo-CVD. The temperature has been controlled from room temperature to about 70^oC. The fractional numbers indicate the volume ratios of mixing gases defined as $C_2H_2/Si_2H_6 = B_2H_6/Si_2H_6$.

Fig. 4 shows that the optical transmittances and optical bandgaps of p-type a-SiC films deposited by photo-CVD were measured as a function of operating temperature. As the operating temperature increased from room temperature to 70^oC, the transmittance decreased, which indicated an increase in the optical absorption coefficients, and the optical bandgaps also decreased.

The temperature coefficient A, defined as $E_{opt}(T) = E_{opt}(0) - A\times T$, of various p-type a-SiC films are summarized in Table 3. The temperature coefficients of the optical bandgaps for p-type a-SiC films were about $8\sim9\times10^{-4}$ eV/K, while that of crystalline silicon was reported to be about 2.8×10^{-4} eV/K.[4] In the case of a-Si:H films, we previously reported that the temperature coefficients of optical bandgaps were about $4\sim6\times10^{-4}$ eV/K.[5]

Table 3. Temperature coefficients of optical bandgaps for various p-type a-SiC films deposited by photo-CVD.

a-SiC samples	# SC-1	# SC-2	# SC-3	# SC-4	# SC-5
E_{opt}(RT)	1.84	2.02	2.02	2.07	2.19
A[$\times10^{-4}eV/K$]	8.85	8.46	8.90	9.60	8.62

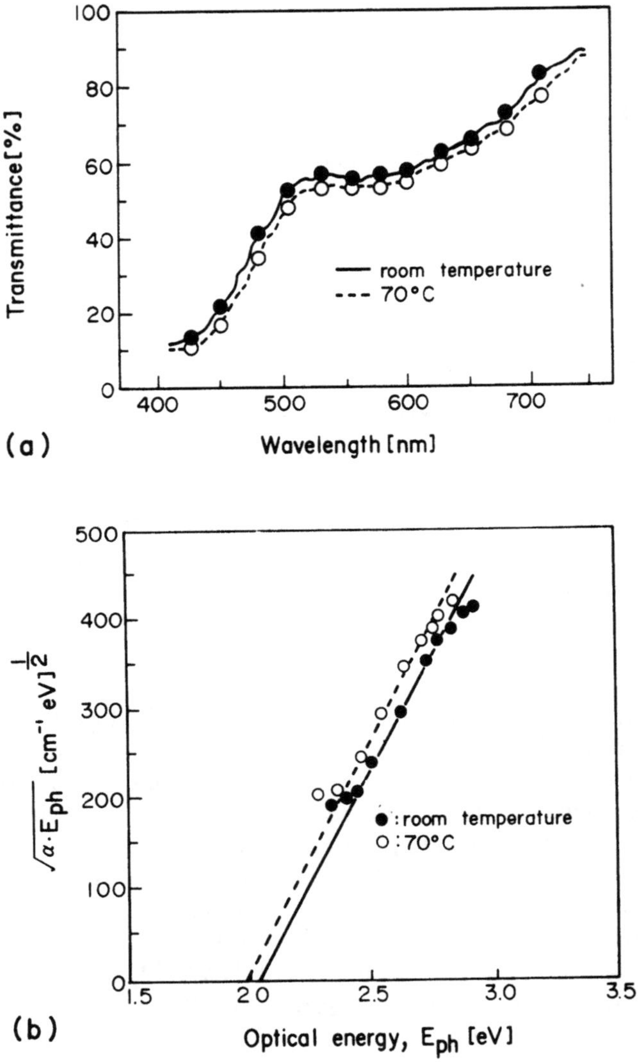

Fig. 4 Variations in optical transmittances (a) and bandgaps (b) of p-type a-SiC deposited at 250^oC with varying operating temperatures.

4. Conclusions

Boron doped p-type a-SiC films were deposited by low temperature photo-CVD utilizing the low-pressure mercury lamp without mercury sensitization. The optical bandgaps and the electrical conductivities were characterized quantitatively, changing the operating temperatures from room temperature to about $70^{o}C$. When the volume ratios of $C_2H_2/Si_2H_6 = B_2H_6/Si_2H_6$ were about 4 ~ 5, the larger optical bandgaps and the higher photoconductivities of p-type a-SiC films were obtained. This result was evidenced by analyzing the activation energies of p-type a-SiC films obtained from the temperature dependences of the dark conductivities.

The temperature coefficients of the optical bandgaps for p-type a-SiC films were about $8 \sim 9 \times 10^{-4}$ eV/K which were larger than that of crystalline silicon and that of boron doped a-Si films. Our experimental results regarding the effects of operating temperatures on optical and electrical properties of p-type a-SiC films may be useful to optimize and characterize the performances of a-Si solar cells and photodetectors.

Acknowledgement. This work has been supported by the Ministry of Energy and Resources, the Republic of Korea as the national photovoltaic project.

References

[1]. Y.Tawada, T.Yamaguchi, S.Nonomura, S.Hotta, H.Okamoto, and Y.Hamakawa : Jpn. J. Appl. Phys., 21, Suppl., 21-1, pp.297 (1982).

[2]. S.Morozumi, H.Kurihara, T.Takeshida, H.Oka, and K.Hasegawa : IEEE Electron Device, ED-32 (8), pp.1546 (1985).

[3]. K.Sasaki, S.Furukawa, and M.M.Rahman : 1985 Int. Electron Devices Meeting, Washington D.C., Tech. Dig., pp.294.

[4]. C.D.Thurmond : J. Electrochem. Soc., 122, pp.1133 (1975).

[5]. J.S.Park and M.K.Han : J. Appl. Phys., 65 (11), pp.4456 (1989).

Polycrystalline Silicon-Silicon Carbide Thin Films Produced by Plasma Enhanced CVD

Y. Onuma[1], *K. Kamimura*[1], *Y. Nagura*[1], *K. Koike*[1], and *S. Yonekubo*[2]

[1]Department of Engineering, Shinshu University, 500 Wakasato, Nagano 380, Japan
[2]Nagano Precision Industry Research Institute, 9959 Okaya-shi, Nagano394, Japan

In order to obtain stable and reliable thin films for electronic devices (sensor elements) which can be operated at high ambient temperature, silicon-silicon carbide thin films were produced by plasma enhanced CVD. Structural and electrical properties were investigated by X-ray diffraction, ESCA profiles and Hall effect measurement. It was found that by varying the ratio of CH_4/SiH_4, polycrystalline silicon-silicon carbide thin films with optical band gap from about 1.2 eV to 2.6 eV and high sensitivity to strain were obtained.

1. Introduction

Many studies have been done on polycrystalline silicon and silicon carbide thin films[1-3], because of many potential applications to electronic devices, such as sensors[4], thin film transistors[5], etc. It is expected that polycrystalline silicon-silicon carbide thin films have some advantages, that is, higher electron mobility than the films with single silicon carbide phase and higher temperature-resistant properties than the films with single silicon phase. In this paper, we deal with the preparation and characterization of silicon-silicon carbide thin films deposited by plasma enhanced CVD and the possibility of its application to a strain sensor.

2. Experiment

The apparatus used for preparation of polycrystalline thin films was a capacitively coupled plasma CVD system which is described elsewhere[1-2]. The mixture of gases SiH_4 (10%) and CH_4 (5%) diluted with H_2 was introduced into the stainless reaction chamber. The flow rate of CH_4/SiH_4 was varied from 2.5 to 10 [non dimension]. The substrate temperature was kept at 800 ℃ which was the optimum temperature for preparing polycrystalline silicon carbide thin films as reported in the reference[2]. Thermally oxidized silicon and vycor glass were used as insulating substrates. The pressure in the reactor was 1 Torr during deposition. The film thickness was about 0.5 μm.

3. Results and Discussions

X-ray diffraction analysis was used to investigate the structure of silicon-silicon carbide thin films. At a flow rate of CH_4/SiH_4 of 10 [non dimension] and with an input power of 80 watt, polycrystalline SiC thin films were formed as reported earlier[1]. Decreasing the CH_4 gases resulted in peaks corresponding to the silicon phase tending to appear among diffraction peaks and the diffraction peaks of silicon becoming larger and sharper. The grain sizes were evaluated from the half width of X-ray diffraction peaks.

Dependence of grain size on the flow rate of CH_4/SiH_4 is shown in Fig.1. These values are calculated at the diffraction angles 2θ of 28.4 (for SiC), 35.6 (for Si),

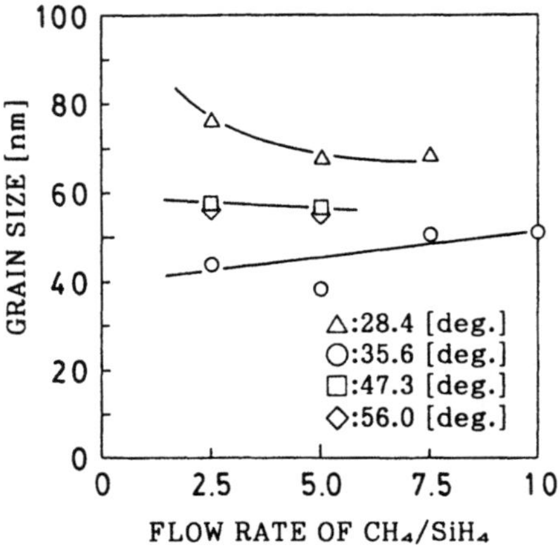

Fig.1. Dependence of grain size on the flow rate of CH_4/SiH_4.

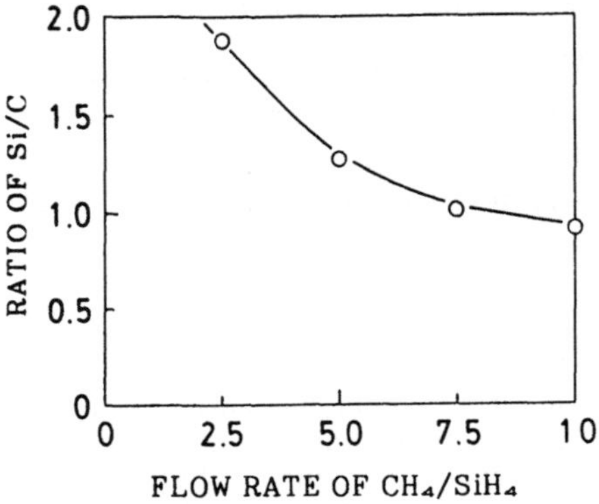

Fig.2 Dependence of composition ratio (Si/C) on the flow rate of CH_4/SiH_4.

47.3 (for Si) and 56.0 (for Si).

Dependence of composition ratio of Si/C in the films on the flow rate of CH_4/SiH_4 was examined by electron probe micro analysis , as shown in Fig.2. The composition ratio of Si/C was clearly dependent on the CH_4/SiH_4 ratio.

Stoichiometry (Si/C) in the films deposited at the flow rate of CH_4/SiH_4 of 10 [non dimension] was about 1:1.02, by Auger electron spectroscopy. Auger profiles of the silicon carbide surface thus obtained are shown in Fig.3 (a) and (b). In Fig.3 (a), silicon, carbon and oxygen were detected. However, after etching of the surface (depth of 500 Å), oxygen was not observed at all.

Chemical bondings for Si-SiC thin films were investigated by using ESCA spectrometer (Shimadzu ESCA-1000, target: Mg ($K\alpha$), accer. volt.: 10 kV, current: 10 mA).

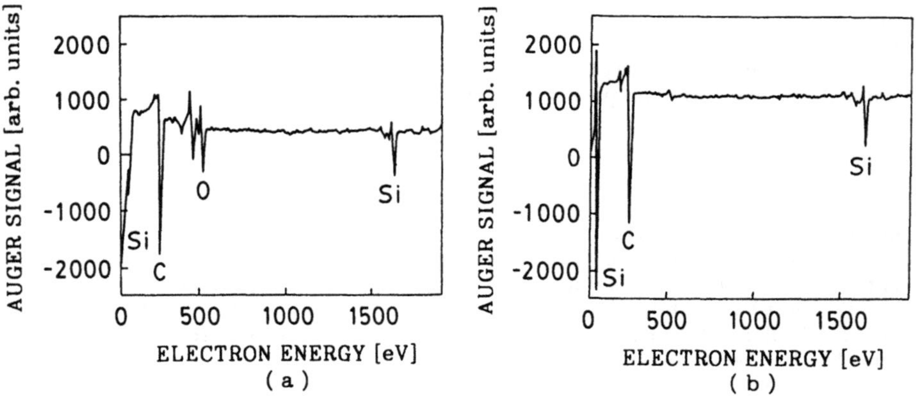

Fig.3. Auger electron spectroscopy of the SiC surface, as grown (a), sputter-etched (depth of 500 Å) (b).

Figure 4 shows the Si2p spectrum for the sample (flow rate of CH_4/SiH_4, 10 [non dimension]) sputter-etched (depth of 500 Å) by Ar ions. The spectrum was resolved into three peaks. These peaks at 100.3, 101.9 and 99.5 eV correspond to Si-C, Si-O and Si-Si bonds, respectively. The surface area calculated from the spectrum for peaks at 100.3 was about 8 times larger than the peak at 101.9.

Figure 5 shows the C1s spectrum for the same sample of Fig.4. The spectrum was resolved into two peaks. Broken line curves 1 and 2 correspond to the peaks at 283.2 eV and 284.8 eV, respectively. The secondary peak at 284.8 eV corresponds to C-O bonds.

Figure 6 (a) and (b) show the Si2p and C1s spectra, respectively, for the samples prepared at various flow rates of CH_4/SiH_4, together with the spectra for a reference α-SiC crystal and single crystal silicon. With regard to Si2p spectrum, a small amount of chemical shift was observed between the peak for α-SiC and that for single crystal silicon. The spectrum varied continuously from

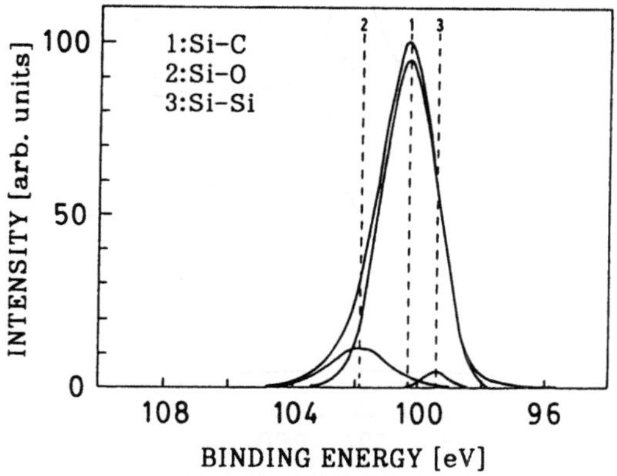

Fig.4. XPS profile of Si2p in silicon carbide prepared with the flow rate of CH_4/SiH_4 of 10 [non dimension].

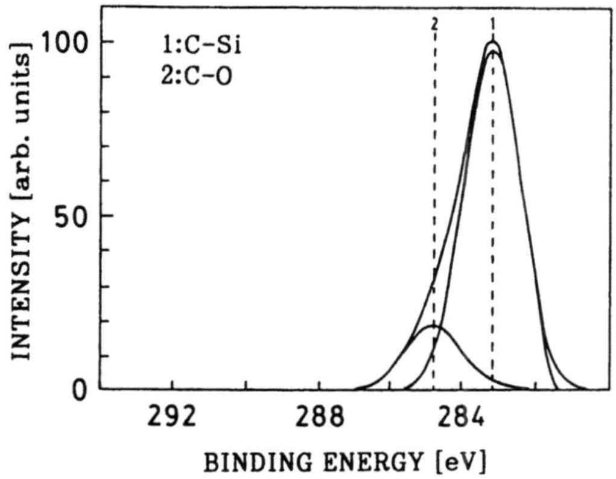

Fig.5. XPS profile of C1s in silicon carbide prepared with the flow rate of CH_4/SiH_4 of 10 [non dimension].

100.4 eV (corresponding to Si-C) to 99.1 eV (corresponding to Si-Si). It is thought that (SiC)-Si bonds were formed in the films.

The band gap was estimated from the photo-absorption measurement. On the $(\alpha\hbar\omega)^{1/2}$ versus $(\hbar\omega)$ characteristics, the linear portions were extrapolated to zero absorption to get the optical band gap[6]. Figure 7 shows the dependence of the optical band gap on the flow rate of CH_4/SiH_4. Optical band gaps were in the range from 1.2 eV to 2.6 eV depending on the flow rate of CH_4/SiH_4.

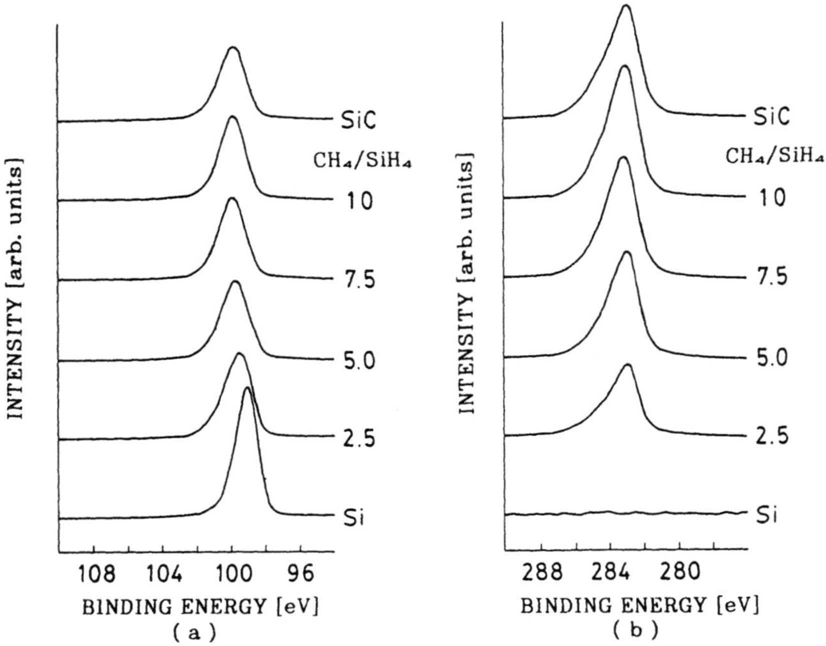

Fig.6. XPS profile of Si2p (a) and Cls (b).

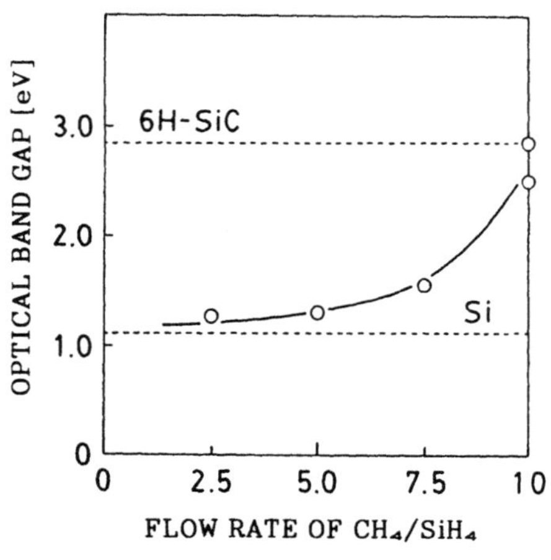

Fig.7. Dependence of optical band gaps on the flow rate of CH_4/SiH_4.

Figure 8 shows carrier concentration and hole mobility of the polycrystalline Si-SiC thin films. The hole mobility slightly increased with decreasing flow rate of CH_4/SiH_4 in the mixture gases. This is due to the existence of the silicon phase. Carrier concentrations decreased with decreasing flow rate of CH_4/SiH_4. The reason for this result is not clear, however, it is considered that grain growth reduced dangling bonds and defects around the grain boundary.

For application to sensors, such as pressure and strain sensitive elements, piezoresistive properties of polycrystalline Si-SiC thin films were measured. The electrical resistance changed linearly against strain, as shown in Fig.9. Sensitivity against strain, gauge factor (determined by $\Delta R/R\varepsilon$, where $\Delta R/R$ is fractional change in resistance and ε is strain), was about 22~33, as calculated from Fig.9. The largest values were obtained for the polycrystalline SiC-Si thin films with large grain size and high hole mobility. These values were higher than that of polycrystalline silicon carbide thin films, as reported earlier[2].

4. Summary

Polycrystalline Si-SiC thin films were prepared by varying the flow rate of CH_4/SiH_4. The optical band gap varied from 1.2 eV to 2.6 eV, depending on the flow rate of CH_4/SiH_4. The chemical shift was about 1.3 eV between the Si2p transition for polycrystalline silicon carbide and that for single crystal silicon. It is thought that a silicon-silicon carbide system was formed. The piezoresistive effect of polycrystalline silicon-silicon carbide thin films was measured. Gauge factor, sensitivity to strain, was 22~33 and was larger than that of homogeneous silicon carbide. It is expected that Si-SiC thin films will be useful materials for heat resistant devices, especially as piezoresistive sensor elements which are needed in the sensor technological field.

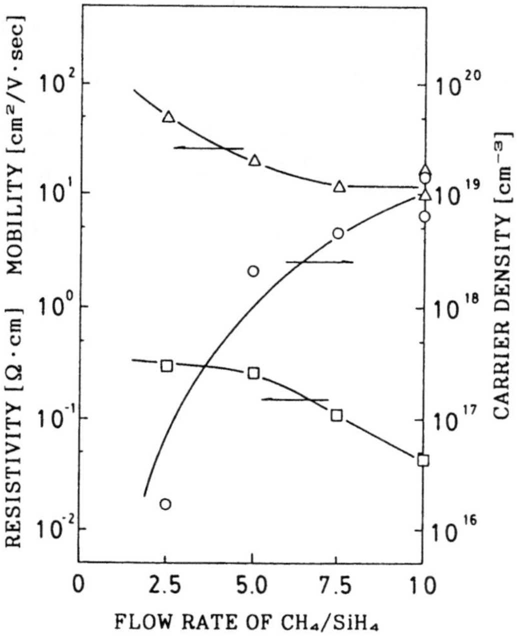

Fig.8. Electrical properties of polycrystalline Si-SiC thin films.

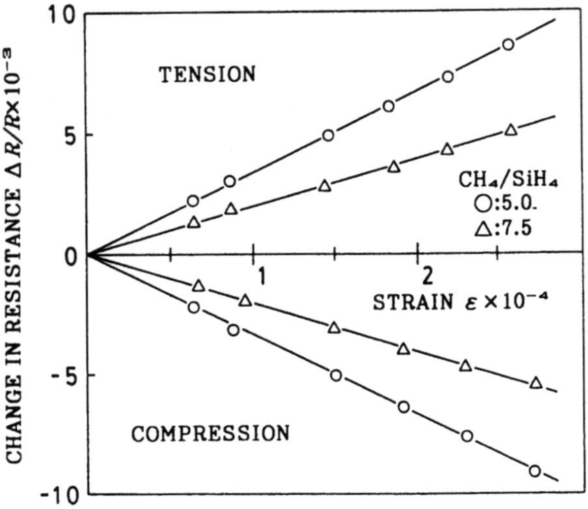

Fig.9. Fractional change in resistance versus strain.

References

1. Y.Onuma, K.Kamimura, F.Nagaune, H.Seki, M.Kiuchi and H.Y.Cai : *The Transactions of The Institute of E.E. of Japan*, 110-A, 101 (1990)
2. Y.Onuma, K.Kamimura, F.Nagaune, H.Seki and K.Tezuka : *The Transactions of The Institute of E.E. of Japan*, 109-A, 219 (1989)
3. Y.Onuma, F.Nagaune and K.Kamimura : in *Amorphous and Crystalline Silicon Carbide*, ed. by G.L.Harris and C.Y.Yang ; Springer Proceedings in Physics 34 (Springer-Verlag, Berlin, Heidelberg 1989) p.142
4. Y.Onuma, H.Seki, K.Kamimura, T.Homma and C.H.Yi : *Technical Digest of the 8th Sensor Symposium*, Japan, 41 (1989)
5. Y.Onuma, S.Miyashita, Y.Nishibe, K.Kamimura and K.Tezuka : in *Amorphous and Crystalline Silicon Carbide* II, ed. by M.M.Rahman, C.Y.Yang and G.L.Harris ; Springer Proceedings in Physics 43 (Springer-Verlag, Berlin, Heidelberg 1989) p.212
6. J.Tauc : in *Amorphous and Liquid Semiconductors*, ed. by J.Tauc (Plenum, New York 1974)

Modeling of Wide Bandgap Microcrystalline Silicon Carbide/Crystalline Silicon Heterojunctions

M.K. Han[1], *J.S. Park*[1], *Y. Mastumoto*[2], *H. Okamoto*[2], *and Y. Hamakawa*[2]

[1]Department of Electrical Engineering, College of Engineering, Seoul National University, Kwanak-Ku Shinrim-Dong, Seoul 151-742, Korea

[2]Department of Electrical Engineering, Faculty of Engineering Science, Osaka University, Toyonaka, Osaka 560, Japan

Abstract. The current-voltage characteristics of wide bandgap B doped microcrystalline silicon carbide/crystalline silicon heterojunction solar cells have been studied in order to investigate the carrier transport mechanisms and junction properties of heterostructures. Electron beam deposited indium tin oxide electrode / electron cyclotron resonance plasma deposited boron doped microcrystalline silicon carbide / N-type polycrystalline silicon / aluminum contact devices have been fabricated and characterized. Capacitance-voltage, current-voltage-temperature, short circuit current, and open circuit voltage measurements have been systematically performed. The results of analysis show that an abrupt heterojunction model and a multi-tunneling capture-emission transport mechanism are found to be valid. The direct deposition of μc-SiC:H on the c-Si causes the poor junction properties, probably due to high power plasma bombardments. A substantial improvement of the junction quality has been achieved by a low power plasma deposition of less than 5 nm thick *a*-SiC:H as a buffer layer and a heterostructure solar cell, of which AM1 efficiency exceeds 15%, is realized.

1. Introduction

Crystalline silicon (c-Si)/amorphous silicon (*a*-Si) heterostructures have attracted considerable interest due to potential device applications such as low cost solar cell[1] and the heterojunction bipolar transistor (HBT).[2] A wide bandgap material is required as a window layer of solar cells to enhance the spectral response of the ultraviolet (UV) region. It is also reported that the back injection of minority carriers is strongly suppressed by utilizing an *a*-Si hetero-emitter.[2] The interface physics and carrier transport mechanisms of c-Si/*a*-Si heterostructures have been studied.[3] However, most of the previous works regarding the carrier transport mechanisms have been limited to *a*-SiC/c-Si structures where an undoped *a*-Si or a relatively small bandgap *a*-SiC is deposited on crystalline silicon.

The device physics dictating characteristics of wide bandgap μc-SiC/c-Si heterostructures has not been investigated yet. In this work, junction properties and transport mechanisms of wide bandgap ECR plasma deposited μc-SiC/poly silicon heterojunction are discussed. We also report light induced degradation effects on micro-crystalline silicon carbide.

Springer Proceedings in Physics, Volume 56 **Amorphous and Crystalline Silicon Carbide III**
Editors: G.L. Harris, M.G. Spencer, and C.Y. Yang

2. Sample Preparation

2.1 Micro-Crystalline Silicon Carbide Films

Wide bandgap μc-SiC films are successfully deposited by an ECR plasma system. The deposition conditions are summarized in Table 1.

Table 1. Deposition conditions of μc-SiC

microwave power	320 watt at 2.54 GHz
base pressure	$< 10^{-7}$ Torr
deposition pressure	7×10^{-4} Torr
H_2 flow rate	45 sccm
SiH_4 flow rate	40 sccm (10 % diluted with H_2)
CH_4 flow rate	20 sccm (10 % diluted with H_2)
B_2H_6 flow rate	75 sccm (0.05 % diluted with H_2)
substrate temp.(T_s)	180 ~ 300 $^{\circ}C$

The various material properties of deposited μc-SiC films are fairly good. The optical bandgap (E_{opt}) of 2.2~2.6 eV , conductivity of $10^{-1} \sim 10^{-2}$ $(\Omega\cdot cm)^{-1}$ and activation energy of 0.1 ~ 0.2 eV have been observed. These electrical and optical properties clearly demonstrate the advantages of the ECR plasma deposited μc-SiC as a window material of solar cells and an emitter material for heterojunction bipolar transistors.

Table 2. Variation of resistivities ($\Omega\cdot cm$) after 8 hours AM1 illumination.

samples	T_s [^{o}C]	power [watt]	E_{opt} [eV]	$\sigma_{bef.}$	$\sigma_{aft.}$
A	300	320	2.20	4.62×10^{2}	1.31×10^{3}
B	180	320	1.95	1.29×10^{2}	8.23×10^{2}
C	200	200	2.00	2.39	2.80

$\sigma_{bef.}$ and $\sigma_{aft.}$ refer to the resistivities before and after illumination, respectively.

It is a well-known fact in hydrogenated amorphous silicon (a-Si:H) films that the light induced degradation of the conductivity, the so-called Staebler-Wronski effect, is critical for the stability of devices. The light induced degradations in μc-SiC materials have not been investigated or reported yet. The conductivity was measured before and after exposure to a light source. The illumination and measurement methods were standard stability measurement techniques well established in a-Si films. The conductivity of μc-SiC, after 8 hours of AM1 illumination, was indeed decreased. As seen in Table 2, the conductivity decreased by about 20 % after illumination. However, the activation energy of μc-SiC films, evaluated by conductivity measurements at the different temperatures, is almost unchanged.

2.2 Fabrication of μc-SiC/poly-Si Heterojunction Solar Cells

The heterostructures employed in this work are heterojunction solar cells, as shown in Fig.1, having the structure of an electron beam deposited indium tin oxide(ITO)/boron doped μc-SiC/N poly Si/Al ohmic contact. A 50 *nm* thick μc-SiC layer has been deposited on a chemically precleaned 2 ~ 4 Ω·cm Wacker N-type polysilicon. About 70 *nm* thick indium tin oxide layer was deposited by an electron beam evaporator and the deposition temperature was about $280°C$. The area of the device is 16 mm^2.

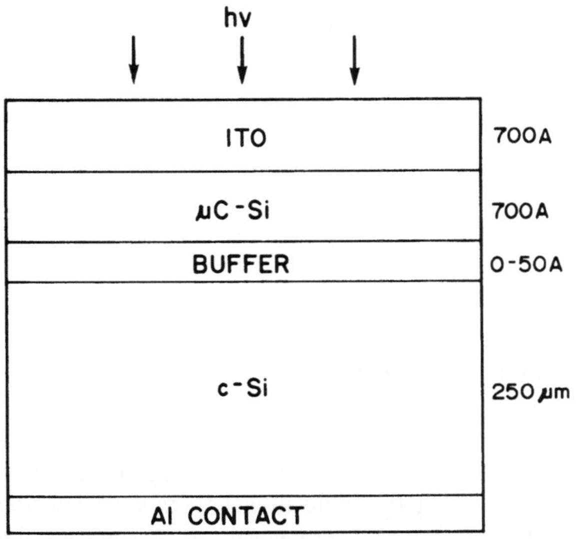

Fig. 1 Structure of μc-SiC/poly Si heterojunction solar cells.

3. Experimental Results and Discussion

3.1 Characteristics of μc-SiC/poly-Si Solar Cells

It is found that the direct deposition of μc-SiC on polysilicon substrates may severely damage the surface of c-Si and poly-Si, probably due to the ion bombardments from high ECR power. A less than 5 *nm* thick boron doped *a*-SiC layer has been deposited on the polysilicon substrates, prior to μc-SiC deposition, by ECR plasma of which the power is less than 100 watt. The deposition conditions of the *a*-SiC buffer layer are identical to those at a temperature less than 150°*C*.

Fig.2 shows the characteristics of μc-SiC/poly-Si solar cells with a buffer layer and without a buffer layer. The cell without a buffer layer, where the μc-SiC layer is directly deposited on the poly-Si, shows a very poor fill factor while the fill factor and short circuit current density of the cell with a buffer layer are substantially improved.

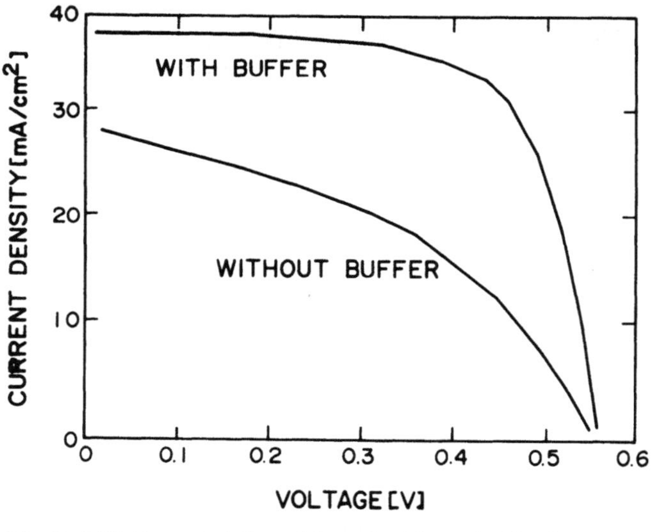

Fig. 2 I-V characteristics with and without buffer layer.

3.2 Multi-tunneling Capture Emission Process

The dark current-voltage characteristics of a μc-SiC/c-Si heterojunction as a function of temperature have been measured and are shown in Fig. 3. The most interesting aspect of the I-V-T data is that the multi-tunneling capture-emission process is the dominant carrier transport mechanism as evidenced by $I(V) = I_o \exp(AV)$ where A is a temperature independent constant.

A multi-tunneling capture-emission transport mechanism in hetrojunctions has been also observed.[4] The carrier transport mechanisms under the illumination conditions may be estimated from V_{oc}-J_{sc}-T data. V_{oc}-J_{sc} under the various illumination levels and temperatures is shown in Fig.4.

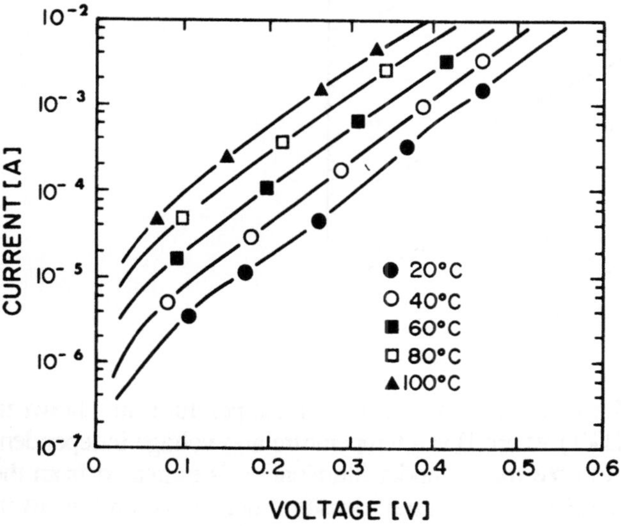

Fig. 3 I-V-T characteristics of solar cells.

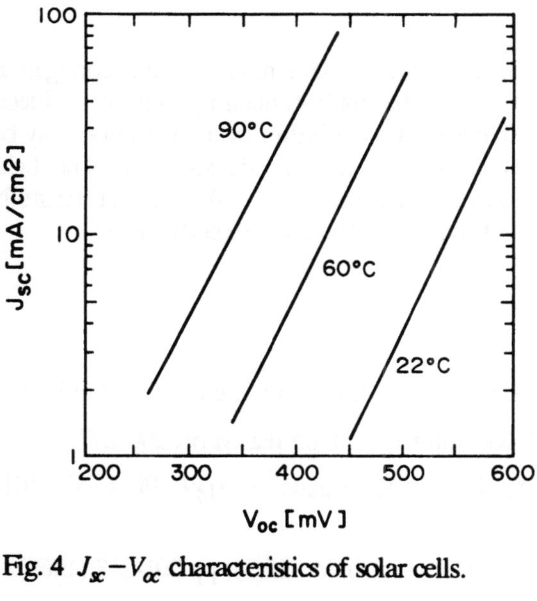

Fig. 4 $J_{sc}-V_{oc}$ characteristics of solar cells.

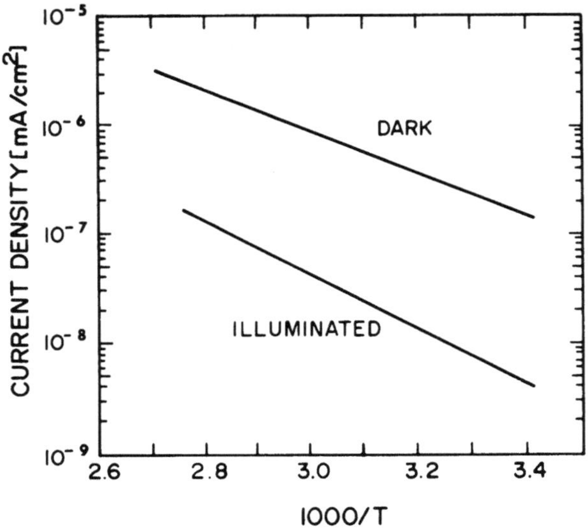

Fig. 5 I_o in the dark and under illumination.

The pre-exponential factors I_o as a function of temperature are shown in Fig.5. I_o varies with exp(B/kT) where B is a temperature and voltage independent constant. It is interesting to note that I_o under illumination is estimated from the open-circuit current by employing a neutral density filter to control incident illumination, and that of the dark condition is not identical as shown in Fig.5 probably due to the illumination.

4. Conclusion

It has been found that the dominant transport mechanism in wide bandgap B doped μc-SiC/poly-Si heterostructures is the multi-tunneling capture emission process rather than a diffusion mechanism. Also, a buffer or a passivation may be required to form a quality heterojunction of μc-SiC/ c-Si structure. The light induced effects on μc-SiC are found to be negligible. We also successfully fabricated μc-SiC/poly Si solar cells whose AM1 efficiency exceeds 14 %.

References

[1] Y. Tawada, H. Okamoto, Y. Hamakawa : Appl. Phys. Lett. 39, 237 (1981).

[2] K. Sasaki, S. Furukawa and M. M. Rahman : 1985 IEDM pp. 294-297.

[3] H.Mastura, T. Okuno, H. Okushi and K. Tanaka : Appl. Phys. 55, 1012 (1984).

[4] H. J. Hovel and A. G. Milnes : Int. J. Electronics, 25(3), pp 201-218 (1968).

Hydrogen Depth Profile Measurement in a-$Si_{1-x}C_x$:H Films by Elastic Recoil Detection

V.Kh. Kudoyarova, G.M. Gusinsky, L.A. Rassadin, and I.V. Kudryavtsev

A.F. Ioffe Physical Technical Institute, Academy of Sciences,
194021 St. Petersburg, Russia

Hydrogen content and depth profiles in a-$Si_{1-x}C_x$:H films have been measured by elastic recoil detection (ERD). It is shown that hydrogen content changes from 15 to 50 at.% as carbon content x grows. At $x < 0.4$, hydrogen content grows mainly due to the increase of the Si-CH_3 contribution, while at $x > 0.4$ it is due to C-H bonds. Comparing the hydrogen density values, defined by IR spectra, with the total hydrogen content, defined by ERD, allowed us to determine the constant A value for the Si-CH_3 rocking oscillation mode.

1. Introduction

Amorphous hydrogenated silicon carbide (a-$SI_{1-x}C_x$:H) is attracting much attention as a possible material for electrophotographic photoreceptors, large multi-colour screens, as a wide gap window for optical devices, and as a subject of purely scientific interest as well.

The structure of these films is not yet understood, especially such details as the method of hydrogen incorporation and the defect nature. The electrical and optical properties of a-$Si_{1-x}C_x$:H depend on carbon and hydrogen total densities and on the nature of hydrogen-silicon and hydrogen-carbon bonds. In practice, it is often useful to know which of these factors really governs the electrical properties. We ran into these problems while studying the optical and electrical properties of a-$Si_{1-x}C_x$:H films for electrophotographic photoreceptors [1].

Infrared (IR) spectroscopy provides a means of determining the nature of hydrogen bonding to silicon and carbon (Si-H, Si-H_2, Si-CH_3 or C-H). Hydrogen content can be measured if the proportionality coefficients A are known. These coefficients correlate the number N of corresponding bonds with absorption intensity $I = \int \alpha(\sigma)/\sigma d\sigma$, i.e. $N = A \times I$. Here α is the absorption coefficient in the corresponding absorption band and σ is the wavenumber.

In the case of a-$Si_{1-x}C_x$:H the IR value at $630\,cm^{-1}$ allows us to estimate the density of Si-H bonds, because the constant A is known for this band [2]. The A constant values for the other bonds are either unknown or, as in the case of the

Springer Proceedings in Physics, Volume 56 **Amorphous and Crystalline Silicon Carbide III**
Editors: G.L. Harris, M.G. Spencer, and C.Y. Yang

C–H stretching mode, vary greatly: $A_{C-H} = 8.7 \times 10^{20} cm^{-2}$ and $1.7 \times 10^{21} cm^{-2}$ [3]. Independent measurements of hydrogen density are needed to determine the constants. SIMS, proton magnetic resonance (H-NMR) and nuclear resonance reaction (NRA) techniques may be such alternative methods.

More powerful, in our opinion, is the elastic recoil detection (ERD) method, as it has certain advantages over the above-mentioned techniques: it is nondestructive; it allows one to determine the total hydrogen content; and it provides depth profiling of hydrogen at a fixed energy of incident particles.

The present paper reports ERD measurements of the hydrogen distribution profile and the total hydrogen content in a-$Si_{1-x}C_x$:H films ($x \leq 0.5$).

2. The Experimental Method

a-$Si_{1-x}C_x$:H films prepared by HF decomposition of silane-methane mixtures of different compositions, diluted with argon and hydrogen, were studied. The substrate temperature was kept constant (T_s = 230°C), providing good adhesion on the one hand and appropriate hydrogenation on the other. The structure and compositions of the films were studied using Auger spectrosopy and X-ray microstructure analysis.

Recoil proton spectra were taken, using a heavy nitrogen ion ^{14}N beam at 10.5 MeV. The experimental arrangement is schematically represented in Fig. 1.

Accelerated ^{14}N ions are incident at an angle $\alpha = 15°$ to the surface, scattered by the heavier silicon atoms and detected with a surface barrier detector D_1. The spectrum of nitrogen ions ^{14}N scattered by silicon atoms is further used for precise measurement of the number of incident particles. When the accelerated nitrogen ions ^{14}N interact with hydrogen, the latter is elastically scattered and recorded with a second surface barrier detector D_2.

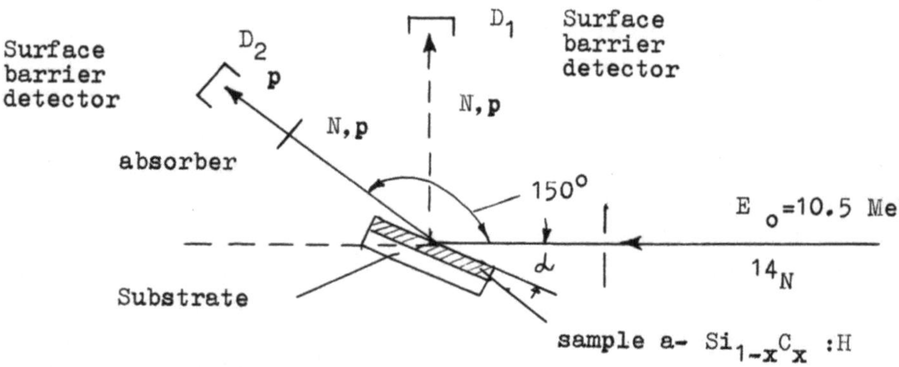

Fig. 1. Schematic representation of the experiment

To stop ^{14}N ions, scattered by silicon, a therylene filter is located in front of detector D_2, transparent for recoil protons (hydrogen). In the experiment energy spectra of elastically scattered recoil protons are observed.

In order to correlate the energy of the outgoing particles (recoiled protons) with the depth at which the elastic scattering took place, the deceleration energy loss and range were calculated for both the bombarding particles (^{14}N ions) and the recoiled protons in the target materials (a-$Si_{1-x}C_x$:H) and in the therylene filter. In the calculations we took into account that the deceleration losses and the stopping power were carbon-content (i.e. "x") dependent.

The resolution of the technique was 150 Å and the maximum probing depth 1 μm. The hydrogen detection limit was 0.1 at.%.

3. Experimental Results

Typical hydrogen depth profiles in a-$Si_{1-x}C_x$:H films with different compositions are shown in Fig. 2. As in the case of a-Si:H [4], layers depleted of hydrogen are observed near the free film surface and at the interface between the a-$Si_{1-x}C_x$:H film and the crystalline substrate. On the whole, using a mixture diluted by hydrogen and the substrate temperature T_s = 230°C, a uniform hydrogen distribution over the film can be obtained.

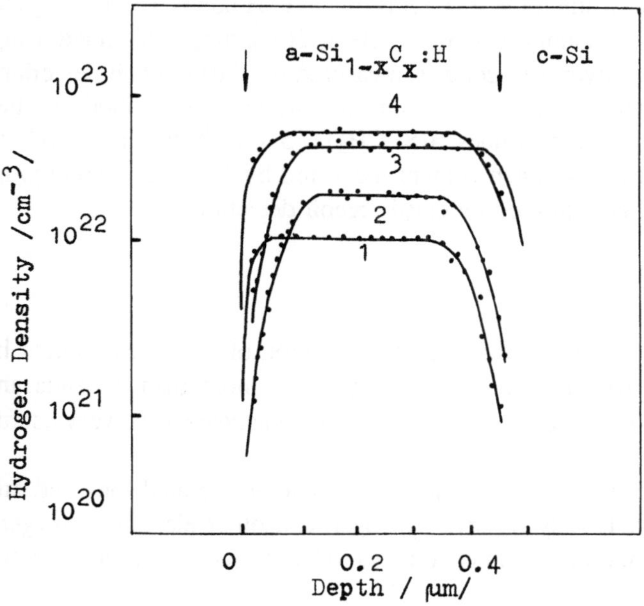

Fig. 2. Hydrogen density depth profile in a-$Si_{1-x}C_x$:H films for different carbon contents: 1 : x = 0.06, 2 : x = 0.088, 3 : x = 0.16, 4 : x = 0.21

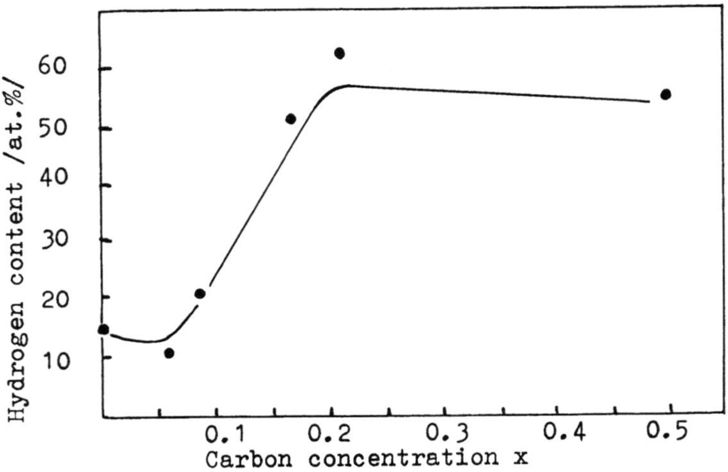

Fig. 3. Hydrogen content as a function of carbon concentration

In Fig. 3 the hydrogen density is shown as a function of carbon concentration. A rise of hydrogen density up to ~ 50 at.% is observed as x increases to $x = 0.2$ and a saturation occurs at $x > 0.2$. Note that addition of small amounts of carbon to a-Si:H increases hydrogen content from 15 at.% to 50 at.%. IR spectra analysis suggests [1] that at $x < 0.4$ hydrogen content grows mainly due to an increase of the Si-CH_3 contribution, while for $x > 0.4$ it is due to C-H bonds. Comparing the density of bonded hydrogen atoms (Si-H, Si-CH_3, C-H bondings), obtained using IR spectra, with the total hydrogen content, measured by elastic recoil detection, allowed us to determine the constant A value for the vibrational mode of the Si-CH_3 bond. If one accepts the values A(Si-CH_3) = 5×10^{19}cm^{-2} and A(C-H) = 1×10^{21}cm^{-2}, the total hydrogen content measured by IR spectroscopy is in good agreement with that measured by elastic recoil detection.

4. Conclusion

In conclusion, it should be noted that the determination of hydrogen content in a-$Si_{1-x}C_x$:H from IR data is rather difficult, because more than 4 bands are observed in the absorption spectra. The ERD technique appears to be very useful in this case.

The ERD has wide possibilities: using appropriate accelerated ions and energies, one may simultaneously observe hydrogen and other elements (oxygen, nitrogen, carbon) in amorphous semiconductors. The results of such studies are to be published.

Acknowledgement. The authors wish to express their thanks to I.N. Trapesnikova for help in IR measurements.

References

1. V.Kh. Kudoyarova, E.I. Terukov, I.N. Trapesnikova, G.N. Violina, N.V. Elkina, R.A. Kavalyauskas: Lith. J. Phys. (1990), to be published
2. H. Shanks, C.J. Fang, L. Ley, M. Cordona, E.J. Demond, S. Kalbitzer: Phys. Status Solidi (b) **100**, 43 (1980)
3. K. Nakazawa, S. Ueda, M. Kumeda, A. Morimoto, T. Shimizu: Jpn. J. Appl. Phys. **21**, L 176 (1982)
4. V.Kh. Kudoyarova, G.M. Gusinsky, L.A. Rassadin, I.A. Kudryavtsev: Proc. Int. Conf. Non-Crystalline Semiconductors-89, Uzhgorod, Vol. 3 (1989) p. 108

Infrared, Raman and ESR Studies of a-SiC Prepared by the Polymer Route

C.-J. Chu, S.-J. Ting, and J.D. Mackenzie

Department of Material Science and Engineering, University of California, Los Angeles, CA 90024, USA

Abstract. The structural evolution of Amorphous SiC by the polymer route was studied using various spectroscopic methods. Infrared studies of this material provide information about bonding and local structure. Raman studies show graphite cluster which was also confirmed by the X-ray and TEM studies. The dominating defects in this new a-SiC were identified as C-dangling bonds by ESR.

1. Introduction

The hydrogenated amorphous silicon carbides (a-SiC:H) have attracted wide interest due to the capability of tailoring electronic, optical and structural properties for optoelectronic device applications. One of the applications is to use a-SiC:H as the window material in p-i-n amorphous silicon solar cells[1]. This will not only reduce the interface reflections and surface absorptions of incident light, but also greatly improve the transmissivity of the p layer in the solar cells. The efficiency will increase from 8% to more than 10%. By introducing microcrystalline phase into the a-SiC to obtain the higher doping efficiency, the efficiency of the solar cells can be increased to above 12%[2].

Because of the difference in chemical properties of constituent elements, it is difficult to form defect-free amorphous network structure, resulting in the low efficiency of the dopants. Understanding the defect structures becomes very important in improving the properties of the a-SiC:H. Electron spin resonance (ESR) is one of the powerful tools that have been used to identify defects in these materials.

The pyrolysis of metal-organic polymers to produce some of the advance ceramics like carbides and nitrides has been proved as another promising method. One of the most widely studied examples is to produce SiC by polycarbosilane[3]. The original objective of this process is to produce polycrystalline SiC fibers. But before the crystallization termperature, and after unreacted organics are removed, an inorganic amorphous phase is formed. Almost no study was done on the structure and properties of this amorphous phase.

The polymer route provides a completely different approach to processing amorphous materials from the traditional amorphous material formed by rapid cooling of the melt, such as oxide glasses. Using this unique polymer route, a-SiC films have first been obtained by authors' group. Some of the electrical and optical properties were studied [4]. Structral evolution has

Springer Proceedings in Physics, Volume 56 Amorphous and Crystalline Silicon Carbide III
Editors: G.L. Harris, M.G. Spencer, and C.Y. Yang

also been studied by the NMR and X-ray diffraction[5]. In this paper, Infrared, Raman and ESR studies of this material have been carried out and related to the other studies, which may give a clearer insight of the structure.

2. Experimental

The starting material, polycarbosilane (PC), is commerically available from Dow Corning (X9-6348). Polycarbosilane can be dissolved in n-hexane or toluene to make a polymer solution.By dipping, spinning or spraying, polymer films can be coated on different substrates like fused quartz, silicon and Al_2O_3. The samples were fired in Ar at different temperatures using a high temperature furnace (Thermal Technology Inc. Astro 1000A). The structural evolution was studied using X-ray diffraction and TEM/SAED. FTIR was used to study the structural evolution of the films that were coated on intrinsic Si wafer with double sides polished. Raman scattering was performed at room temperature using the 5145Å line of an Argon ion laser. The Raman spectra were measured in the backscattering configuration and recorded at the power of ~40mW. Electron spin resonance (ESR) was also used to study defect concentration.

3. Results and Discussions

Chemical analysis of PC shows its composition to be $SiC_{2.0}H_{4.1}$. From the NMR study, polycarbosilane can be assumed to form chains of Si-CH_2 bonds with two kinds of Si units, -(CH_2)-Si(CH_3)$_2$-(CH_2)- and-(CH_2)-SiH(CH_3)- (CH_2)- in approximately one to one ratio[6]. From the ideal composition fomular SiC_2H_5 of PC, a simple presumption is SiC_2H_5-- SiC + CH_4 + 1/2H_2. But from gas chromatography results, the release of CH_4 is small compared with that of H_2[7]. This appears to account for the excess carbon in the a-SiC by the polymer route. Chemical analysis of the samples fired above 800°C shows a C/Si ratio close to 1.5. Figure 1 shows the possible condensation reaction of polycarbosilane chains with evolution of H_2 and CH_4, and lead to the formation of rings of Si and C atoms. The rings may have various sizes instead of the identical 6-member rings in the crystalline ß-SiC. NMR results show no evidence of the Si-Si bonds. The imperfection of the structure is mainly due to C-C bonds. The breaking of the C-H bonds will introduce defects like C=C double bonds and C-dangling bonds at higher firing temperatures.

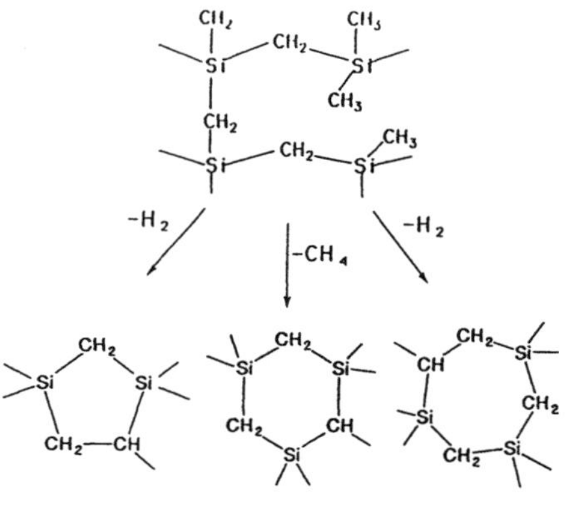

Fig.1 Possible condensation reaction of polycarbosilane chains.

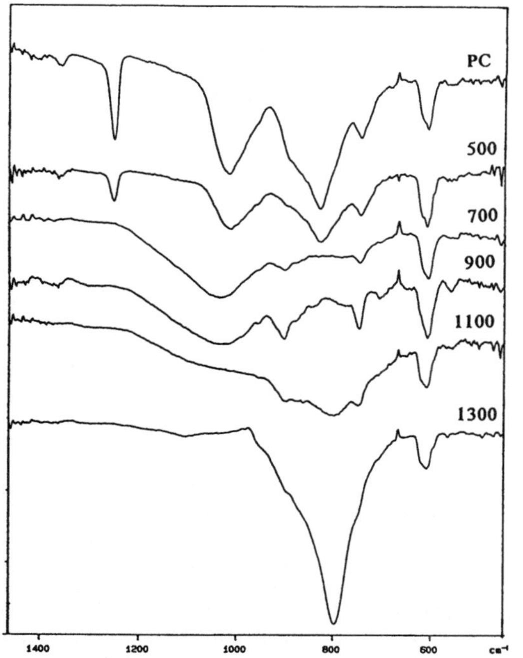

Fig. 2 IR spectra for samples fired at different temperatures

Figure 2 shows the IR spectra from 450 cm^{-1} to 1450cm^{-1} for samples fired at different temperatures. The absorption peak at 820 cm^{-1} of the PC has been assigned to Si-CH_3, which disappears for samples fired above 700°C, due to the breaking of C-H bonds. The peak at 800cm^{-1} corresponds to the crystalline SiC which starts to appear for samples fired above 1000°C. DTA analysis of PC heat treated in N_2 atmosphere shows an exothermic peak at 1000°C, which may be due to the crystallization. X-ray diffraction data also indicate microcrystalline structure for the samples above 1000°C. The crystallite size for the 1000°C sample was found to be around 20Å by the TEM/SAED studies[8]. The peak at 1025 cm^{-1} was assigned to the Si-CH-Si bonds, which broadened and finally disappeared at 1200°C. This also indicates the breaking of the C-H bonds, as a result increasing the crystallization. The peak at 1270 cm^{-1} was assigned to the CH_3 bonds which disappeared at temperatures higher than 500°C as the C-H bonds broke.

X-ray diffraction pattern shows a small peak around $2\theta=26°$. The intensity of this peak reaches maximum for the 1200°C sample and decreases with increasing temperature. It could be assigned to the (002) line of carbon, which may be due to the presence of small clusters of graphite in the amorphous phase. Figure 3 shows the preliminary result from Raman scattering for the sample fired at 900°C. The broad peaks around 1350 cm and 1620 cm have also been assigned as small cluster of graphite. From structure study by HRTEM[9], the free aromatic carbon coexists

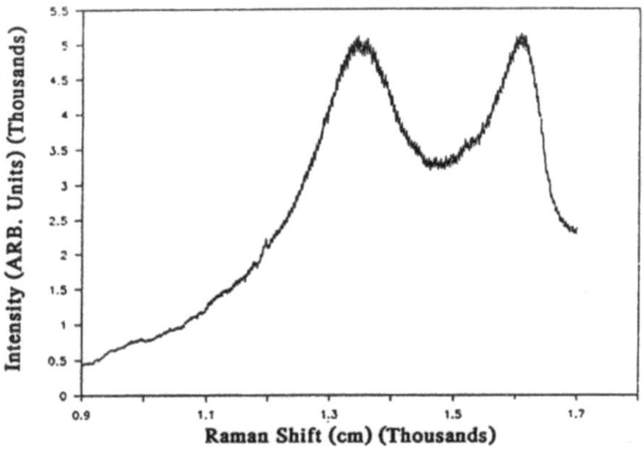

Fig. 3 Raman scattering for the samples fired at 900°C

with SiC microcrystals. Carbon tend to form oriented rims along particle edges, and could slow the rate of crystal growth.

Figure 4 shows the ESR spectra of some samples fired at different temperatures. All the samples show the same g-value to be 2.003, which has been assigned to the C-dangling bonds. The a-SiC:H prepared by glow-discharge with the same C/Si ratio (C/Si=1.5) showed similar result that C-dangling bonds were the dominate defect[10]. Figure 5 compares the spin density of the samples fired at different temperatures. The spin density around $10^{19}cm^{-3}$ is in agreement with the number of defects found in the a-SiC:H by glow-discharge. The density of the C-dangling bonds reaches a maximum at 1200°C, which correlates well with the X-ray diffraction results of the maximum formation of graphite clusters at 1200°C.

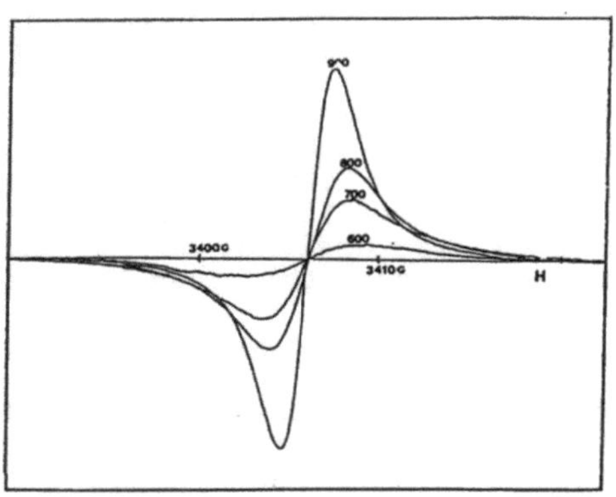

Fig. 4 ESR spectra of some samples fired at different temperatures.

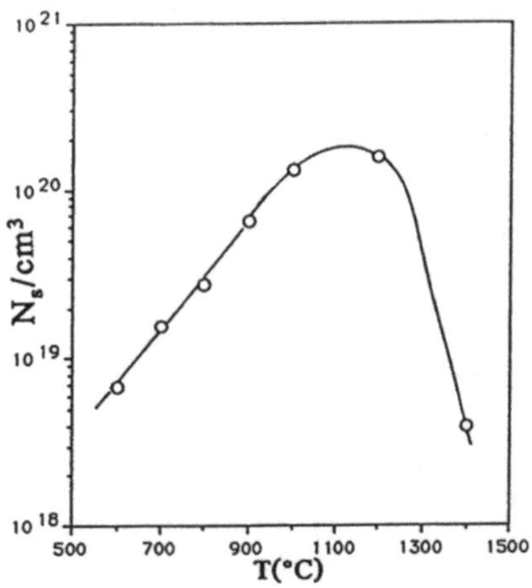

Fig. 5 Spin density of the samples fired at different temperatures.

Based on the reaction scheme shown in Figure 1, samples fired in H_2 atmosphere instead of Ar atmosphere, should decrease the evolution of H_2 and increase the release of CH_4. Defects in the SiC may also decrease. Some of the samples were prepared in 100% H_2 atmosphere and studied by ESR, FTIR and NMR. The results show good match with the theorical approach, and will be published separately.

4. Conclusions

In the structural evolution, graphite clusters show a dominating effect. The aromatic carbon not only prevent or slow the the rate of SiC crystal growth, but also relate with the C-dangling bond formation. Heat treatment in H2 may provide the possibilities of lowering the defect density and obtaining more ordered structure.

5. Acknowledgements

The authors would like to thank Jame Strouse for the ESR and FTIR experiments, T-S Ho for the Raman studies. This research was funded by NSF under grant No. DMR 87 06379.

6. References

[1]. Y. Tawada, M. Kondo, H. Okamoto and Y. Hamakawa, Solar Energy Mater. 6, 299 (1982).

[2]. Y. Hattori, D. Kruangam, T. Toyama, H. Okamoto and Y. Hamakawa, J. Non-Cryst. Solids 97&98, 1079 (1987).

[3]. S. Yajima, J. Hayashi and M. Omori, Chem. Lett. 931 (1975).

[4]. C-J. Chu, G.D. Soraru, F. Babonneau and J.D. Mackenzie, proceeding of the 2nd International Conference on Amorphous and Crystalline Silicon Carbide and Related Materials (ICACSC, 88), Santa Clara, CA, U.S.A.(1988), p64.

[5]. G.D. Soraru, F. Babonneau and J.D. Mackenzie, Presented at the 7th International Symposium on Ceramics, Dec. 14-16, 1988, Bologna, Italy: in press in the proceeding.

[6]. G.D. Soraru, F. Babonneau and J.D. Mackenzie, J. Non-Cryst. Solids 106, 256 (1988).

[7]. S. Yajima, Y. Hasegawa, J. Hayashi and M. Iimura, J. Mat. Sci. 13, 2569 (1978).

[8]. Y. Sato, M. Kamo and N. Setaka, Carbon 16, 279 (1978).

[9]. J. Ayache, S. Bonnamy, X. Bourrat, A. Deurbergue, Y. Maniette, A. Oberlin, E. Bacque, M. Birot, J. Dunogues and J.P. Pillot, J. Mat. Sci. Lett. 7, 885 (1988).

[10]. S. Liedtke, K. Jahn, F. Finger and W. Fuhs, J. Non-Cryst. Solids 77&78, 849 (1985).

Amorphous SiC/Si Heterojunctions Prepared by the Polymer Route

C.-J. Chu, S.-J. Ting, and J.D. Mackenzie

Department of Material Science and Engineering, University of California, Los Angeles, CA 90024, USA

Abstract. The amorphous SiC thin films were obtained by depositing polycarbosilane solution on a variety of substrates. The electrical and optical properties of this new a-SiC thin film have been studied. P-N control of this material can also be achieved by the polymer route. The doped a-SiC thin films were coated on the Si wafer to make a-SiC/Si heterojunctions, which show typical I-V characteristics of the p-n junction device. Photovoltaic effect was also studied.

1. Introduction

The hydrogenated amorphous silicon carbide (a-SiC:H) is a wide bandgap semiconductor, which has attracted much interest due to the potential applications in electronic device. One of the most significant applications is the a-SiC/a-Si heterojunction solar cells, achieving a 12% efficiency[1]. The other application is to use the a-SiC:H as an emitter of a Si heterojunction bipolar transistor(HBT). Theorically, the a-SiC emitter HBT has a higher current gain than a conventional Si homojunction transistor[2]. Some other applications like semi-transparent a-SiC solar cells, a-SiC thin film light emitting diode, and electrophotographic receptors, have been reported or proposed[3].

Recently, a polymer route has been investigated for the production of some advanced ceramic materials like carbides and nitrides. One of the advantages of this polymer process is that the polymer can be easily fabricated into different shapes, likes fibers and films. The most well known example is the fabrication of polycrystalline SiC fiber from polycarbosilane precursor[4]. By firing polycarbosilane to a certain temperature so that the organics are removed before the crystallization temperature is reached. The amorphous SiC can be obtained in this temperature range, which can not be obtained by the traditional melting process. The studies of the structure and properties of this new amorphous SiC has just begun. Amorphous SiC films can also be easily fabricated from polycarbosilane.

2. Experimental

Polycarbosilane can be dissolved in many organic solvents. In this study, n-hexane was chosen as the solvent for its low boiling point (69°C) and for safety considerations. By dipping, spinning or spraying, polymer films were coated on the substrates. The polymer films were then pyrolyzed in argon at

500°C to 1400°C for one hour in a high temperature furnace(thermal Technology Inc. Astro 1000A). The heating rate is 10°C/min. The new a-SiC thin films have been successfully coated on a variety of substrates, like silicon, quartz glass, Al_2O_3 and MgO for different characterization purposes.

Film thickness was measured by Dektak, ellipsometer and SEM. SEM was also used to study the structural morphology. TEM/SAED (Jeol STEM 100CX)was used to follow the structural evolution. Infrared spectroscopy (Perkin Elmer 1600 FTIR) and UV visible spectroscopy (Perkin Elmer 330 spectrophotometer)were used to characterize the optical properties of the films. DC conductivity was measured by the four-probe method.

The doped a-SiC was prepared by first reacting dimethyl -dichlorosilane with appropriate boron or phosphorus compound. After polymerization, the doped polycarbosilane was obtained. Infrared and chemical analysis proved that dopants existed before and after firing. The electrical properties were also studied.

Amorphous SiC/Si heterojunctions were fabricated for preliminary study on device application. N-type a-SiC (p-doped) thin films were coated on p-type crystalline Si substrates(ρ=0.01-0.02). C-Si was also used as an electrode. Al or ITO(indine tin oxide) thin films were also coated on the top of the a-SiC films as another electrode. The p-type a-SiC on n-type c-Si was obtained via the same method. The I-V characteristic was studied using the semiconductor parameter analyzer(HP-4145A). The photovoltaic effects of these devices were measured under the standard condition for AM1.5 with incident intensity of 100mW/cm^2.

3. Results and Discussions

Film thickness was determined in the dipping technique by the concentration of the polycarbosilane solution, the dipping speed, and the firing temperature. A single dip can create thickness variation from a few hundred Å to about one µm. Thicker than one µm films can be achieved by multiple dipping.

Absorption coefficient measurements were made at wavelengths between 190 to 1500 nm on the films coated on fused quartz glass or Al_2O_3 substrates. The plot of $(\alpha h\upsilon)^{1/2}$ versus $h\upsilon$ in Fig. 1 follows the Tauc approximation[5]. The ΔE_{opt} can be derived from the slope of each curve. Fig. 2 shows the room temperature DC conductivity, the activation energy for conduction, and the optical gaps of the samples fired at different temperatures. The activation energies for electrical conduction follow the trend of one half of the optical gaps for the samples fired below 900°C. But for the samples fired above 1000°C, the activation energy dropped to about 0.1 eV. This drop has been assigned to the formation of microcrystalline phase[6]. The X-ray diffraction and TEM/SAED results confirm this structure evolution.

DC conductivity of the doped a-SiC was measured. Figure 3 shows a comparison of the conductivity of some of the doped a-SiC and undoped a-SiC films prepared under the same conditions. The results show a significant increase in conductivity for both B and P doped samples; therefore p-n control of this new a-SiC from the polymer route becomes

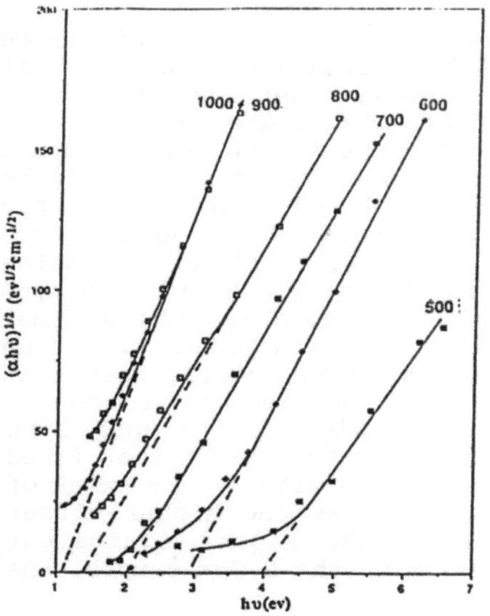

Fig. 1 Optical absorption of a-SiC fired at different temperatures.

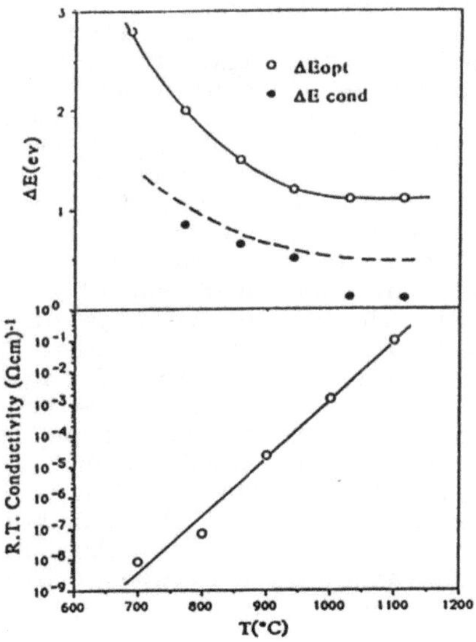

Fig. 2 D.C. conductivity, the activation energy for conduction, and the optical gap of the samples fired at different temperatures.

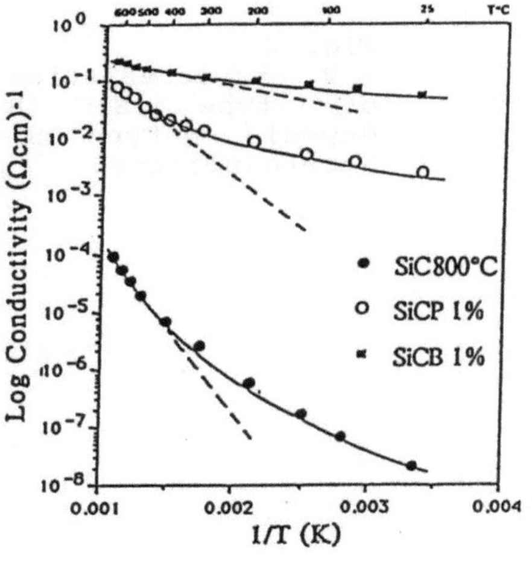

Fig. 3 Conductivities of undoped a-SiC, SiC with 1% B dopping, and SiC with 1% P dopping.

possible. This result also highlights the possibility of device application. The structure and electrical and optical properties of these doped SiC is currently under investigation.

The a-SiC/c-Si heterojunctions were fabricated with various thickness of a-SiC thin films at different firing temperatures. Most of the samples show typical I-V characteristics as the p-n junction diodes, as shown in Fig. 4 and 5. In Fig. 4 the dashed line represents the junction fired at 800°C, whereas the solid line represents the sample fired at 700°C. Both samples have the same thickness of 0.3 µm for the p-type a-SiC on n-type c-Si. The offset voltage of the 800°C sample is smaller than that of the 700°C sample. This result indicates that the equivalent resistance of the 800°C sample is smaller than that of the 700°C sample, which may be due to the increased conductivity of the p-type layers. In Fig. 5, the dash line represents the sample with 0.2 µm of the n-type a-SiC that fired at 1100°C, and the solid line is the sample with 0.3 µm thick of the n-type a-SiC. The thicker sample shows the higher offset voltage , which may also be due to the higher equivalent resistance. In the reverse bias region of the 0.2µm sample, the

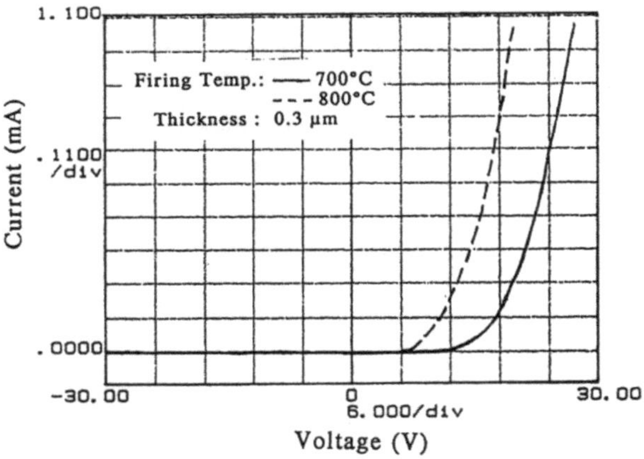

Fig. 4
I-V characteristics of P-type a-SiC (B doped)/ N-type Si heterojuctions.

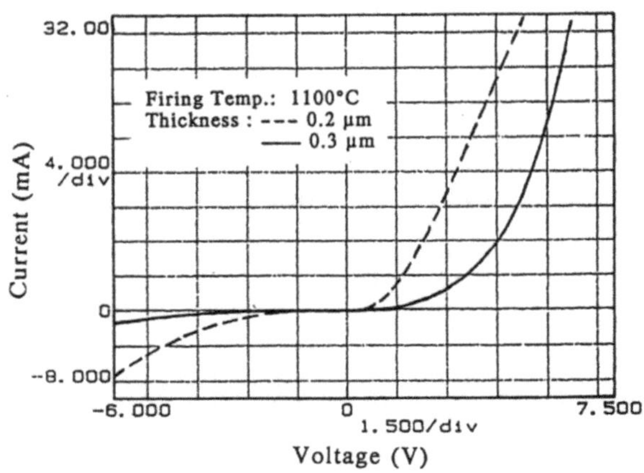

Fig. 5
I-V characteristics of N-type a-SiC (P doped)/ P-type Si heterojuctions.

reverse current increases with the square root of reverse-bias voltage, which represents the carrier generation within the transition region.

Photovoltaic study of the junctions shows that the maxium output voltage is about 0.3 V, and the maxium output current is about 0.57 mA/cm2. The fill factor is only about 0.3. This small value of the fill factor should be due to the tunnel and recombination current through the junction. This can be improved by adding an intrinsic layer to form the depletion layer which can suppress the generation of tunnel current[7]. The study on the p-i-n structure of the a-SiC by polymer route is in progress.

4. Conclusions

P-N control of this new a-SiC by the polymer route has been achieved, which provides the possibility of device applications. The a-SiC/Si heterojunctions showed I-V characteristics as a typical p-n junction device. The heterojunctions also showed photovoltaic effect with low fill factor, which may be improved by adding an intrinsic layer.

5. Acknowledgements

The authors would like to thank T-S Ho and Mary Colby for thickness measurements by Dektak and SEM. S-W Hsiao for I-V measurements. This research was funded by NSF under grant No. 87 06379.

6. References

[1]. Y. Tawada, M. Kondo, H. Okamoto and Y. Hamakawa, Solar Energy Mater. 6, 299 (1982).

[2]. K. Sasaki, T. Fukazawa ans S. Furukawa : 1987 Intl. Electron Devices Meeting, Washington D.C., Tech. Dig. p 186.

[3]. Y. Hamakawa, proceeding of the 2nd International Conference on Amorphous and Crystalline Silicon Carbide and Related Materials (ICACSC, 88), Santa Clara, CA, U.S.A.(1988), p164.

[4]. S. Yajima, J. Hayashi and M. Omori, Chem. Lett. 931 (1975).

[5]. E.A. Davis and N.F. Mott, Phil. Mag., 22, 903 (1970).

[6]. Y. Hattori, D. Kruangam, T. Toyama, H. Okamoto and Y. Hamakawa, J. Non-Cryst. Solids, 97&98, 1079 (1987).

[7]. D.E. Carlson, C.R. Wronski, J.I. Pankove, P.J. Zanzucchi and D.L. Staebler, RCA Review, 38, 211 (1977).

Part III

Materials and Device Properties

Crystal Growth of β-SiC on Si and Its Application to MOSFETs

A. Suzuki, K. Furukawa, Y. Fujii, M. Shigeta, and S. Nakajima

Central Research Laboratories, Corporate Research and Development Group, Sharp Corporation, 2613-1 Ichinomoto, Tenri, Nara 632, Japan

Abstract. Heteroepitaxial CVD growth and defect control of β-SiC films on Si are described. A new approach to reduce stacking faults in the films is presented using novel off-axis substrates. Important device processes of thermal oxidation and ion implantation for the films are studied and fundamental characteristics of inversion-type MOSFET's are examined.

1. Introduction

New power devices and IC's operative in high-temperature environments where Si devices cannot be used are strongly desired for various electric power sources in the future. High-frequency operations are also important for these devices. Among wide bandgap semiconductors, which are necessary for the conductivity control at high temperatures, both α-SiC and β-SiC are outstanding candidates because of their extremely stable material properties as represented by their high Debye temperatures. Especially, electric breakdown fields about one order larger than in Si [1] are quite favorable for high-voltage power devices.

Very high performances will be required for these devices because severe operative conditions such as high voltage, large current, and high temperature are imposed and Si power devices are their competitors. Therefore, in addition to high quality of SiC crystals, various device fabrication techniques have to be developed. From this point of view, SiC is very promising because several techniques similar to those for Si devices have already been attained: for example, CVD growth and doping for p- and n-types, ion implantation, thermal oxidation, dry etching, Schottky and ohmic contacts [2]. Especially, MOS structures with thermal oxides present excellent properties comparable to those of Si, which is advantageous for future SiC power devices considering recently developed Si MOS power IC's.

The authors have investigated heteroepitaxial growth of β-SiC on Si and its application to MOS devices. Large-area single crystal films of 2 inches diameter by CVD growth similar to Si CVD are very useful and attractive for mass production. Various device fabrication processes for SiC MOSFET's have been studied using these films. At the present stage, however, the films contain a lot of stacking faults (~ 10^7 cm^{-2} for 20-μm-thick film), which have to be decreased drastically for practical devices.

In the present paper, CVD growth and defect control of SiC films on Si are described. Important device processes of thermal oxidation for gate oxides and ion implantation for source and drain regions as well as

Springer Proceedings in Physics, Volume 56 Amorphous and Crystalline Silicon Carbide III
Editors: G.L. Harris, M.G. Spencer, and C.Y. Yang

fundamental characteristics of inversion-type MOSFET's are presented. Some problems to be solved are pointed out.

2. Thin Film Growth and Defect Control

2.1 Film Growth and Doping

Heteroepitaxial CVD growth on Si has been investigated by many researchers [1-3]. The CVD growth system in the present work is similar to their systems using a horizontal quartz reaction chamber with an inner diameter of 12 cm. The source gases are SiH_4 and C_3H_8 in a H_2 carrier gas. The growth temperature is about 1350 ℃. The growth rate is about 3 μm/h. Prior to CVD, the Si surface is carbonized with C_3H_8. The substrate orientation is (100). Films on (111) substrates show severe wafer warpage and cannot be used for device fabrication.

Non-doped n-type films show about 3×10^{16} cm^{-3} for carrier concentration and 600 cm^2/Vsec for electron mobility for 10-μm-thick films. Decrease in carrier concentration to lower than 10^{14} cm^{-3} is necessary in the future. Concentrations of 10^{16} - 10^{18} cm^{-3} can be controlled by the addition of N_2 gas for n-type and TMA (trimethylaluminum) for p-type. Because of a deep Al acceptor level, the number of electrically unactivated Al impurities at room temperature is two orders of magnitude larger than the hole concentration [2], and the concentration drastically increases with temperature. This may give some problems for crystal quality and temperature dependence of device characteristics. Doping with B_2H_6 brings about only a highly-resistive p-type layer (higher than 100 Ωcm) and the resistivity control will be difficult.

2.2 Defect Control

Generally in SiC films, stacking faults are easily introduced due to quite a low energy of stacking fault (2.5 mJ/m^2) compared with other compound semiconductors (55 mJ/m^2 for GaAs). The authors have used electrolytic etching with HF acid to observe stacking faults and anti-phase boundaries (APB's) [3]. Film surfaces are converted to p-type to be easily etched. Figures 1(a) and 1(b) are SEM images of etched films on on-axis (100) and off-axis (100) tilted toward [011] by 2°. Stacking faults crossing each other at 90° and APB's are observed in the on-axis (100) films, while only elongated stacking faults perpendicular to the off-direction [011] are observed in the off-axis (100) films. Stacking fault densities are shown in Fig. 2 as a function of film thickness. They decrease with film thickness exponentially. But stacking faults of larger than 10^7 cm^{-2} exist even in thick films of 25 μm.

The stacking fault planes are along {111} faces. The authors speculate that when stacking faults meet each other during the growth, they disappear at a certain rate[3]. Figure 3(a) shows speculated surface steps for the off-axis (100) films tilted toward [011]. Stacking faults along [011] are probably confined within surface terraces during the growth and disappear

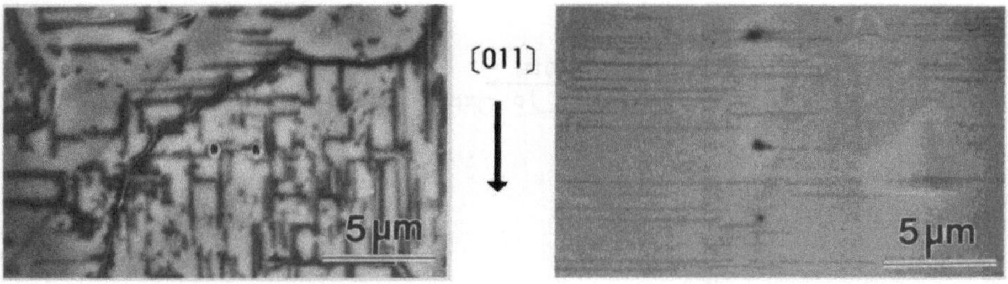

a b

Fig. 1 SEM images of etched films on (a) on-axis and (b) off-axis Si(100) substrates

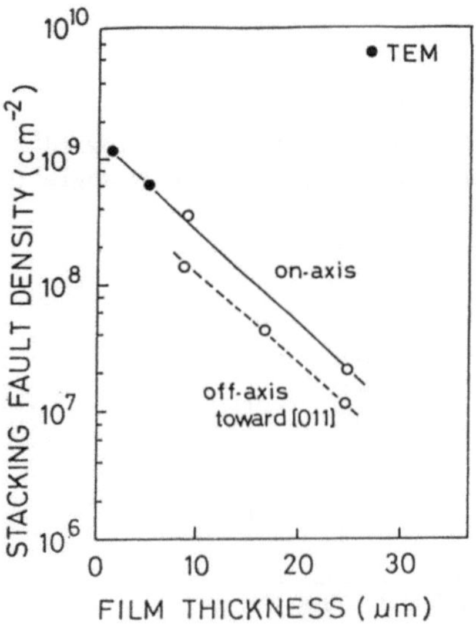

Fig. 2 Dependence of stacking fault density on film thickness

due to their increased energies. Only elongated stacking faults along surface steps $[01\bar{1}]$ remain. The result that the densities for off-axis films in Fig. 2 are about half of those for on-axis films supports the above speculation.

Poor electrical characteristics, especially large reverse leakage currents and soft breakdown at lower than 5 V, of p-n junction diodes of SiC on Si have a strong relation to these stacking faults or other defects combined with stacking faults. Since a p-n junction is one of the most important elements in all device structures, a fundamental study to decrease the stacking faults has been done. Novel off-directions deviated by θ from the conventional [011] are used as shown in Fig. 3(b). Speculated surface steps become smaller terraces. Therefore, elongated stacking faults along $[01\bar{1}]$ will be again confined by the steps along [011] and disappear. As an example, distributions in stacking fault density as a function of stacking fault length are shown in Fig. 4 for 0° and 30° as θ. The off-angle ϕ is

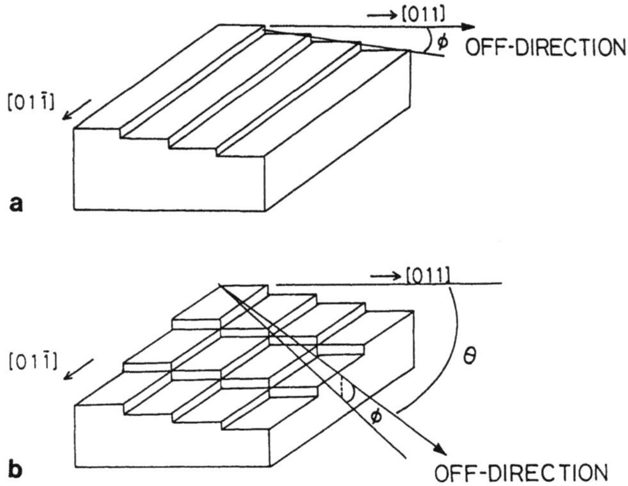

Fig. 3 Speculated schematical off-axis (100) surfaces tilted (a) toward [011] and (b) toward the direction deviated from [011] to [01$\bar{1}$] by angle θ

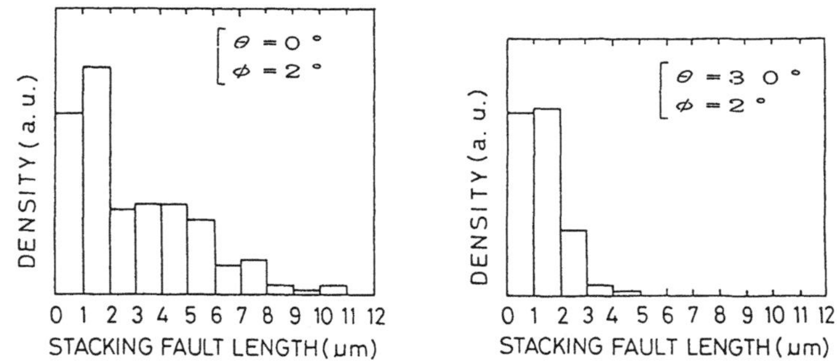

Fig. 4 Distributions of stacking fault density as a function of stacking fault length

fixed at 2°. Stacking faults longer than 5 μm disappear and those shorter than 3 μm mainly remain for θ = 30°. These results show the possibility of control of stacking faults of SiC on Si by surface steps [4].

3. Fabrication Techniques for MOSFET's

3.1 Thermal Oxidation for Gate Oxides

Thermally oxidized SiO_2 layers of SiC on Si at 1000 - 1200 ℃ have sufficient resistivities and electric breakdown fields comparable to those of Si. Interface properties of SiO_2-SiC have been studied for wet and dry oxidations for on-axis and off-axis films [5]. Wet oxidation is carried out

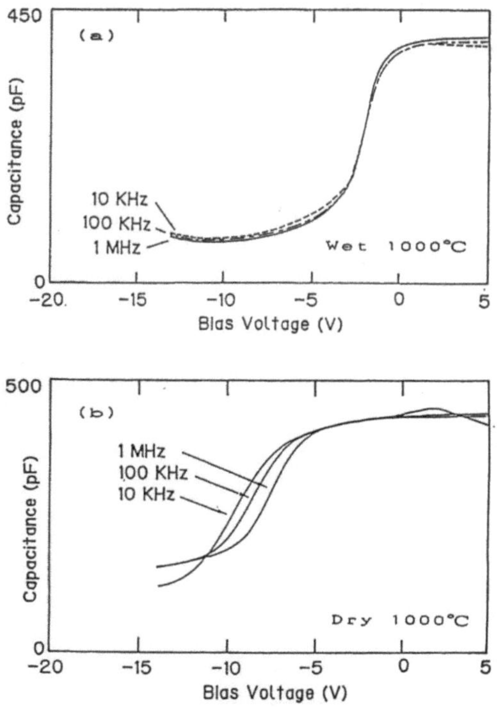

Fig. 5 C-V characteristics of MOS structures by (a) wet and (b) dry oxidations

by the reaction of H_2 with O_2. Figure 5 shows typical results of C-V characteristics of MOS structures on on-axis films oxidized at 1000 ℃. The oxide thicknesses are about 35 nm. The MOS structure by wet oxidation shows excellent characteristics with a small shift of flat-band voltage without frequency dispersion. Fixed-charge density calculated from the flat-band voltage is $1\text{-}2 \times 10^{12}$ cm^{-2}. The minimum interface state density is very low and comparable to Si MOS structures. Bulk trap density in the depletion layer is one order lower than that by dry oxidation. SiC films on off-axis substrates have no APB's, but MOS characteristics of off-axis films are similar to those of on-axis films. This implies that APB's and stacking faults do not affect MOS characteristics significantly.

3.2 Ion Implantation for Source and Drain Regions

Selective doping by ion implantation is very important for SiC devices because diffusion doping cannot be used due to an extremely small diffusion coefficient at practical process temperatures. Ion implantations of P, N, B, and Al into SiC on Si have been reported by several researchers [6,7]. Hot implantation at 350 - 750 ℃ is effective for the electrical activation of implanted species. In the present work, annealing effects on the activation

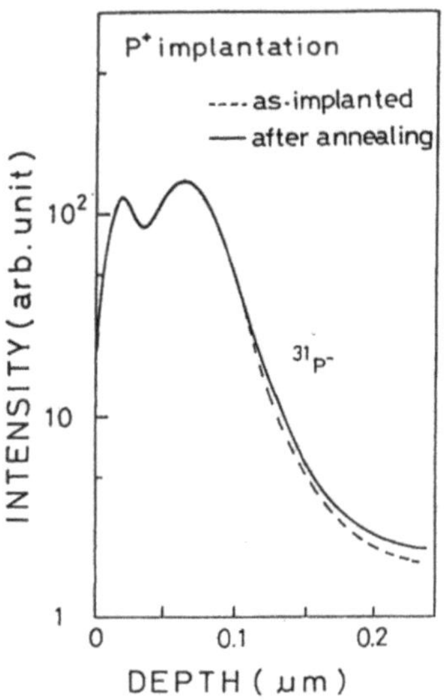

Fig. 6 SIMS depth profiles of P^+ ion-implanted SiC film

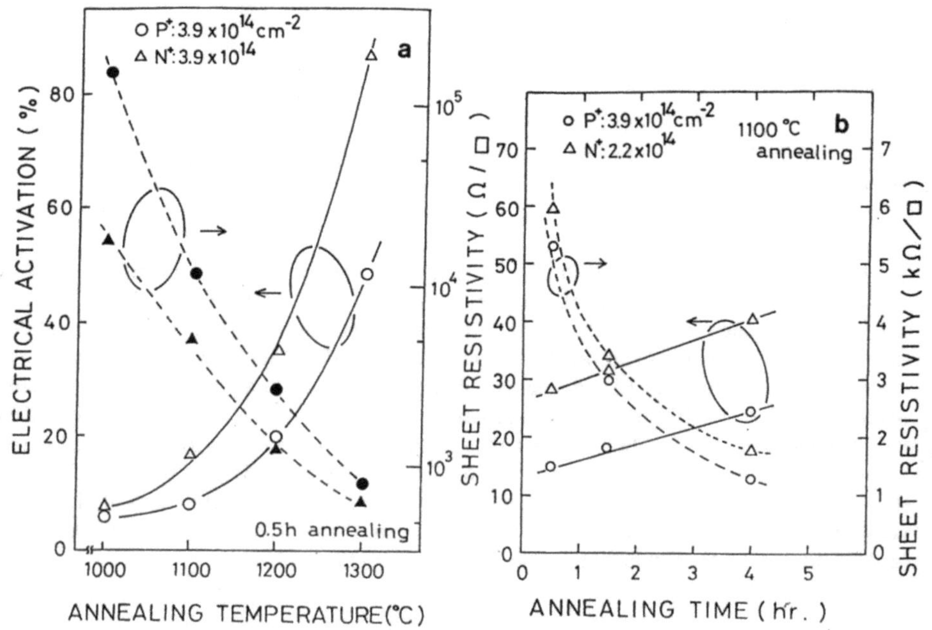

Fig. 7 Sheet resistivity and electrical activation of implanted SiC layers as a function of (a) annealing temperature (0.5 h annealing) and (b) annealing time (1100 ℃ annealing)

after room-temperature implantation are studied. Figure 6 shows an example of SIMS depth profile of implanted atoms for as-implanted and annealed samples at 1300 ℃ for 0.5 h. P^+ ions are dually implanted with accelerating voltages of 100 kV (1.0×10^{15} cm^{-2} dose) and 30 kV (3.0×10^{14} cm^{-2} dose). Even after annealing at 1300 ℃, the depth profile does not change due to the extremely small diffusion coefficient. This is quite a different feature from Si.

Figure 7 shows dependences of electrical activation and sheet resistance on annealing temperature for P^+ and N^+ ions (0.5 h annealing). Activation rates are higher than 50 % at 1300 ℃. For the fabrication process of MOSFET's, annealing temperatures should be lower than or, at most, comparable to thermal oxidation temperatures of 1000 - 1100 ℃ in order to prevent the interface reaction of field oxide and/or gate oxide with SiC surface. But at these temperatures activation rates are lower than 30 % after 0.5 h annealing. Annealing-time dependences at 1100 ℃ are shown in Fig. 7(b). Activation rates increase gradually with annealing time, but rather a long time is necessary for a sufficient activation at 1100 ℃.

Characteristics of p-n junctions implanted with P^+ and N^+ into Al-doped p-SiC on Si are generally poor with remarkable reverse leakage currents. The characteristics for P^+ and N^+ implanted junctions change differently by annealing. This implies that stacking faults or defects combined with stacking faults react differently with different implanted species, but details remain unknown.

3.3 Fabrication of Inversion MOSFET's

Inversion-type MOSFET's of β-SiC on Si using thermal oxidation for gate oxides and ion implantation for source and drain regions have already been reported [8,9]. SHIBAHARA et al.[8] reported about 200 µS/mm for transconductance g_m for the gate length of 20 µm, and PALMOUR et al.[9] obtained about 500 and 200 µS/mm for the gate lengths of 5 and 20 µm, respectively. In the present work, two kinds of procedures have been examined to fabricate inversion-type MOSFET's as shown in Fig. 8. Figure 8(a) shows self-aligned type using poly-Si gate metal. After thermal oxidation for gate oxides at 1100 ℃ for 5 h, N^+ ions are implanted and annealed at 1100 ℃ for 0.5 h. The p-type SiC film is doped with B_2H_6 and highly resistive (larger than 100 Ωcm). The gate length is 20 µm. On the other hand, in Fig. 8(b) ion implantation is carried out before thermal oxidation for gate oxides. Gate metal is aluminum. MOSFET's fabricated by the latter process show sufficient characteristics as shown in Fig. 9. Those fabricated by the former one show very small drain currents. These are probably explained from the sufficient annealing after implantation during thermal oxidation for the latter process.

The dependences of threshold voltage V_t, effective mobility μ_{eff}, and transconductance g_m on the ionized impurity of SiC film have been examined. The amount of the ionized impurity will be controlled by the doping amount of B_2H_6. In Fig. 10, drain current - gate voltage characteristics are shown for three different amounts of B_2H_6. The oxide thickness is 76 nm. The value of V_t can be controlled by the amount of B_2H_6. About 100 cm^2/Vsec is

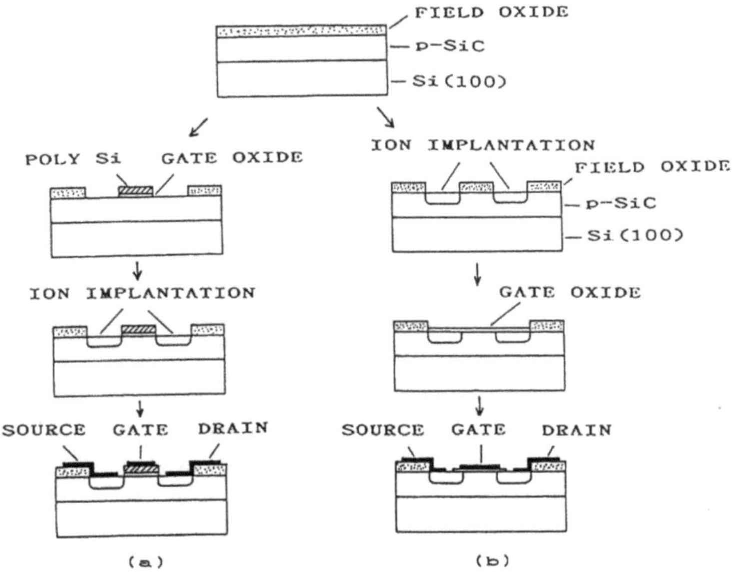

Fig. 8 Fabrication procedures of inversion-type MOSFET's

Fig.9

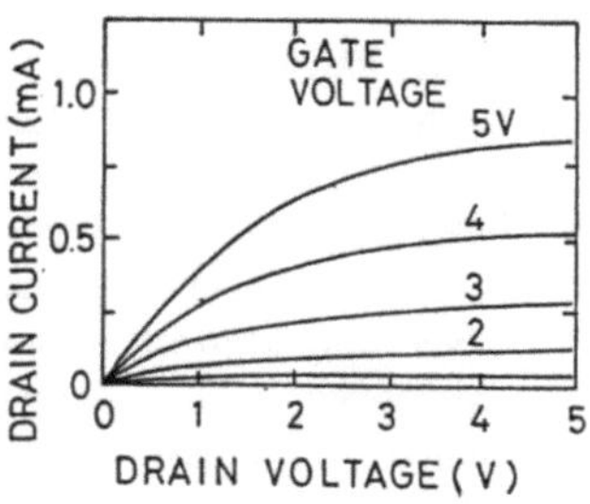

Fig.10

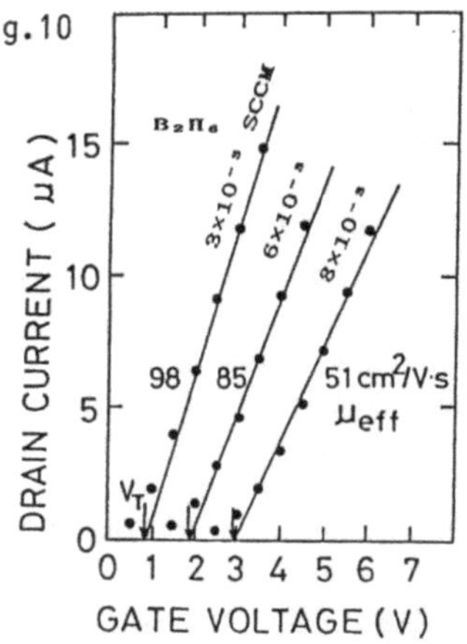

Fig. 9 Typical drain characteristics of an inversion-type MOSFET

Fig. 10 Drain current - gate voltage characteristics of MOSFET's as a function of B_2H_6 flow rate

obtained for μ_{eff}. In Fig. 11, g_m is found to depend on the oxide thickness and B_2H_6 amount. About 500 μS/mm for gate length of 20 μm is obtained. From these results, V_t is found to be controlled and μ_{eff}, g_m to be affected by the doping amount of B_2H_6.

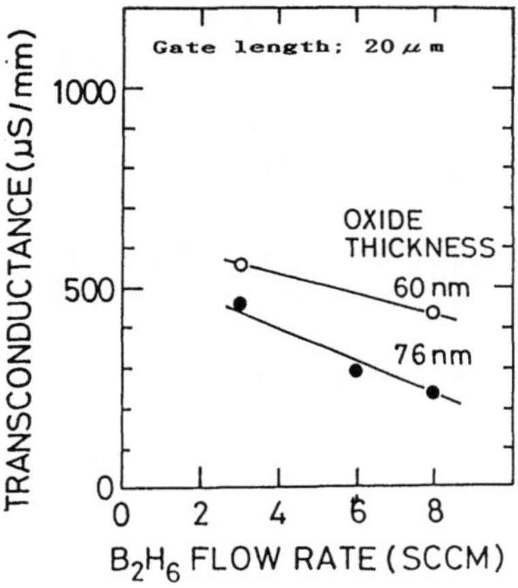

Fig. 11 Tansconductances of MOSFET's as a function of B_2H_6 flow rate and oxide thickness

4. Summary

Our recent work on crystal growth of β-SiC on Si and its application to MOSFET fabrication has been reviewed from the standpoint of developing various processes similar to Si processes. A new approach to reduce the stacking faults in the films was presented for the crystal growth. Several important device processes for MOSFET's were examined. MOS interface characteristics are quite excellent, while there remain some problems to be solved for ion implantation. The authors consider the various fundamental problems for the fabrication of MOSFET's already solved. As the next stage, using SiC films of much higher quality including homoepitaxial films to get excellent p-n junction interfaces, high-power MOS devices will be designed and realized.

Acknowledgment

The authors would like to thank Prof. Y. Yasuda, Dr. S. Zaima, and K. Onoda of the Department of Crystalline Materials Science, Nagoya University, for their valuable cooperation in the study of MOS structures. They are also much indebted to Y. Okamoto for SIMS analysis.

References

1. K. Shibahara, S. Nishino, and H. Matsunami, J. Cryst. Growth, 78, 538 (1986).
2. For example, R. F. Davis, Z. Sitar, B. E. Williams, H. S. Kong, H. J. Kim, J. W. Palmour, J. A. Edmond, J. Ryu, J. T. Glass, and C. H. Carter, Jr., Mater. Sci. Eng. B1, 77 (1988).
3. M. Shigeta, Y. Fujii, K. Furukawa, A. Suzuki, and S. Nakajima, Appl. Phys. Lett. 55, 1522 (1989).

4. K. Furukawa, Y. Fujii, A. Suzuki, and S. Nakajima, to be published in Epitaxial Heterostructures, Proceedings of Materials Research Society Symposia, 1990, San Francisco.
5. K. Onoda, Y. Fujii, S. Zaima, Y. Koide, K. Furukawa, A. Suzuki, S. Nakajima, and Y. Yasuda, in Extend. Abst. 50th Autumn Meet. Jpn. Soc. Appl. Phys. 1989, p. 240.
6. J. A. Edmond, K. Das, and R. F. Davis, J. Appl. Phys. 63, 922 (1988).
7. K. Shibahara, T. Takeuchi, S. Nishino, and H. Matsunami, Jpn. J. Appl. Phys. 28, 1341 (1989).
8. K. Shibahara, T. Saito, S. Nishino, and H. Matsunami, IEEE Electron Device Lett. EDL7, 692 (1986).
9. J. W. Palmour, H. S. Kong, and R. F. Davis, J. Appl. Phys. 64, 2168 (1988).

Sensor Properties of n-Type β-SiC

J.S. Shor, D. Goldstein, and A.D. Kurtz*

Kulite Semiconductor Products, Leonia, NJ 07605, USA
*Also: Columbia University Microelectronics Sciences Laboratories, New York, NY 10025, USA

In this paper, the properties of n-type ß-SiC for piezoresistive pressure and strain sensors are reported. Namely, the gauge factor (GF) and temperature coefficient of resistivity (TCR) of unintentionally doped and nitrogen doped ß-SiC are examined over a range of temperatures. The maximum GF of high resistivity unintentionally doped SiC is -31.1. For the nitrogen doped samples, this GF decreases to -25.9. The TCR of the unintentionally doped samples (0.2-0.7 Ω-cm.) was 0.72 %/°C over the range 25-650°C, and the donors were fully ionized at room temperature. For the nitrogen doped samples (0.02 Ω-cm.), the donors were not fully ionized until ≈ 250°C. Thus, below 250°C, the TCR was negative. Between 300-550°C, the TCR was positive and approximately constant with a value of 0.09 %/°C. A decrease in resistance due to intrinsic carrier generation was not observed for any of the samples up to the limits of temperatures studied. The TCR's of these samples are similar to those of other semiconductors, and the gauge factors are much higher than those of metallic foil gauges (GF≈2), demonstrating that SiC is a potentially useful material for high temperature microsensors.

Introduction

Semiconductor materials, such as silicon, have proven to be very useful for pressure and strain sensors due to their large piezoresistive coefficients [1]. Currently, there are semiconductor pressure sensors available for operation at temperatures up to 500°C. However, there is a demand for sensors to operate at 500-800° C in areas such as wind tunnels and jet engines. A promising semiconductor material for this temperature range is SiC. Its wide bandgap, high thermal conductivity and high melting point make SiC a potentially useful material for high temperature environments. In this paper, we will evaluate the feasibility of SiC for sensors by examining its piezoresistive and sensor properties, namely its gauge factor (GF) and temperature coefficient of resistivity (TCR). These parameters will be reported for undoped n-type and nitrogen doped n+ ß-SiC. The application of ß-SiC for high temperature sensors will also be discussed.

The piezoresistive properties of SiC have not been fully characterized in the literature. There was some early Russian work done on the piezoresistance of α-SiC [2,3]. The piezoresistive coefficients of these crystals are reasonably large (e.g. high resistivity n-type 6H-SiC has $\pi_{11}=-142\times10^{-12}$ cm^2/dyne at 273°K [2]). There was only one study done on the piezoresistive properties of ß-SiC [4], and that reported only the hydrostatic pressure coefficient, $\pi_{11}+2\pi_{12}$, of n-type material.

Editors: G.L. Harris, M.G. Spencer, and C.Y. Yang

Experimental

Epitaxial n-type ß-SiC films were grown on Si by CVD at the NASA Lewis Research Center [5]. These films were of the <100> orientation and were 10 μm thick. Samples of different resistivities were tested to determine the effect of carrier concentration on the sensor properties. Unintentionally doped samples had resistivities ranging from 0.2-0.7 ohm-cm while nitrogen doped films had resistivities of 0.02 ohm-cm. The samples were metallized and patterned, and the silicon substrates were removed with HF:HNO_3, leaving a self supported rectangular SiC strip with contacts and leads at both ends. For all TCR measurements, Kelvin contacts were used.

Since piezoresistance is a tensor property, several gauge configurations were tested (fig. 1). By applying a stress along the axis of the strips in these configurations and measuring the change in resistivity, the three piezoresistive coefficients can be calculated. The quantity measured in these experiments was the gauge factor which is defined as:

$$GF = \delta R / \epsilon R \quad \text{where } \epsilon = \text{applied strain.}$$

A positive gauge factor would correspond to an increase in resistance with tensile strain, while a negative GF would correspond to a decrease. The gauge factor of a given configuration is related to the piezoresistive coefficient by Young's Modulus and some dimensional corrections, which are relatively small [6]. To measure the gauge factor, the SiC strips (strain gauges) were mounted to a steel cantilever beam with epoxylite 8121 strain gauge cement, and tensile strain was applied by bending the

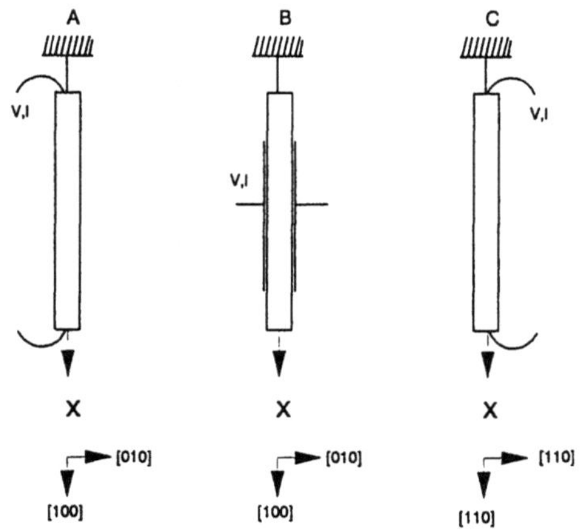

$$\frac{\delta\rho}{X\rho} = \Pi_{11} \qquad \Pi_{12} \qquad \frac{1}{2}(\Pi_{11}+\Pi_{12}+\Pi_{44})$$

Fig. 1: Piezoresistance measurements and the coefficients measured. X=applied stress. Stress/strain = Young's modulus Gauge factor measurements express these quantities in terms of strain instead of stress.

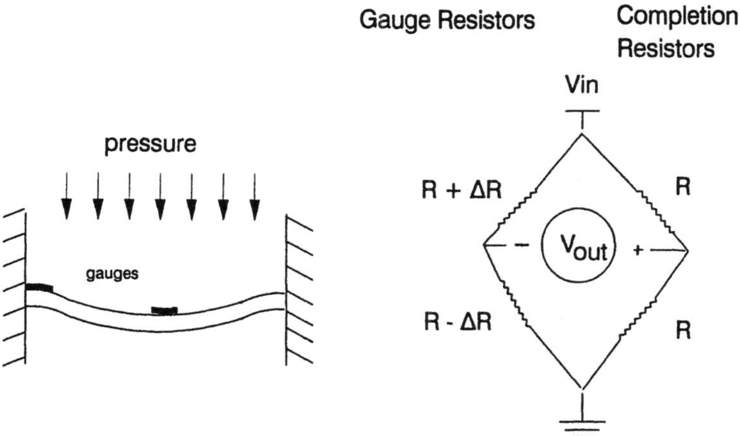

Fig 2: a) Strain gauges mounted on a circular diaphragm   b) Half bridge configuration

beam. Control p-Si strain gauges were used to determine the experimental error introduced by mounting in each experiment, which is estimated to be within 10%.

In order to simulate the operation of a pressure transducer, two n-type (0.7 ohm-cm) ß-SiC strain gauges of the "A" configuration were mounted onto a thin circular steel diaphragm in a wheatstone half bridge configuration [7], as is shown in figures 2a and 2b. The mounting agent used was Corning 7555 fritted glass. When a pressure is applied to the diaphragm, causing it to deform, the strain gauge in the center of the diaphragm will be subjected to a compression, while the one at the edge will be placed in tension. In this configuration, the bridge will yield an output:

$$V_{out} \propto GF \times \text{Pressure}.$$

By plotting V_{out} vs. temperature, the temperature dependency of the gauge factor can be measured. This was done over the temperature range 25-350°C.

Results

The room temperature gauge factors of n-type ß-SiC are listed in table 1. For the undoped samples, all three configurations of figure 1 were measured. Configuration "A" has the highest magnitude with GF=-31.8. The gauge factors of silicon that appear in the table are those listed in the literature. The gauge factor of n+ SiC in configuration "A" (GF=-26.6) is slightly lower than the undoped samples, and both of these are considerably lower than the GF of n-Si which is -133.

In figure 3, the variation of bridge output with temperature is plotted for gauges of the "A" configuration. The output shows a steady decrease over the temperature range 25-200°C, but begins to level off beyond this point. At 350°C, the output is almost completely constant with temperature, at approximately 54% of its room temperature value.

Table 1: Room temperature gauge factors of ß-SiC

Material	resistivity (ohm-cm)	GF_A	GF_B	GF_C
n ß-SiC	0.7	-31.8	+19.2	-3.7
n^+ ß-SiC	0.02	-26.6		
n Si	11.0	-133.0	+68.3	-52.0

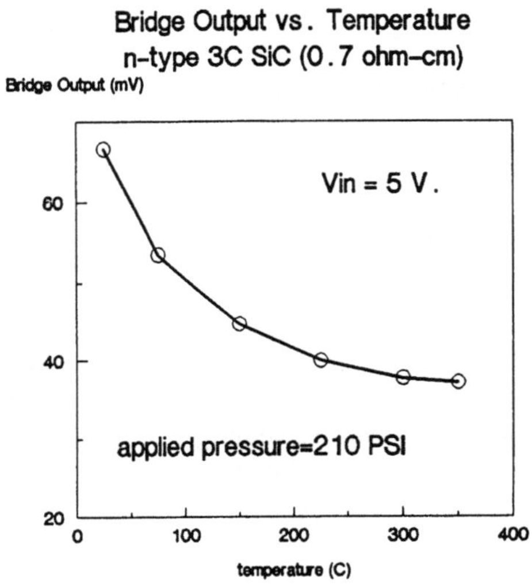

Fig. 3: Output of half bridge circuit. Sensing elements are two n-type 3C-SiC strain gauges mounted on a circular steel diaphragm with fritted glass.

In figure 4, the resistance of undoped and nitrogen doped samples is plotted against temperature for calculation of the TCR which is defined as:

$$TCR=\{R(T)-R(T_{ref})\}/\{T-T_{ref}\}.$$

The values of resistance have been normalized to their room temperature values so that various resistivities could be compared on a linear scale. The undoped samples (0.2-0.7 ohm-cm) show an initial decrease in resistance with temperature over the range -150°C to -75°C. Above -50°C, the resistance increases with temperature and above room temperature this increase is approximately linear. The TCR of these samples between room temperature and 650°C is 0.72 %/°C, which is similar to Si samples of comparable carrier concentration ($\approx 10^{17}$ /cm^3). The resistance of the nitrogen doped films varies much less with temperature, as can be seen in figure 4a. The TCR of n+ SiC is shown on an enlarged scale in fig. 4b. These samples exhibit a negative TCR up to 200°C. Only after 350°C does the resistance increase linearly with temperature, yielding a TCR of 0.091 %/°C between 350-550°C.

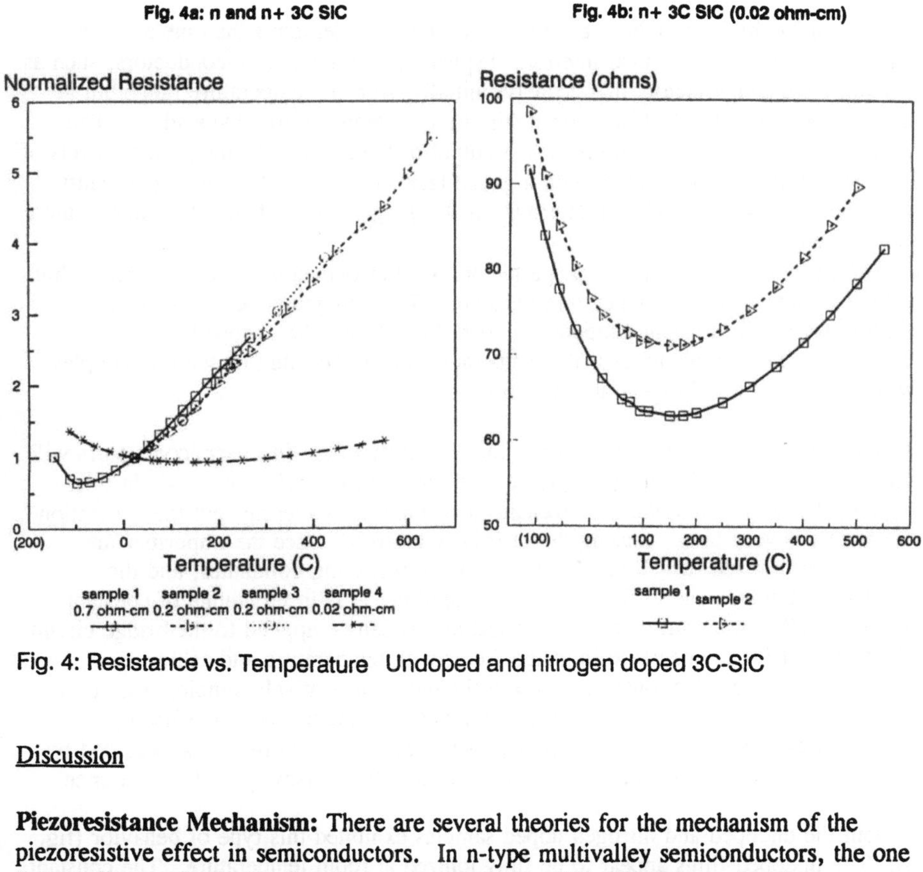

Fig. 4: Resistance vs. Temperature Undoped and nitrogen doped 3C-SiC

Discussion

Piezoresistance Mechanism: There are several theories for the mechanism of the piezoresistive effect in semiconductors. In n-type multivalley semiconductors, the one that applies most often is the electron transfer effect [1]. This theory has been accepted for n-Si. The theory states that under certain strain configurations, electrons will redistribute themselves among different multivalleys in momentum space, causing a change in the population of carriers which have mobilities parallel or perpendicular to the major axis of the valley. If the mobility is anisotropic within the valley, this will result in a change in resistivity under strain. Electron cyclotron resonance experiments have determined that ß-SiC has its multivalley minima in the <100> direction [8]. This is also the case for n-Si. For such a semiconductor, the theory predicts that π_{11} will be the largest coefficient with $\pi_{12} \approx -\frac{1}{2}\pi_{11}$ and $\pi_{44} = 0$. Deviations from this behavior can be due to other, smaller effects. The data in table 1 shows that for n-type ß-SiC, $\pi_{12} \approx -0.6\pi_{11}$ and that π_{44} is the smallest coefficient of the three. This implies that the electron transfer effect is most likely the dominant mechanism for the piezoresistance of n-type 3C-SiC, with other effects playing a small role as well.

Gauge Factor: The room temperature gauge factor of -31.8 exhibited by ß-SiC is considerably larger than those of conventional foil gauges, which typically have GF ≈ 2. At the temperatures of interest for SiC, 500-800°C, other semiconductor strain gauges which may have comparable gauge factors to SiC are not readily available.

Thus if SiC exhibits a higher gauge factor at these temperatures than metallic foil gauges, it will be of practical interest. Typically, low doped semiconductors, such as silicon, have gauge factors that decrease initially over the temperature range but at high temperatures level off and eventually become temperature independent. This behavior is exhibited in figure 3 for the output of the SiC half bridge, which levels off with temperature at 300-350°C with a gauge factor of ≈ -15. Further experiments over the range 400-800°C will determine if the gauge factor retains this value at these temperatures.

At higher doping levels, the gauge factors of semiconductors usually decrease, but exhibit a more constant temperature behavior. The nitrogen doped films have a slightly lower GF at room temperature (16% lower) than the undoped ones. Preliminary measurements over the temperature range indicate that the n+ samples will level off to a GF of -14 at 350-425°C.

TCR: In general, at low temperatures, the resistivity of extrinsic semiconductors will decrease with increasing temperatures due to the ionization of impurities. In this regime, the dominant scattering mechanism is impurity scattering, but the generation of carriers is very large, causing the resistivity to drop. Once the impurities are almost fully ionized (usually at T ≈ 0°C), lattice scattering dominates, and the resistivity increases due to the decrease in mobility. It is in this range of constant, positive TCR that circuit compensation techniques can be applied to the bridge circuit [7] to correct for the decrease in gauge factor with temperature and achieve a temperature independent output. These techniques are relatively simple, since they only entail placing appropriate resistors in series with the bridge. At very high temperatures (300-1000°C), the impurity carrier concentration becomes swamped by the thermal generation of intrinsic carriers, causing the resistivity to decrease once again.

Both the undoped and nitrogen doped samples exhibited this type of behavior (fig. 4). The undoped films appear to be fully ionized at room temperature. The constant TCR over the range 25-650°C makes these samples potentially useful within this temperature range. The nitrogen doped films are not fully ionized until ≈ 250°C, and the curve is linear only in the range 350-550°C, limiting the usefulness of the n+ samples to higher temperatures only. The TCR of the n+ samples is much lower than that of the undoped ones, which is typical for highly doped semiconductors. Intrinsic carrier generation was not observed for any of the samples up to the temperatures studied. Investigations of carrier concentration vs. temperature by other authors indicate that no intrinsic carriers appear up to 1000°K [9].

Conclusion

We have tested the GF and TCR of n-type and n+ ß-SiC. The maximum gauge factor of n-type material is -31.8, while the n+ samples show a GF=-26.6. These gauge factors are substantially higher than those of conventional foil gauges. The gauge factor decreases over the temperature range 25-200°C, but levels off to a value of ≈ -15 at 350°C. The n-type unintentionally doped samples are fully ionized at room temperature and exhibit a constant TCR = 0.72 %/°C over the range 25-650°C, making them potentially useful at these temperatures. The n+ samples have a much lower TCR (0.092 %/°C over the range 350-550°C), but are not fully ionized until

250°C. These low temperature studies indicate that SiC has good sensor properties. Future studies of the gauge factor and TCR over the range 400-800°C will further determine the applicability of SiC at these temperatures.

Acknowledgements

We would like to thank H. Bernstien and R.A. Weber for their useful discussions as well as A. Boseman and S. Goodman for their technical assistance. We would also like to thank Tony Powell from NASA Lewis for his helpful suggestions and for generously supplying the samples used in this study.

References

1. R.W. Keyes: *Solid State Physics* **11**, 149 (Academic Press, NY 1960)
2. I.V. Rapatskaya, G.E. Rudashevskii, M.G. Kasaganova, M.I. Iglitsin, M.B. Reifman, and E.F. Fedotova : Sov. Phys. - Solid State **9**, 2833 (1968)
3. G.A. Lomakina: Sov. Phys. - Solid State **17**, 1808 (1976)
4. G.N. Guk, N. Ya. Usol'tseva, V.S. Shadrin, and N.K. Prokop'eva: Sov. Phys. Semicond. **10**, 83 (1976)
5. J.A. Powell, L.G. Matus, and M.A. Kuczmarski: J. Electrochem. Soc. **134**, 1558 (1987)
6. P.J. French and A.G.R. Evans: Solid State Electron. **32**, 1 (1989)
7. U.S. Patent # 3,245,252: D.J. First, A.D. Kurtz, J.A. Pugnaire (1966)
8. R. Kaplan, R.J. Wagner, H.J. Kim, R.F. Davis: Solid State Comm. **55**, 67 (1985)
9. A. Suzuki, A, Uemoto, M.Shigeta, K. Furukawa, and S. Nakajima: Appl. Phys. Lett. **49**, 450 (1986)

Charge Trapping in Cubic Silicon Carbide MIS Capacitors

J.J. Kopanski[1] *and R.E. Avila*[2]

[1]Semiconductor Electronics Division, National Institute of Standards and Technology, Gaithersburg, MD 20899, USA
[2]Department of Electrical and Computer Engineering, University of Puerto Rico at Mayaguez, Mayaguez, Puerto Rico 00709

The charge trapping properties of the insulator and the insulator-SiC interface of cubic SiC metal-insulator-semiconductor (MIS) capacitors have been studied. The interface trap level density, D_{it}, was determined by the high-frequency capacitance-voltage and the conductance-voltage techniques. The number of active interface traps increases sharply in the range from room temperature to 260°C. SiC MIS capacitors exhibit a slow-trapping instability when subjected to a stress voltage. Both the bulk oxide trap density, N_{ot}, and the D_{it} are seen to increase during a voltage stress. The conduction mechanism in thermal oxide layers on SiC is limited by Fowler-Nordheim emission with a barrier height of about 2.9 eV.

1. Introduction

A crucial component of metal-oxide-semiconductor field-effect-transistors (MOSFETs) is the MOS capacitor, which is used to control the conductivity of the channel between the oppositely doped source and drain regions. MOS capacitors on both cubic (3C) [1-5] and hexagonal (6H) [6-7] SiC, as well as operating MOSFETs, have been reported. These reports dealt with devices fabricated using a limited number of oxidation conditions. They demonstrated that unpinned MOS capacitors on SiC could be made, i.e., the surface of the underlying SiC could be inverted.

The promise of SiC devices is operation at high temperatures and electric fields. If devices incorporating MIS capacitors are to function under these conditions, their stability with respect to temperature and electric field stress must be established. In the present study, the effect of oxidation conditions on the electrical properties of MOS capacitors was determined. The capacitance-voltage (C-V), conductance-voltage (G-V), and current-voltage (I-V) behavior of the capacitors was analyzed to determine the material properties of the SiC and the properties of the insulator-SiC interface. All four charge densities commonly present in MIS capacitors (oxide fixed, N_f, oxide trapped, N_{ot}, interface trapped, N_{it}, and mobile ionic, N_m, charge densities) were quantified.

2. Capacitor Fabrication

MIS capacitors were fabricated on 3C SiC heteroepitaxially grown [8] on (001) silicon substrates that were either on-axis or tilted 0.5° to 1° off-axis. SiC grown on on-axis Si contains inversion-domain-boundaries (IDBs), while SiC grown on off-axis Si does not [9]. The SiC layers had n-type conductivity with electron carrier densities in the low 10^{16} cm^{-3} range.

Springer Proceedings in Physics, Volume 56 Amorphous and Crystalline Silicon Carbide III
Editors: G.L. Harris, M.G. Spencer, and C.Y. Yang

Thermal oxides were grown at temperatures of 1050, 1100, 1150, and 1200°C in hydrogen and oxygen. Oxidation time was adjusted to produce an oxide thickness of about 600 Å [10]. Capacitors were also fabricated with chemical-vapor-deposited silicon dioxide as the insulating layer. Sputter-deposited Al (with 1.5% Si) was used to make 305-μm-diameter field plates and top-side ohmic contacts through contact openings in the insulating layers. Capacitors with thermal oxide insulating layers were microalloyed in 10% hydrogen, 90% nitrogen at 450°C after Al contact deposition. Additional details of the fabrication process are available elsewhere [3].

3. Interface Charge Density

Values of N_f between 4 and 9×10^{11} cm^{-2} were determined from the difference between the measured and theoretical values of the flatband voltage. The microalloying step was critical in reducing N_f. D_{it} was determined from the high-frequency C-V characteristics using the Terman technique [11]. The interface traps affected the capacitance most strongly for ac measurement frequencies, ω, less than 400 kHz. Both N_f and D_{it} were found to have no clear dependence on the oxide growth temperature.

C-V curves measured at 1 MHz at temperatures between 23 and 260°C for a SiC capacitor with a thermal oxide insulator are shown in Fig. 1. Interface trap profiles, calculated from these C-V curves [11], are shown in Fig. 2. As the measurement temperature is increased, greater numbers of interface traps become active, increasing the stretchout in the C-V curve. This change in stretchout is not permanent, but is solely dependent on the measurement temperature. The limitations of this technique when applied to SiC are such that the determined D_{it} is less accurate near the conduction band edge and below mid-gap. The interface trap capacitance has not been taken into account in this calculation, leading partly to the dip of D_{it} near the conduction band edge, E_c. Also, at the sweep rate used (100 mV/S) and at low temperature, traps deep in the band gap may not follow the dc signal, leading to the apparent drop of D_{it} after the peak located about 0.6 eV below E_c. Notice that the origin of the energy axis in Fig. 2 is at E_c, which decreases slightly with rising temperature.

D_{it} has also been calculated from G-V measurements [12] at frequencies between 5 and 100 kHz. The values of D_{it} were calculated from $< G_p/\omega > /0.4q$, where q is the electron charge and G_p is the conductance parallel to the depletion capacitance,

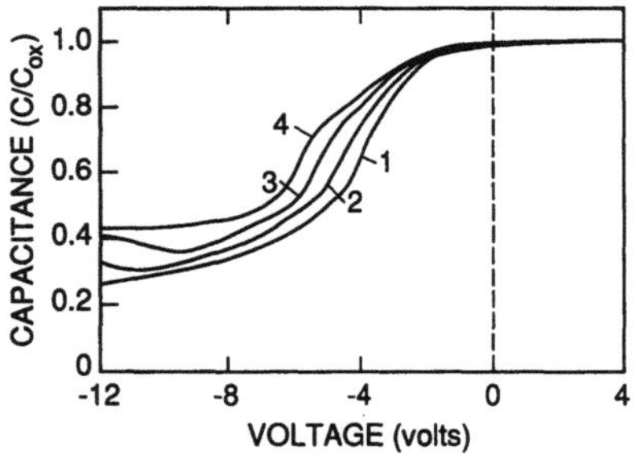

Fig. 1. C-V characteristics of a SiC MIS capacitor at 1) 23°C; 2) 115°C; 3) 175°C; and 4) 260°C

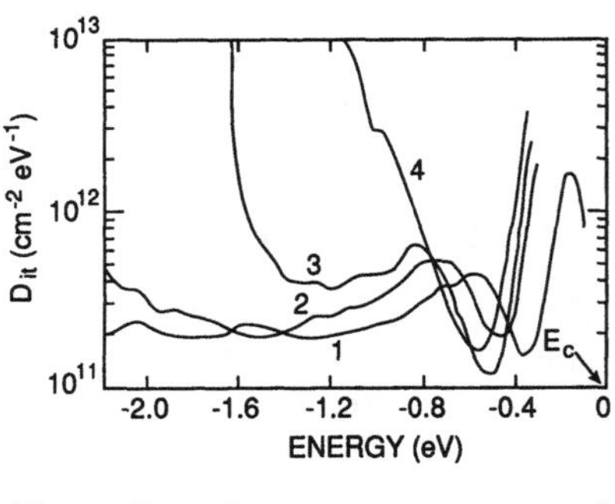

Fig. 2. D_{it} values vs. temperature calculated from the C-V curves in Fig. 1

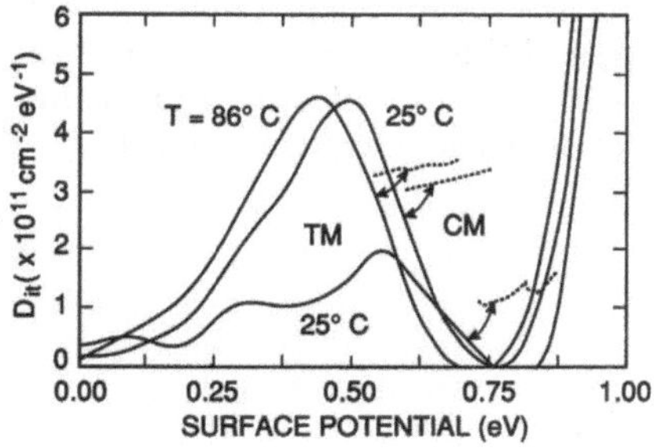

Fig. 3. Interface trap density calculated using G-V measurements (CM) and the Terman method (TM). The higher curves are from the same device, the lower curve from another

after eliminating the series resistance and taking the oxide capacitance into account. Typical results are depicted in Fig. 3. Here the energy axis is referenced to mid-gap (Surface Potential = 0). Band bending fluctuations were not taken into account in this calculation. For comparison, the Terman method calculation is also shown in Fig. 3. D_{it} calculated from conductance is regarded as more reliable than that from the Terman method. A preliminary calculation of D_{it} using capacitance vs. frequency measurements agrees with the rise of the Terman method result near the conduction band edge. This rise is consistent with what is usually found in silicon.

4. Oxide Conduction Mechanism

The I-V characteristics of a typical capacitor are shown in Fig. 4. The oxide resistivity, ρ, was estimated from the current seen at low voltages as $\rho = (V/t_{ox})/(I/A)$, with t_{ox} the insulator thickness and A the capacitor area. The average ρ of thermal oxides on SiC was about 10^{16} Ω-cm. The breakdown voltage was evident from the I-V measurements as a sharp increase in current. The breakdown electric field, E_{bd}, for all oxides on as-grown SiC was tightly distributed around 3.3×10^6 V/cm. Thermal oxide layers on SiC containing IDBs had slightly lower E_{bd}'s than on SiC without IDBs.

Similarly to SiO_2 thermally grown on Si, the electronic conduction mechanism in thermal SiO_2 on SiC was determined to be limited by Fowler-Nordheim (F-N) emission [13]. F-N emission is the tunneling of electrons from near the electrode Fermi level through the forbidden energy gap into the conduction band of the oxide. F-N conduction is identified by a J/F^2 dependence of the form [14]:

$$J/F^2 = Ce^{-\beta/F}, \tag{1}$$

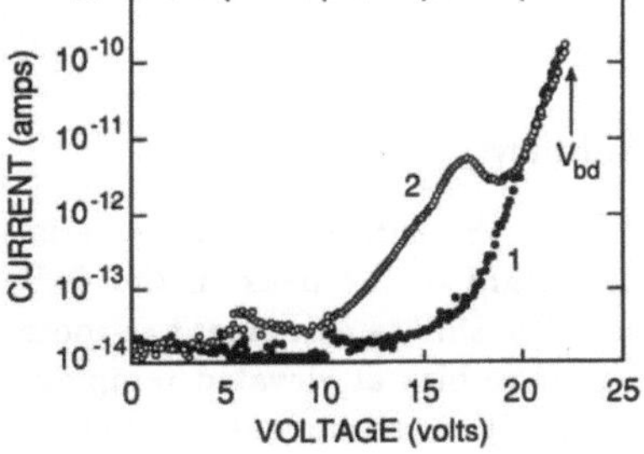

Fig. 4. Typical I-V characteristics for a SiC MOS capacitor in the 1) forward (accumulation) and 2) reverse (depletion) direction

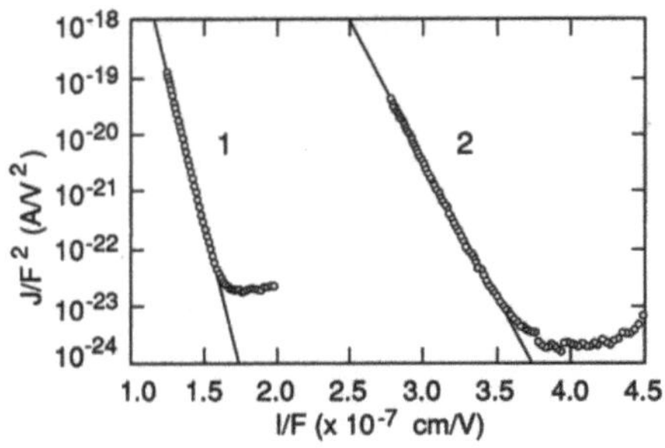

Fig. 5. Fowler-Nordheim plot from I-V characteristics of 1) Si MOS capacitor and 2) as-grown-SiC MOS capacitor showing the linear relation between $\log(J/F^2)$ and $1/F$

where J is the current density through the insulator and F is the electric field applied to the insulator. The pre-exponential term, C, and β are constants dependent on the effective mass, m_{ox}, of an electron in the insulator, and ϕ_B, the barrier between the emitting electrode and the insulator [13]. A plot of $\log(J/F^2)$ vs. $1/F$ of a capacitor limited by F-N emission will be a straight line with slope β and y-intercept C from which m_{ox} and ϕ_B can be determined.

All SiC capacitors with thermal oxides displayed I-V characteristics resembling Fowler-Nordheim tunneling when a positive voltage was applied. Positive voltages imply that electrons are tunneling from the substrate into the oxide. F-N plots showing the expected linear behavior for MOS capacitors on silicon and as-grown-SiC are shown in Fig. 5. The SiC-SiO_2 barrier height was determined by fitting the F-N formula to the observed I-V characteristics using an m_{ox} of 0.5 the mass of the electron in free space [14]. The oxide thickness was estimated from the oxide capacitance.

The apparent barrier height between as-deposited 3C SiC and SiO_2 was determined as approximately 1.8 eV, compared to 2.9 eV for Si [14]. This is substantially less than the predicted value of about 3.1 eV, based on a SiC electron affinity of 4.3 eV [15]. The rough surface of the as-deposited SiC layers results in thin spots in the oxide or regions of enhanced electric field near sharp protrusions. These non-idealities result in current flowing at lower fields and an apparent lowering of the SiC-SiO_2 barrier height.

Subsequential I-V measurements were made using MOS capacitors fabricated on SiC mechanically polished with 1-μm diamond paste and then oxidized and etched to remove the damaged layer. Capacitors on this substrate had an oxide thickness of 560 Å, and E_{bd} from 7 to 8×10^6 V/cm. The barrier height determined from I-V measurements of capacitors on polished SiC varied from about 2.7 to 2.9 eV. While the surface of the SiC appears nearly featureless (even under ×70 magnification), some microscopic roughness may remain. Thus, the value of 2.9 eV should be regarded as a lower limit of the actual barrier height that would be measured with an ideal, atomically abrupt SiC-SiO_2 interface.

5. Bulk Oxide Trap Density and Capacitor Stability

A negative (depletion) voltage stress results in a shift of the entire C-V characteristic to more negative voltages and an increase of the height of the peak in the G-V curve, indicating that both N_{ot} and D_{it} were increased. The shift is permanent at room temperature, but could be reversed somewhat by a positive bias at elevated temperature. No corresponding voltage shift was observed for a positive (accumulation) voltage

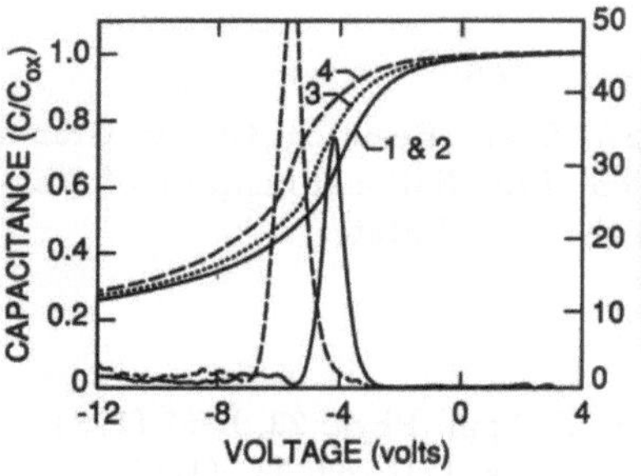

Fig. 6. C-V and G-V response at 10 kHz for a SiC MOS capacitor showing the "slow trapping" instability 1) before any voltage stress; and after 10-min, room-temperature voltage stresses at 2) -10 V; 3) -15 V; and 4) -20 V

stress. Typical shifts in the C-V and G-V characteristics as a result of voltage stresses are shown in Fig. 6. This type of response to an electric field stress is similar to the "slow trapping" instability reported for MIS capacitors on Si [16-17] and other semiconductors. The sense of the shifts indicate that positive (hole) traps have been created or charged in the oxide.

The changes in capacitor N_{ot} and D_{it} following a voltage stress were deduced from the relative shifts of the C-V curve at the flatband, mid-gap, and inversion potentials [18]. A negative voltage stress increased N_{ot} by as much as 5×10^{11} cm^{-2} and increased D_{it} by as much as 5×10^{10} cm^{-2} eV^{-1} (about 25% greater than its initial value).

There is a distinct dip in the I-V curve measured in the depletion region (Fig. 4) roughly correlated with the previously observed recovery of the capacitance from deep depletion [2-3]. Negative voltages imply that electrons are tunneling or are injected from the gate. Similar ledges or dips have been reported in the I-V characteristics of Si MOS capacitors when the current reaches a level at which bulk SiO_2 traps start to fill significantly [19].

For Si under high negative fields, it has been suggested that hot holes are generated by impact ionization by electrons entering the Si from the oxide conduction band [20]. If hot holes were generated in such a manner in SiC, they could then tunnel from the SiC valence band into the oxide where they would become trapped. This phenomenon is an attractive explanation as it accounts for both the generation of an inversion layer in the SiC through a process that creates holes in the SiC and the positive charge creation in the oxide. Electron-hole pairs may also be created within the oxide via impact ionization by electrons injected into the SiO_2 by Fowler-Nordheim tunneling [21].

The extent of contamination with mobile ionic charge, N_m, was determined from bias-temperature drift experiments [22]. Shifts in the C-V curve of the sense expected for mobile ionic contamination were observed. The minimum N_m observed was 5×10^{10} cm^{-2}.

6. Summary

The oxide charges in a SiC MOS capacitor with a thermal oxide were insensitive to oxide growth temperatures between 1050 and 1200°C. The number of active interface traps was seen to increase with temperature. SiC MOS capacitors were seen to exhibit a slow trapping instability when subject to a voltage stress. A voltage stress created/charged as many as 5×10^{11} cm^{-2} bulk oxide traps and D_{it} was increased by as much as 25%. Fowler-Nordheim tunneling has been identified as the oxide conduction mechanism, with a ϕ_B of about 2.9 eV with $m_{ox} = 0.5$. The I-V characteristics showed a region of decreasing current with voltage when measured in the depletion direction.

Acknowledgments

The authors acknowledge the support of NASA Lewis Research Center Grants #C-30007-K and #C-30018-M. Official contribution of the National Institute of Standards and Technology, not subject to copyright in the United States.

References

1. K. Shibahara, S. Nishino, H. Matsunami: Jpn. J. Appl. Phys. 23, L862 (1984).
2. R. E. Avila, J. J. Kopanski, C. D. Fung: Appl. Phys. Lett. 49, 334 (1986).
3. J. J. Kopanski and D. B. Novotny: Extended Abstracts of the 176th Electrochemical Society Meeting 89-2, Hollywood, Fla., 1989, p. 722.
4. M. I. Chaudhry and W. B. Berry: J. Mater. Res. 4, 1491 (1989).
5. S. M. Tang, W. B. Berry, R. Kwor, M. V. Zeller, L. G. Matus: J. Electrochem. Soc. 137, 221 (1990).
6. R. C. A. Harris: Solid State Electron. 19, 103 (1976).
7. A. Suzuki, H. Ashida, N. Furui, K. Mameno, H. Matsunami: Jpn. J. Appl. Phys. 21, 579 (1982).
8. S. Nishino, J. A. Powell, H. A. Will: Appl. Phys. Lett. 42, 460 (1983).
9. J. A. Powell, L. G. Matus, M. A. Kuczmarski, C. M. Chorey, T. T. Cheng, P. Pirouz: Appl. Phys. Lett. 51, 823 (1987).
10. C. D. Fung and J. J. Kopanski: Appl. Phys. Lett. 45, 757 (1984).
11. L. M. Terman: Solid State Electron. 5, 285 (1962).
12. E. H. Nicollian and A. Goetzberger: Bell Syst. Tech. J. 46, 1055 (1967).
13. M. Lenzlinger and E. H. Snow: J. Appl. Phys. 40, 278 (1969).
14. Z. A. Weinberg: J. Appl. Phys. 53, 5052 (1982).
15. M. A. Kadyrov, N. V. Kunina, A. S. Saidov, Kh. A. Shamuratov, A. Ya. Shik, and Yu. V. Shmartsev: Sov. Phys. Semicond. 14, 942 (1980).
16. S. R. Hofstein: Solid-State Electron. 10, 657 (1967).
17. D. Lu, G. A. Ruggles, and J. J. Wortman: Appl. Phys. Lett. 52, 1344 (1988).
18. P. S. Winokur, J. R. Schwank, P. J. McWhorter, P. V. Dressendoefer, D. C. Turpin: IEEE Trans. Nucl. Sci. NS-31, 1453 (1984).
19. D. J. DiMaria and D. W. Dong: J. Appl. Phys 51, 2722 (1980).
20. Z. A. Weinberg, W. C. Johnson, M. A. Lampert: J. Appl. Phys. 47, 248 (1976).
21. H. E. Boesch and J. M. McGarrity: IEEE Trans. Nucl. Sci. NS-26, 4814 (1979).
22. E. H. Snow, A. S. Grove, B. E. Deal, C. T. Sah: J. Appl. Phys. 36, 1664 (1965).

Development Prospects for SiC Bipolar Transistors and Thyristors

V.E. Chelnokov

A.F. Ioffe Physical-Technical Institute, Academy of Sciences, Polytechnicheskaya, 26, 194021 St. Petersburg, Russia

Abstract. This review describes our investigations on SiC-based bipolar-type devices including: diodes, transistors, and thyristors. The structure and characteristics of the SiC dynistor and bipolar transistor will be discussed.

1. Introduction

We consider discrete SiC-based bipolar-type devices (relying on minority charge carriers) including: diodes, transistors, and thyristors.

The basic element in the structure of these devices is a contact for injecting minority carriers, which is a homo- or heterojunction. We deal with an abrupt nondegenerate p-n junction which is uniform across the junction area, its major parameters being: admissible operating temperature T_w and limiting (maximum) temperature T_m, maximum reverse (breakdown) voltage V_m, admissible forward current density j_m, speed of operation.

These major parameters determine the useful switched power of the devices and the operating frequency range with an upper limiting frequency f_m.

These basic parameters are considered for a p-n junction in SiC-6H [1]. For comparison, we referred to silicon p-n junctions as the most commonly available, though junctions in III-V compounds could also be used for this purpose. It was assumed that the ratio W/L_d could be varied in the same ranges as in the case of silicon (L_d is the minority carrier diffusion length and W the thickness of the junction base). As the result of our investigation, we have demonstrated that silicon carbide could be the material for producing injection devices such as diodes [2-6] and thyristors [7,8], in which specific switched power and switching speed exceed those in silicon devices by more than an order of magnitude, with operating temperatures up to 1000°C.

2. Experimental Results

2.1. Dynistor

Structure

The area of the top of the mesa was $0.15 \times 0.15 mm = 2.2 \cdot 10^{-4} cm^2$. It is seen that in addition to the usual four layers the structure

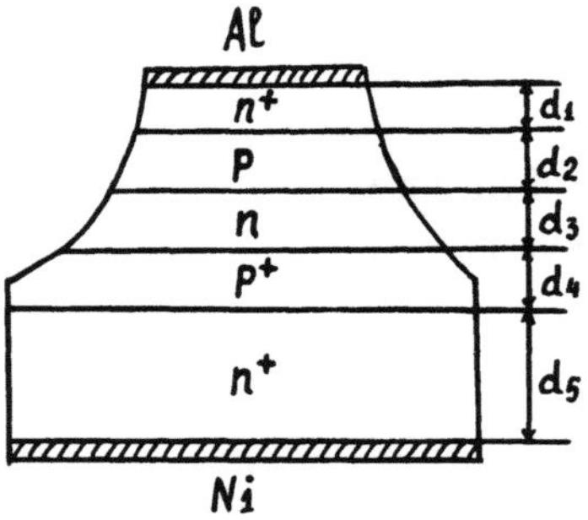

Fig.1 Schematic cross-section of SiC dynistor:

d_1=2-3 μm $N_D-N_A=10^{19}cm^{-3}$,
d_2=3 μm $N_A-N_D=10^{18}cm^{-3}$,
d_3=1 μm $N_D-N_A=5\cdot10^{17}cm^{-3}$,
d_4=20 μm $N_A-N_D=10^{19}cm^{-3}$,
d_5=300 μm $N_D-N_A=3\cdot10^{18}cm^{-3}$.

has a fifth layer with an extra junction (p^+n^+) contributing to the structure resistance in the forward direction. The structure illustrated in Fig.1 was fabricated by V.A.Dmitriev [7] using the method of container-free liquid-phase epitaxy.

Static current-voltage characteristics

Fig.2 shows forward and reverse current-voltage characteristics of the dynistor and may be summed up as follows. The reverse current-voltage characteristic of the dynistor for which the emitter p^+-n junction is responsible is that usual for a diode, as described in Ref.6. The value of V_m is 50-100 volt.

The forward current-voltage characteristic in a low current range is given in Fig.3.

V_{sw}=50-60 V and is essentially independent of T up to 200°C; j_{sw}=3-5 A/cm² and rises markedly as T is increased; V_h=2.5-5 V,

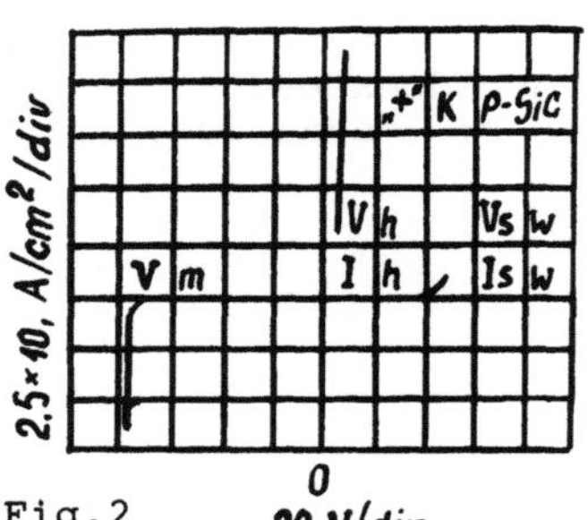

Fig.2

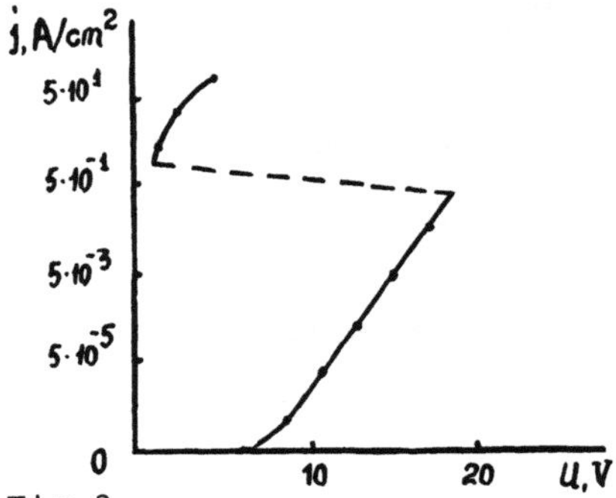

Fig.3

Fig.2 Static current-voltage characteristic of SiC dynistor.
Fig.3 Forward current-voltage characteristic of SiC dynistor in a low current range.

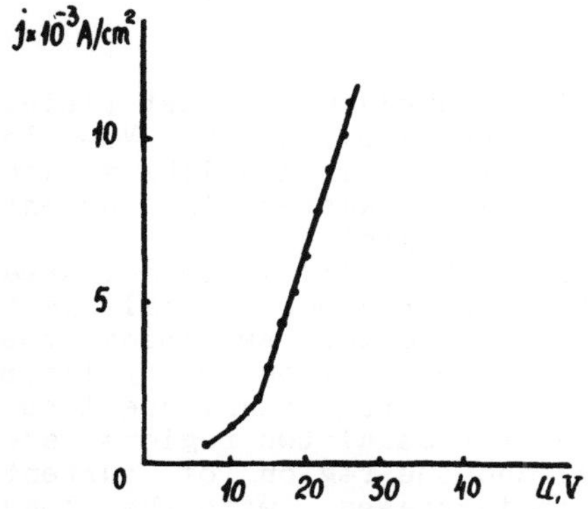

Fig.4 Forward current-voltage characteristic of SiC dynistor in the on-state.

a value that indicates the sign inversion of voltage across the p-n collector; V_h rises rather strongly as T is increased; j_h=2-10 A/cm² and rises with temperature dramatically.

Differential resistance r_{diff} in a current range $j > j_{diff}$ is $(2\text{-}2.5)\cdot 10^{-2}$ Ohm•cm² and drops with increasing j until it becomes constant above $j \approx 5\cdot 10^3$ A/cm².

In Fig.4 we give a forward current - voltage characteristic of the dynistor in the on-state. In a current range $j > 5\cdot 10^3$ A/cm² it is linear:

$$j=(V-V_{cut-off})/R_g$$

where $V_{cut-off}$=13 V, $r_{diff}=1.4\cdot 10^{-3}$ Ohm•cm².

Comparing this characteristic with those for diodes, one notes their identity concerning the value of r_{diff} at high currents and substantially larger value of $V_{cut-off}$ for the former [2-6].

The above data on the static forward current-voltage characteristic of the dynistor permit us to draw the following conclusions: 1) The current-voltage characteristic of the dynistor is explained by commonly accepted concepts of the theory of such devices, with major parameters of the characteristic being within theoretical predictions. In particular, the fact that the minimal value of V_h in the forward direction equals the voltage drop across of one p-n junction of the dynistor proves the occurrence of sign inversion of the voltage on the collector p-n junction. This observation is also supported by the value of r_{diff} at high currents. 2) A decrease of r_{diff} with increasing j observed above j_h can be explained by an increase in the effective contact area of the dynistor after switching-on (see below). 3) The large cut-off voltage $V_{cut-off}$ can be explained by introduction in the structure of the fourth junction p^+-n^+. 4) Anomalous temperature dependences of j_{sw}, V_h and j_h have not yet been fully explained. In particular, account should be taken of a decrease with temperature of coefficients for injection into the base regions of the emitter n^+-p junction and the collector n-p junction.

Speed of operation

The dynistor was switched on either by applying a fast-rising voltage dV/dt not exceeding V_{sw} or by applying $V > V_{sw}$. In visual observations the initial contact area looked like a ring at the sample periphery of about 3μm in diameter. The current density in this area amounted to 10^4-10^5 A/cm².

With a further increase in current, the initial contact area widens and above $j > 300$ A/cm² extends over a substantial part of the structure surface. A critical charge for switching the dynistor on amounts to 10^{17}cm^{-3}, while for silicon or gallium arsenide thyristors it is one to three orders ofmagnitude less.

On a current-time diagram (Fig.5) the usual two regions are seen, viz., the delay region (t_d) and the region of current rise (t_r). As the applied voltage is increased, with the load in an external circuit kept constant, the respective times both decrease to their minimum values, $t_d < 10$ ns, $t_r < 60$ ns. The rise time t_b for current determined with reference to 0.1 and 0.9 levels of j_{max} was 10-40 ns. The time constant τ_o at the initial (exponential) stage of current rise was 1-10 ns.

Switching the dynistor off after passing of the forward current pulse was performed by applying voltage in a reverse direction (i.e., by passing the current in the reverse direction) and the total switching-off time t_{off} was determined using a probing pulse of direct current. t_{off} was 100-200 ns and increased with forward current. The diameter of the current filament observed prior to transition into the off-state was 10μm, which corresponded to a j_h value of 10^2-10^3 A/cm².

Some estimates of the lifetime τ_p in the thin n-base give $\tau_p \approx 40$ns, i.e., $L_p \approx 1$ μm or $d_3/L_p < 1$. On the other hand, it is possible to have $d_2/L_n > 1$.

Thus, conditions for switching the dynistor into the on-state appear to be provided almost exclusively by the thin base. This is substantiated by some of the above data on the static current-voltage characteristic.

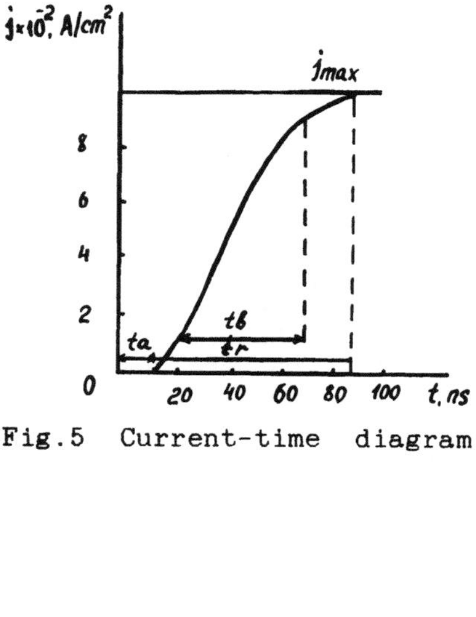

Fig.5 Current-time diagram.

2.2. Bipolar transistor

In this section, we present the first results of our recent research on the possibility of producing a SiC bipolar transistor.

The transistor structure grown by container-free liquid phase epitaxy (perfomed by V.A.Dmitriev and T.P.Samsonova) is analogous to that described by Munch [9]. The difference consisted in that an epitaxial n-layer has been pre-grown onto the n+ substrate, so that the resulting structure had the form n++ -p+ -n -n+. As became apparent, it has two essential disadvantages: very large emitter-base input resistance reaching 10^5Ohm at room temperature, which falls to $6 \cdot 10^3$Ohm at 200°C, and small diffusion length L_n=0.4μm or d_2/L_n=1.2. The latter was determined using the value of the emitter current gain β < 0.55 and assuming for the injection coefficient in the emitter junction a value of τ=1. Both these properties are fairly well explained taking into account the measurement results on p^+ layer properties obtained independently. Very large input resistance precluded measurements in a common emitter circuit, so a common base circuit was employed. Though in some samples the maximum voltage across the p^+-n collector V_m was as high as 100 V, measurements were made at V_{c-b}< 10 V.

3. Conclusion

Preliminary results of experiments aimed at realizing three basic injection-type SiC devices have been presented. Two major conclusions can be drawn from the above study: 1) Preliminary device results indicate that SiC diodes, dynistors, and bipolar transistors promise substantial improvement in power handling capabilities, operating temperature, and speed. 2) These results also suggest that we will be able to raise the power in discrete components and develop new types of transit-time avalanche-like devices as well as SiC integrated circuits. Further work is under way to improve on these and other device results.

References

1. Otblesk A.E., Chelnokov V.E.: Proceedings of the III All-Union Conference on wide-gap semiconductors, LETI, Leningrad, p.197-211 (1979) [in Russian]
2. Anikin M.M., Lebedev A.A., Popov I.V., Sevastjanov V.E., Syrkin A.L., Suvorov A.V., Chelnokov V.E., Shpynev G.P.: Pisma w ZhTF, v.10, N17, p.179-182(1984) [in Russian]
3. Anikin M.M., Vainshtein S.N., Levinshtein M.E., Strel'chuk A.M., Syrkin A.L.: Pisma w ZhTF, v.14, N6, p.545-547(1988) [in Russian]
4. Anikin M.M., Lebedev A.A., Popov I.V., Pjatko S.N., Rasstegaev V.P., Syrkin A.L.,Tsarenkov B.V.,Chelnokov V.E.: Fiz. Techn. Poluprov., v.22, N1, p.133-136(1988)[in Russian]
5. Anikin M.M., Lebedev A.A., Popov I.V., Strel'chuk A.M., Suvorov A.V., Syrkin A.L., Chelnokov V.E.: Fiz. Techn. Poluprov., v.20, N15, p.844-848(1986) [in Russian]

6. Anikin M.M., Lebedev A.A., Popov I.V., Rasstegaev V.P., Strel´chuk A.M., Syrkin A.L., Tairov Yu.M., Tsvetkov V.F., Chelnokov V.E.: Fiz. Techn. Poluprov., v.22, N2, p.298-300(1988) [in Russian]
7. Vainshtein S.N., Dmitriev V.A., Syrkin A.L., Chelnokov V.E.: Sov. Tech. Phys. Lett., v.13, N8, p.413-414(1987) [in Russian]
8. Dmitriev V.A., Levinshtein M.E., Vainshtein S.N., Chelnokov V.E.: Electronics Letters, v.24, N16, p.1032-1033(1988)
9. Munch W.V., Hoeck P.: Solid State Electronics, v.21, p.479-480(1978)

Investigation of the Blue Electroluminescence Efficiency of Epitaxial SiC-6H Structures

L.M. Kogan and Ya.V. Morozenko

A.F. Ioffe Physical-Technical Institute, Academy of Sciences, 26, 194021 St. Petersburg, Russia

Abstract. At present time the most efficient blue electroluminescence in silicon carbide has been obtained from SiC-6H pn junctions doped with nitrogen (donor) and aluminium (acceptor). In the present paper, we report on investigations aimed at raising the efficiency of blue electroluminescence; to this end we searched for optimum concentrations of Al and N in SiC-6H epitaxial layers and pn structures.

1. Epitaxial Structures

Epitaxial layers and pn structures were grown by container-free liquid phase epitaxy of SiC from a Si-C molten solution [1].

For photoluminescence studies the epitaxial layers of two sorts were made. In the first series of samples donor concentration was $0.5 \cdot 10^{18} cm^{-3}$; in the second series $1 \cdot 10^{18} cm^{-3}$. In both series of samples the Al content was varied over a wide range: an amount of aluminium introduced into the Si-C molten solution prior to epitaxy varied from 0 to 14 wt.%.

2. Photoluminescence Studies of SiC-6H Epitaxial Layers

Studies of PL spectra of the n-type epitaxial layers with low aluminium content (N_{Al} = 0.1 wt.%) has shown that at low temperatures (80 K) a major contribution into PL comes from donor-acceptor recombination of N-Al pairs (E_{max} = 2.68 eV). weak bands of the free exciton annihilation (E_{max} = 2.92 eV), and the exciton bound at neutral nitrogen (E_{max} = 3.0 eV) line was observed in the spectra [2]. As temperature is increased the bands of donor-acceptor recombination and bound exciton annihilation are strongly quenched; the band of free exciton is not affected up to 350 K and at room temperature dominates the spectrum.

At higher Al contents (0.25-2 wt.%) in low temperature spectra alongside the donor-acceptor band a new band of unknown origin is observed that peaks at 2.56 eV. In samples with maximum compensation of conductivity this new band is most pronounced. With rising temperature this band becomes noticably quenched and at 200 K the band with E_{max} = 2.56 eV is hardly observable. Quenching of the donor-acceptor band in this sample is spelling that in samples with low Al content.

As temperature is raised the maximum of the donor-acceptor band shifts to shorter wavelengths, mainly due to a contribution from conduction band - acceptor (Al) recombination. At T = 300-350 K the luminescence spectrum gives

Editors: G.L. Harris, M.G. Spencer, and C.Y. Yang

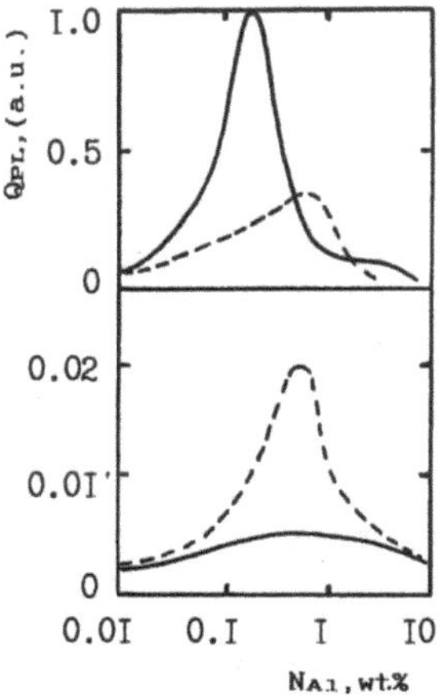

Fig.1

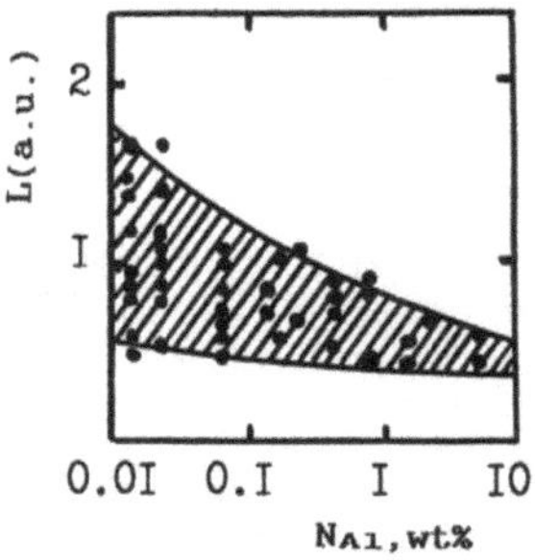

Fig.2

Fig.1. A dependence of the quantum efficiency of 470 nm - photoluminescence (Q_{PL}) on Al concentration(N_{Al}): (a) T=80 K, (b) T=295 K. Continuous lines - series 1 samples, dotted lines - series 2 samples.

Fig.2. A dependence of the intensity photoluminescence due to free exciton annihilation(L) on Al concentration(N_{Al}). T=295 K.

positions of the bands of free exciton annihilation, donor-acceptor recombination in N-Al pairs and of conduction band - Al acceptor recombination. Further increase of Al content (2-14 wt.%) leads to an appearence in the spectrum of the green luminescence band due to defects (E_{max} = 2.46 eV) [2].

To determine optimum concentrations of Al and N, we studied quantum yield of blue PL(Q_{PL}) as a function of Al content in samples of series 1 and 2 (Fig.1) in a temperature range 100-300 K. It has been found that at low temperatures (100 K) a curve for $Q_{PL}(N_{Al})$ has a clearly defined maximum at Al concentration approximately corresponding to the point conductivity compensation. Maximum values of Q_{PL} for series 1 samples (nitrogen concentration $0.5 \cdot 10^{18} cm^{-3}$) greatly exceed (by a factor of 5) Q_{PL}(max) in series 2 samples (N_N = $1 \cdot 10^{18} cm^{-3}$). Rising temperature quenches PL, in the region of maximum Q_{PL} quenching is much stronger in series 1 samples. Thus, at room temperature maximum value of Q_{PL} in samples with high nitrogen content turns out to be noticeably lower than Q_{PL} (max) in samples with low nitrogen.

To estimate an effect on concentrational quenching of doping level, we determined the dependence of the free exciton luminescence intensity on aluminium concentration in epitaxial layers (Fig.2). We found that the free exciton luminescence falls off monotonically and rather weakly as Al concentration is increased. Therefore, we suppose that the reduction of PL with rising Al concentration is not due to increased concentration of nonradiative recombination centres at high doping levels.

We have also studied the effect on the shape of luminescence spectrum of excitation level in samples with different Al concentrations. It was found that in n- and p-layers at low temperature no appreciable changes in the spectral shapes occur when excitation intensity is changed by a factor of 3. In compensated layers, though, the same increase in excitation intensity brings about noticeable shift of the donor-acceptor recombination to higher energies.

We suppose that the appearance of a maximum on the $Q_{PL}(N_{Al})$ curve is explained by the fact that conductivity compensation causes formation of potential relief of the bands. An occurrence of band relief formation is evidenced by: 1) approximate correspondence of the maximum position in $Q_{PL}(N_{Al})$ curves to conductivity compensation points in samples with different donor concentration; 2) an appearance in nearly compensated samples of a new emission band (E_{max} = 2.56eV) shifted to lower energies relative to the donor-acceptor band; 3) an observed considerable shift of the donor-acceptor band in compensated samples to short wavelengths with increasing excitation intensity

In view of the fact that formation of the potential relief is accompanied with partial localisation of nonequlibrium carriers in bottom parts of the relief where the carriers are isolated from nonradiative centers an increase in compensation degree should lead to an increase in the quantum efficiency of luminescence.

3. Electroluminescence of SiC-6H Epitaxial pn-Structures

We have investigated an influence of doping p-layer with aluminium on the quantum yield of pn structures. pn structures were produced by epitaxial growth of p-layers with different Al content on SiC-6H monocrystalline n-type wafers containing $3 \cdot 10^{18} cm^{-3}$ of uncompensated donors. The PL spectrum of these wafers at a low temperature had an appearance typical of donor-acceptor luminescence of N-Al pairs; the wafers having near values of the quantum yield of blue PL were chosen. The hole concentration in epitaxial p-layers measured by Hall method was in a range $1 \cdot 10^{18} cm^{-3}$ to $5 \cdot 10^{19} cm^{-3}$.

At low temperatures (≈100 K) the electroluminescence spectrum of the pn structures was similar to PL spectra of p-layers. As temperature is increased (> 200 K) the direction of predominant carrier injection changes and the electroluminescence spectrum is found to coincide with n-layer PL spectrum. So, at room temperature the main contribution to emission from the pn structures studied comes from the n-region.

A dependence of the electroluminescence quantum yield (Q_{EL}) on hole concentration in p-emitter in these structures was determined at T = 300 K (Fig.3). Initially, Q_{EL} rises with the hole concentration and then at hole concentrations higher than $10^{19} cm^{-3}$ falls drastically. We suppose that this rise is due to an enhancement of hole injection into n-region where radiation comes from. Subsequent decrease of Q_{EL} appears related to formation of nonradiative recombination centers in n-layer due to strains arising from lattice in the heavity doped p-layer. This was confirmed in X-ray investigations of heavily doped (Al) layers grown by LPE [3].

Studies of electroluminescence pn structures incorporating a p-layer optimally doped with Al and n-layer in which Al, N

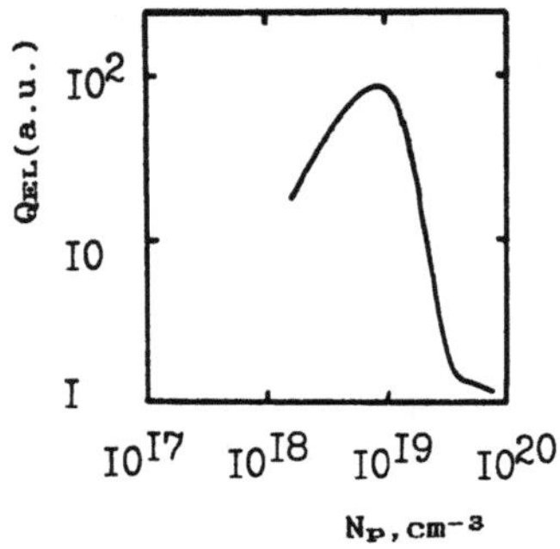

Fig.3. A dependence of the quantum yield of electroluminescence (Q_{EL}) on concentration of holes in p layers(N_P). T=295 K.

concentrations were varied have shown that at T = 300 K the quantum yield of blue electroluminescence depends on Al concentration in a way analogous to the dependence of $Q_{PL}(N_{Al})$ in epitaxial layers. Therefore, the results on optimization of doping in epitaxial layers obtained using data on PL (alongside with studies of pn structure electroluminescence) were utilized in realization of an optimal blue-emitting pn structure. In the optimized pn structure the electroluminescence spectrum depends on temperature in the following way:

1) in a temperature range 77-400 K the spectrum is dominated by the band of donor-acceptor recombination of nitrogen-aluminium pairs;

2) further increase of temperature (> 400 K) leads to appearence in the spectrum of the band due to recombination between the conduction band and an acceptor (Al); at 470 K this band dominates the spectrum.

Using optimized pn structures, we have fabricated blue SiC-6H light-emitting diodes. The luminous intensity of the LED's equals 9 mcd (at 20 mA current, the divergence angle 10°), the wavelenth in the maximum region of electroluminescence is approximately 470 nm.

4. References

1. V.A.Dmitriev, P.A.Ivanov, I.V.Korkin, Ya.V.Morozenko, I.V. Popov, T.A.Sidorova, A.M.Strel'chuk, V.E.Chelnokov: Sov. Tech.Phys.Lett., Vol.11, 98 (1985)
2. Y.M.Tairov, Y.A.Vodakov: Topics in Appl.Phys., Vol.1, 1 (1977)
3. V.A.Dmitriev, I.V.Popov, V.E.Chelnokov: in 7th Conf. on Crystal Growth, 9-13 June 1986, Novosibirsk, Tesises, Vol.1, 51,1986)

Spectroscopic Studies of Donors in 3C-SiC Films

J.A. Freitas, Jr.[1], *W.E. Carlos*[2], *and S.G. Bishop*[3]

[1]Sachs-Freeman Associates, Landover, MD 20785, USA
[2]Naval Research Laboratory, Washington, DC 20375, USA
[3]Center for Compound Semiconductor Microelectronics and Department of Electrical and Computer Engineering, University of Illinois, Urbana, IL 61821, USA

The evidence from photoluminescence spectroscopy for the existence of a 54 meV donor in both undoped n-type and Al-doped cubic SiC is reviewed along with investigations of the three-line electron paramagnetic resonance spectra which reveal the presence of nitrogen donors in the same material. The basis for identifying isolated, substitutional nitrogen as the 54 meV donor in cubic SiC is then discussed.

1. Introduction

The development of chemical vapor deposition techniques [1-4] for the growth of thin films of cubic (3C) SiC has triggered a resurgent interest in the potential application of this semiconductor in a variety of electronic and optoelectronic devices. However, realization of the full device performance potential of CVD cubic SiC will require the identification, control, and or elimination of the defect levels which dominate the electrical properties of the thin heteroepitaxial films of this material.

In a previous paper [5], we reviewed in detail the optical and magnetic resonance spectroscopy studies of both Lely-grown crystals and CVD films of cubic SiC and their application to the identification of background donors. In this paper, we focus on spectroscopic studies of CVD cubic SiC alone and attempt to resolve some of the controversy surrounding the interpretation of the photoluminescence (PL) [6-12] and electron paramagnetic resonance (EPR) [13-15] spectra. Of particular interest is the apparent contradiction between the 54 meV donor which manifests itself in the PL spectra of both undoped and Al-doped cubic SiC, and the 15-20 meV donor which is detected in temperature dependent Hall effect measurements [16-19] on as-grown n-type films of CVD cubic SiC. We relieve some of the confusion by initially separating the discussion of the determination of donor binding energies from the issue of donor identification. After reviewing the unequivocal optical evidence for the existence of a 54 meV donor in cubic SiC, and the three-line EPR spectra which reveal the presence of neutral nitrogen donors in the same material, we outline the basis for identifying isolated, substitutional nitrogen as the 54 meV donor.

2. Photoluminescence Spectroscopy of CVD Cubic SiC

2.1 Bound Exciton Spectra

Several workers [6-11] have reported the observation at liquid helium temperature of a characteristic five-line near-band edge (~2.38 eV) PL spectrum in undoped films of CVD-grown cubic SiC. Examples of such spectra obtained by FREITAS et al.[8] are shown in Fig. 1. They are basically equivalent to the five-line PL spectra first observed by CHOYKE et al. [20] in Lely-grown crystals of cubic SiC. HARTMAN and DEAN [21] studied the Zeeman (magnetic field) and uniaxial stress splitting of this spectrum in Lely crystals and concluded that the luminescence spectrum is attributable to the recombination of excitons bound to neutral donors, and that the donors are isolated point defects, not complexes. In 1977, DEAN et al. [22] studied the two electron satellite transitions associated with the five-line donor-bound exciton spectrum and inferred a highly accurate value of 53.6 +/-0.5 meV for the donor.

Springer Proceedings in Physics, Volume 56 **Amorphous and Crystalline Silicon Carbide III**
Editors: G.L. Harris, M.G. Spencer, and C.Y. Yang

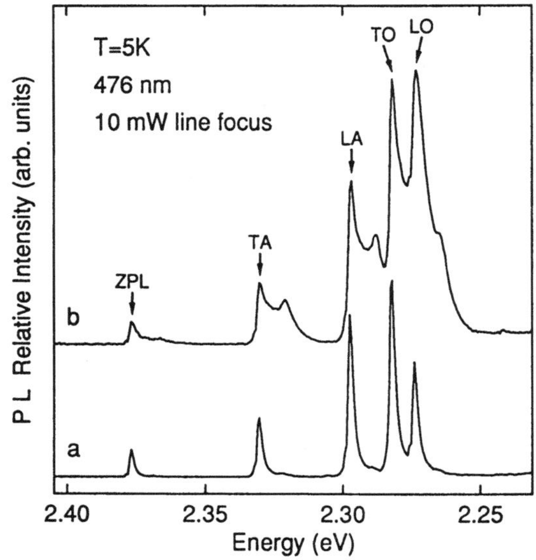

Fig. 1 The 54 meV donor bound exciton PL spectra measured at 5K in two undoped 3C-SiC films grown with two different values of C/Si source gas ratio: a) 2.4; b) 1.2. The zero phonon line (ZPL) and phonon replicas (TA, LA, TO, and LO) are indicated in the spectrum b.

As the quality of CVD films of cubic (3C) SiC has improved, the spectral quality and detail of the five-line bound exciton PL spectra obtained from the films has approached that of the spectra from Lely-grown crystals. Recently, POWELL et al.[11] reported the observation of low temperature bound exciton PL spectra in CVD films of 3C-SiC grown on 6H-SiC crystals which exhibited no strain induced red-shifts or spectral broadening relative to corresponding PL spectra from Lely crystals. The exact equivalence of the five-line bound exciton PL spectra in CVD films and Lely crystals of cubic SiC having been established, it follows that the observation of this characteristic PL spectrum in all of the undoped films which have been studied constitutes strong evidence for the pervasive presence of the ~54 meV donor in these films of cubic SiC.

The bound exciton PL spectra from CVD films of cubic SiC exhibit shoulders on the low-energy side of each of the five lines including the zero phonon line (ZPL) and the four well-resolved phonon replicas (Fig. 1). FREITAS and coworkers [8] have shown that the strength of these shoulders is enhanced with decreasing C/Si source gas ratio, suggesting that the recombination centers responsible for these PL transitions might be associated with nonstoichiometric defects. On the basis of the spectral position of these shoulders, their measured thermal quenching energy [8], and lifetime studies [23], it has been suggested [8,23,24] that the shoulders arise from recombination of excitons bound to donors having a somewhat larger binding energy than the pervasive ~54 meV donor which gives rise to the principal five line bound exciton spectrum. In contrast, there is no evidence in the near-band edge PL spectra for the recombination of excitons bound to a shallower donor.

2.2 Donor-Acceptor Pair Spectra

Evidence for the presence of a 54 meV donor in the CVD fims of cubic SiC has also been inferred from donor-acceptor pair (DAP) PL spectra observed in Al-doped films. Figure 2 shows DAP PL spectra obtained by FREITAS et al.[8] from films which were heavily doped with Al by introducing trimethyl aluminum with the source gases during growth. In the DAP recombination process [25], electrons bound at neutral donors recombine radiatively with holes

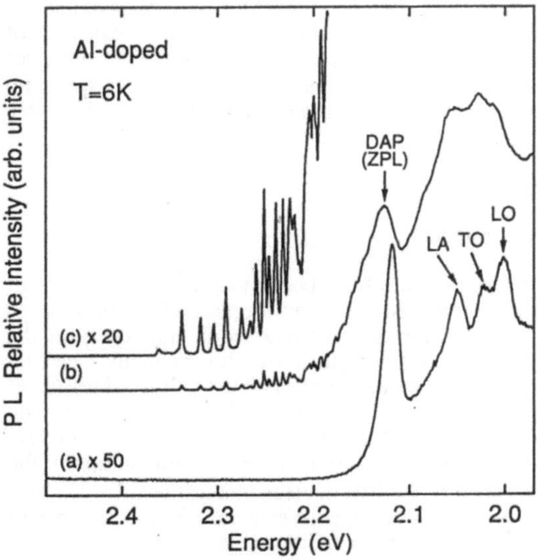

Fig. 2 Low temperature PL spectra of an Al-Doped 3C-SiC film: a) low excitation power (~ 10^{-4} W/cm^2) spectrum showing the DAP and associated phonon replicas; b) high excitation power (1W/cm^2) spectrum with close and distant pair bands; c) detail of sharp-line close pairs spectrum.

bound at neutral acceptors. The emitted photon energy is a function of the spatial separation of the donor and acceptor atoms involved in the recombination. The closest pairs give rise to the discrete, sharp line spectra between 2.38 and 2.2 eV in Fig. 2, while more distant pairs give rise to the broad peak at about 2.12 eV.

These DAP PL spectra provide two independent measurements of the binding energy of the donor participating in the recombination process. First, quantitative analysis of the spectral energies of the sharp-line close pair spectra provides a value of 310 meV for the sum of the aluminum acceptor and unspecified donor binding energies. The binding energy of the deeper Al acceptor can be determined from PL spectra obtained at higher temperatures (>120K) for which all of the shallow donors are ionized and an Al acceptor free-to-bound PL transition is observed [8,24,27], whose energy is approximately equal to the band gap minus the acceptor binding energy. Having thus measured a 257 meV Al acceptor binding energy, subtraction from 310 meV yields a donor binding energy of 53 meV. These values of the binding energies determined by FREITAS et al.[8] in Al-doped films of CVD SiC are in exact agreement with those measured by CHOYKE et al. [26] from DAP spectra in Al-doped Lely-grown crystals of cubic SiC. In addition, Kuwabara et al. [28] reported DAP PL spectra in boron-doped cubic SiC from which they inferred a value of 735 meV for the boron acceptor binding energy and 54 meV for the binding energy of the background donor which paired with the boron acceptors. Clearly, the same background donor seems to be involved in the DAP spectra of both the Al-doped and B-doped SiC.

A second, independent measure of the donor binding energy in Al-doped films of CVD cubic SiC was obtained by FREITAS et al.[8] who studied the thermal quenching of the sharp close DAP bands. Because the donor binding energy is about one fifth that of the 257 meV Al acceptor, the holes remain bound over the temperature range of donor ionization (≤25K) and the PL quenching curve (Arrhenius plot) is characterized primarily by the donor binding energy [25]. The value of the donor binding energy inferred from the slope of the exponential DAP PL quenching curve was 54 meV [8].

We conclude this brief review of the PL studies of cubic SiC by emphasizing the following points: There is conclusive evidence in the PL spectra for the pervasive existence of a ~54 meV donor in undoped and Al-doped films of cubic SiC films grown by CVD techniques and in Lely-grown crystals of cubic SiC. The sharp, five-line bound exciton PL spectrum which appears in all undoped films and crystals of cubic SiC is an unambiguous, unmistakable signature of this 54 meV donor.

3. Electron Paramagnetic Resonance Spectra

The EPR spectrum exhibited by the unpaired spins of electrons bound to donors can provide an identification of the donor species through the interaction of the paramagnetic electrons with the nuclear magnetic moments of the donor atoms. This central hyperfine interaction splits the EPR spectrum into 2I+1 lines, where I is the nuclear spin, providing in many cases an identification of the donor. In a previous paper [5], we have reviewed the published results of EPR spectroscopy as applied to both Lely crystals and CVD films of cubic SiC. From the perspective of the present discussion, the most important result of these studies is that a three-line EPR spectrum has been observed which is attributed to the hyperfine interaction of the paramagnetic electron with the magnetic moment of the ^{14}N nucleus (nuclear spin I = 1). This three-line EPR spectrum has been established as an unambiguous signature for the presence of neutral nitrogen donors and, under the proper experimental conditions, its intensity can provide a measure of the concentration of neutral nitrogen donors.

Examples of low-temperature EPR spectra obtained by CARLOS and coworkers [13] from undoped CVD films of cubic SiC are shown in Fig. 3. The small, isotropic hyperfine interaction which they observed (A = 1.2G) is consistent with expectations for a hydrogenic donor and is in excellent agreement with that reported by ALTAISKII et al. [29] for Lely grown cubic SiC and also by other workers [14-15] in CVD films. CARLOS et al. found that the ability to obtain well-resolved three-line spectra is apparently strongly dependent upon donor concentration in undoped films. At higher concentrations (>>10^{17} cm^{-3}) broadening due to exchange interactions could obscure the relatively small hyperfine splitting [13]. However, studies of films prepared with a variety of growth parameters (e.g. C/Si source gas ratio) have shown that it is also possible that the loss of the hyperfine structure in the EPR spectra of higher concentration samples is caused by the superposition of a broad, structureless spectrum which arises from a second, distinct neutral donor [15,23,24].

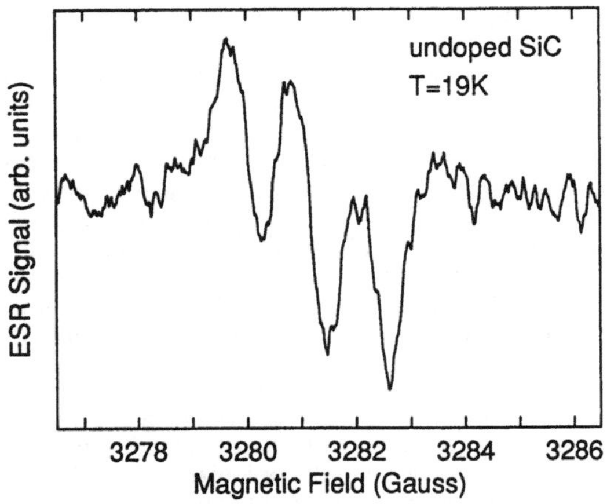

Fig. 3 EPR spectrum of the undoped 3C-SiC CVD film grown with C/Si source gas ratio of 2.4 (PL spectrum shown in Fig. 1a). The microwave power level was ~ 2 microwatts.

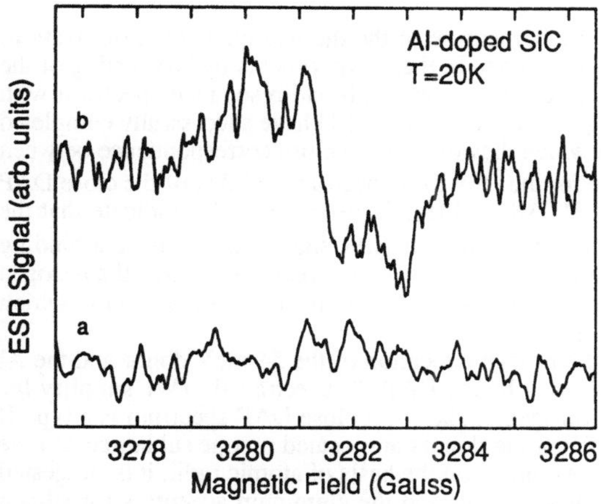

Fig. 4 Low temperature EPR spectra obtained at 9GHz from the Al-doped film (PL spectrum shown in Fig. 2) under two different conditions: a) unilluminated (dark); b) illuminated with 457.9 µm argon laser line with power density of ~ 10^{-4} W/cm^{2}.

The photoinduced EPR spectrum (b) of Fig. 4 indicates that nitrogen donors can also be observed in Al-doped CVD films (FREITAS AND CARLOS, to be published [30]). In the dark (unilluminated) EPR spectrum (a) of Fig. 4, no EPR signal is observed in the spectral range of the donor lines, as might be expected for an Al-doped sample in which most if not all of the donors are compensated by the Al acceptors. Compensated (charged) donors have no bound electrons with unpaired spins and are not paramagnetic centers. However, when an Al-doped, compensated sample is illuminated with above-gap light, a quasi-equilibrium population of photoinduced neutral donors is generated and the three-line neutral nitrogen donor EPR spectrum appears as shown in the illuminated or photoinduced EPR spectrum of Fig. 4.

4. Discussion

Just as the PL results established the existence of a pervasive background donor with a 54 meV binding energy in CVD films of cubic SiC, the EPR spectra demonstrate that nitrogen is a pervasive background donor in these SiC films. In what follows we present the evidence and reasoning which support an association of the ubiquitous 54 meV donor in cubic SiC with isolated nitrogen donors.

The observation of sharp-line donor bound exciton and sharp close DAP PL spectra indicates that the donors participating in these radiative recombination processes are located on lattice sites of high crystalline quality. Optical (PL) spectra associated with donors which are located in highly strained material or in proximity with high concentrations of structural defects would be severely broadened and perhaps be unobservable. The same can be said for donors which give rise to EPR spectra with well resolved hyperfine splitting, such as the three-line ^{14}N EPR spectrum observed in the CVD films of cubic SiC. In addition, studies of the magnetic field orientation dependence of both the five-line donor bound exciton PL spectrum [21] and the three-line ^{14}N EPR spectrum [29] gave no evidence of anisotropy. This means that both the 54 meV donor giving rise to the donor bound exciton PL spectrum and the neutral ^{14}N donor responsible for the three-line EPR spectrum are isolated, point defects.

Measurement of the binding energy of the nitrogen donor which exhibits the three-line EPR spectrum presents an obvious means of resolving the question of the possible connection between the 54 meV donor and the nitrogen donor. CARLOS et al.[13] attempted to accomplish this by measuring the thermal quenching of the ^{14}N EPR spectrum in an undoped n-type film of cubic SiC. However, an exact measure of the thermal quenching (ionization) energy of the neutral nitrogen donors was thwarted in these experiments by broadening of the three-line EPR spectrum and interference of an underlying broad resonance spectrum with increasing temperature. Recently, Nashiyama and coworkers [31] have successfully completed such a temperature dependence study in which they observed a close correspondence between the slope of the thermal quenching curve for the ^{14}N EPR spectrum and that of the close DAP PL spectrum reported previously by FREITAS et al.[8]. These results [31] indicate that the neutral nitrogen donors giving rise to the three-line ^{14}N EPR spectrum also have a binding energy of ~54 meV, and they constitute the most direct evidence yet obtained that the nitrogen donor detected in EPR spectra of cubic SiC and the pervasive 54 meV donor which characterizes the optical (PL) spectra are the same donor.

Some valuable insight concerning the probable location of the 54 meV donor and the Al and B acceptors in the SiC lattice is provided by the DAP PL spectra [5]. (We simplify the discussion by referring to the donor as nitrogen.) The N-Al close DAP spectrum is a type II spectrum [25] which means that the participating donors are located on one sublattice and the Al acceptors on the other sublattice. For example, on the basis of atomic radii, it is suggested that the nitrogen donors substitute for carbon atoms and the aluminum acceptors for silicon atoms. On the other hand, the N-B DAP close pair spectrum is type I [25], meaning that donors and acceptors are located on the same sublattice. An internally consistent description, that is, one which is consistent with the requirements of the type I and II PL spectra and the sizes of the atoms involved, places both the nitrogen donors and the boron acceptors on the carbon sites.

Implicit in the above discussion is the fact that spectroscopic techniques such as PL and EPR automatically single out the high quality regions of the crystalline sample, and the spectral widths, lattice site configurations and symmetries, binding energies, etc. inferred from these spectra are characteristic of these high quality regions only. Conversely, transport measurements such as the Hall effect perform an average over the entire crystal (film) volume. The conductivities of high quality and highly defective regions of the sample are measured in parallel. Obviously, this can lead to some unavoidable ambiguities in any attempt to extract parameters such as donor binding energies and concentrations, compensation ratios, and thermally activated carrier concentrations.

We now discuss the possible relationship or contradiction between the 54 meV donor (apparently nitrogen) and the 15-20 meV donors which dominate the electrical properties [13,16-19] of the n-type CVD films of cubic SiC. It has been observed [25] that optical transitions which depend only upon the ground state energies can remain sharp and unshifted at relatively high impurity concentrations for which thermal activation energies or binding energies for carriers bound to the impurities can be depressed as much as 10-40% because of apparent overlap in the impurity excited states. In addition, SEGALL et al.[17,18] have invoked a concentration dependent reduction in the thermal activation energy to explain the apparent variation in donor binding energies inferred from temperature dependent Hall effect measurements on n-type CVD films of cubic SiC. However, thermal quenching studies of both the DAP PL spectra [8] and the EPR spectra [31] have measured 54 meV binding energies equivalent to those determined from the optical spectral energies. Furthermore, MOORE [32] has shown that if the published temperature dependent Hall effect data for CVD SiC films (from which donor ionization energies ranging from 15-40 meV have been inferred) are analyzed correctly, one consistently obtains activation energies in the range 15-20 meV, whose variation is easily accounted for by uncertainties in the fitting or analysis procedure. This means that there is not a continuous range of donor binding energies varying from the 54 meV measured optically to the lowest (~15 meV) value measured by Hall effect. Instead, the optical measurements detect one donor with a relatively precise value of 53-54 meV for its binding energy, and the Hall effect measurements yield a second shallower donor, with considerable uncertainty in its 15-20 meV binding energy.

Thus it appears that isolated nitrogen donors, which are located on undistorted lattice sites and have a 54 meV binding energy, cannot be responsible for the 15-20 meV donor binding energy inferred from temperature dependent Hall effect measurements. However, this does not mean that nitrogen impurities cannot be associated in some other configuration with this shallow donor. For example, the nitrogen impurities could substitute on either the carbon sublattice or the silicon sublattice in SiC, with different binding energies in the two cases, and the 54 meV binding energy can represent only one of these two cases. In addition, nitrogen impurities as well as other electrically active lattice defects could be located or concentrated in inhomogeneous regions of the crystal. Under these circumstances the nitrogens could occupy highly distorted lattice sites or interstices, or be complexed with other defects and therefore have binding energies quite different from the 54 meV of the isolated, substitutional nitrogen donor.

Acknowledgments

The authors wish to thank P.E.R. Nordquist, Jr. and M.L. Gipe for providing some of the SiC samples, and P.B. Klein and W.J. Moore for many helpful discussions. This work has been partially supported by the Office of Naval Research.

References.

1. S. Nishino, J.A. Powell, and H. A. Will, Appl. Phys. Lett. 42 460 (1983).
2. A. Addamiano and P.H. Klein, J. Cryst. Growth 70, 291 (1984).
3. K. Sasaki, E. Sukuma, S. Misawa, S. Yoshida, and S. Gonda, Appl. Phys. Lett. 45, 72 (1984).
4. H.P. Liaw and R.F. Davis, J. Electrochem. Soc. 131, 3014 (1984).
5. S.G. Bishop, J.A. Freitas, Jr., T.A. Kennedy, W.E. Carlos, W.J. Moore, P.E.R. Nordquist, Jr., and M.L. Gipe, in: Amorphous and Crystalline Silicon Carbide, Springer Proc. in Physics, Vol. 34, ed. G.L. Harris and C.Y.-W. Yang (Springer-Verlag, Berlin, Heidelberg, 1987), p. 90.
6. J.A. Freitas, Jr., S.G. Bishop, A. Addamiano, P.H. Klein, H.J. Kim, and R.F. Davis, Mats. Res. Soc. Symp. Proc., 46, 581, (1985).
7. J.A. Freitas, Jr., S.G. Bishop, J.A. Edmond, J. Ryu, and R.F. Davis, J. Appl. Phys. 61, 2011 (1987).
8. J.A. Freitas, Jr., S.G. Bishop, P.E. R. Nordquist, Jr., and M.L. Gipe: Appl. Phys. Lett. 52, 1696 (1988).
9. W.J. Choyke, Z.C. Feng, and J.A. Powell, J. Appl. Phys. 64, 3163 (1988).
10. J.A. Freitas, Jr. and S.G. Bishop, Appl. Phys. Lett. 55, 2757 (1989).
11. J.A. Powell, D.J. Larkin, L.G. Matus, W.J. Choyke, J.L. Bradshaw, L. Henderson, M. Yoganathan, J. Yang, and P. Pirouz, Appl. Phys. Lett. 56, 1442 (1990).
12. H. Okumura, M. Shinohara, E. Muneyama, H. Daimon, M. Yamanaka, E. Sakuma, S. Misawa, K. Endo, and S. Yoshida, Japan. J. Appl. Phys. 27, L116 (1988).
13. W.E. Carlos, W.J. Moore, P.G. Siebenmann, J.A. Freitas, Jr., R.Kaplan, S.G. Bishop, P.E.R. Nordquist, Jr., M. Kong, and R.F. Davis, in Proc. Mats. Res. Soc. Symp. 97, 253 (1987).
14. I. Nashiyama, H. Okumura, E. Sakuma, S. Misawa, K. Endo, S. Yoshida, H. Itoh, and N. Hayakawa, in: Amorphous and Crystalline SiC II, Springer Proc. in Physics, Vol. 43, eds. M.M. Rahman, C.Y.-W. Yang, and G.L.Harris (Springer-Verlag, Berlin, Heidelberg, 1988), p. 100.
15. I. Nashiyama, H. Okumura, E. Sakuma, K. Endo, and S. Yoshida, proceedings of this conference.
16. A. Suzuki, A. Uemoto, M. Shigeta, K. Furukawa, and S. Nakajima, Appl. Phys. Lett. 49, 450 (1986).
17. B. Segall, S.A. Alterovitz, E.J. Haugland, and L.G. Matus, Appl. Phys. Lett 49, 584 (1986).
18. B. Segall, S.A. Alterovitz, E.J. Haugland, and L.G. Matus, Appl. Phys. Lett. 50, 1533 (1987).
19. M. Shinohara, M. Yamanaka, H. Daimon, E. Sakuma, H. Okumura, S. Misawa, K. Endo, and S. Yoshida, Japan J. of Appl. Phys. 127, L434 (1988).
20. W.J. Choyke, D.R. Hamilton, and L. Patrick, Phys. Rev. 133 A1163 (1964).
21. R.L. Hartman and P.J. Dean: Phys. Rev. B 2, 951 (1970).
22. P.J. Dean, W.J. Choyke, and L. Patrick, J. Lum. 15, 299 (1977).

23. J.A. Freitas, Jr., S.G. Bishop, P.B. Klein, P.E. R. Nordquist, Jr., and M.L. Gipe, in: Amorphous and Crystalline SiC II, Springer Proc. in Physics, Vol. 43, eds. M.M. Rahman, C.Y.-W. Yang, and G.L.Harris (Springer-Verlag, Berlin, Heidelberg, 1988), p. 106.
24. S.G. Bishop and J.A. Freitas, Jr., to be published in J. Crystal Growth (1990).
25. P.J. Dean, in: Progress in Solid State Chemistry, 8, eds. J.O. McCaldin and Y.G. Somarjai (Pergamon, Oxford) 1973, p. 1.
26. W.J. Choyke and L. Patrick, Phys. Rev. B2, 4959 (1970).
27. G. Zanmarchi, J. Phys. Chem. Solids, 29 (1968) 1727.
28. H. Kuwabara, S. Yamada, and S. Tsunekawa, J. Lumin. 12/13, 531 (1976).
29. Yu.M. Altaiskii, I.M. Zaritskii, V.Ya. Zevin, and A.A. Konchits, Sov. Phys.-Solid State 12, 2453 (1971).
30. J.A. Freitas, Jr. and W.E. Carlos, to be published.
31. I. Nashiyama and coworkers, private communication.
32. W.J. Moore, proceedings of this conference.

Electron Irradiation Effects on CVD-Grown 3C-SiC Epilayers

H. Itoh[1], *M. Yoshikawa*[1], *I. Nashiyama*[2], *S. Misawa*[2], *H. Okumura*[2], and *S. Yoshida*[2]

[1]Japan Atomic Energy Research Institute, 1233 Watanuki, Takasaki, Gunma 370-12, Japan

[2]Electrotechnical Laboratory, 1-1-4 Umezono, Tsukuba, Ibaraki 305, Japan

Abstract. Effects of 1 MeV electron irradiation on electrical properties of CVD-grown 3C-SiC epilayers were studied by Hall and electron spin resonance (ESR) measurements. The carrier densities and the Hall mobilities of the epilayers decreased with the electron fluence in the range from $1x10^{17}/cm^2$ to $1x10^{18}/cm^2$. The carrier removal rate and the mobility damage constant were estimated to be $1.4x10^{-2}cm^{-1}$ and $3.4x10^{-22}$ Vs, respectively, which suggests that 3C-SiC epilayers are more tolerant of radiation than Si. Isochronal annealing indicated that the carrier density recovered in the same temperature region that the irradiation induced ESR centers disappeared. The intensity of the N-related hyperfine ESR-signal decreased by the irradiation. The reduction of the carrier density and the N-related ESR-signal can be interpreted by the introduction of deep electron traps by the irradiation.

1. Introduction

Cubic silicon carbide (3C-SiC) has received considerable interest in recent years in connection with its application to electronic devices used in radiation environments such as outer space. To use 3C-SiC epilayers for electronic devices which can operate in a radiation field, it is important to clarify the radiation effects on electrical properties of the epilayers. Regarding radiation induced defects in 3C-SiC epilayers, there exist several investigations[1,2]. However, there has been only a little amount of information about radiation hardness of 3C-SiC epilayers. Knowledge of radiation induced defects is also useful for understanding the origin of residual carriers in the epilayers. In the previous papers[3,4], we have investigated nitrogen (N)-related 3-lines structures by electron spin resonance (ESR) measurements and reported the production of 5-lines ESR centers (T1) by electron irradiation.

In this paper, we report the effects of 1 MeV electron irradiation on the carrier density and the Hall mobility of CVD-grown 3C-SiC epilayers. We also report their recovery process by thermal annealing and the irradiation effects on N-related ESR-signal. The results obtained are all discussed in terms of the introduction of electron traps by the irradiation.

2. Experimental

Single crystalline 3C-SiC layers with thicknesses of 20-40 μm were grown epitaxially on Si (100) substrates by chemical vapor deposition (CVD)

using SiH_4-C_3H_8-H_2 system at 1410°C. The substrates were removed from the epilayers with HF-HNO_3 after the growth. The samples used were non-doped, n-type epilayers with carrier densities of about $1x10^{16}/cm^3$ and Hall mobilities of about $500cm^2/Vs$ at room temperature (RT). These epilayers were irradiated with 1 MeV electrons up to a fluence of $1.5x10^{18}/cm^2$. The temperature of the epilayers was kept below 50°C during the irradiation.

The carrier densities and the Hall mobilities of the epilayers were measured by van der Pauw method at temperatures between 14K and RT. Ohmic contacts were made on four edges of the epilayers by evaporating Ni followed by heating at 950°C for 5 minutes. The epilayers irradiated with electrons were annealed isochronally for 5 minutes in a flowing pure Ar gas. ESR measurements were performed at 30K and RT with an X-band (9GHz) microwave incident upon a TE_{011} cylindrical cavity.

3. Results and discussion

Figure 1 shows the dependences of the carrier densities and the Hall mobilities at RT on the electron fluence for non-doped 3C-SiC and n-type Si. The carrier densities and the mobilities of the as-grown 3C-SiC epilayers were $3.8x10^{16}/cm^3$ and $480cm^2/Vs$ at RT, respectively. Those of the Si sample were $7.4x10^{15}/cm^3$ and $1320cm^2/Vs$, respectively. Both the carrier densities and the mobilities decreased with the fluence. A carrier density n after an irradiation of fluence Φ is represented empirically as

$$n = n_o - K_n\Phi ,$$

where n_o is the initial carrier density and K_n the carrier removal rate[5,6]. On the other hand, the change in a mobility μ is described as

$$\mu^{-1} = \mu_o^{-1} + K_u\Phi ,$$

where μ_o is the mobility before irradiation and K_u the mobility damage constant[6]. The values K_n and K_u for the 3C-SiC epilayers were estimated to be $1.4x10^{-2}cm^{-1}$ and $3.4x10^{-22}Vs$, respectively, from the result shown in Fig.1. On the other hand, the values K_n and K_u for the

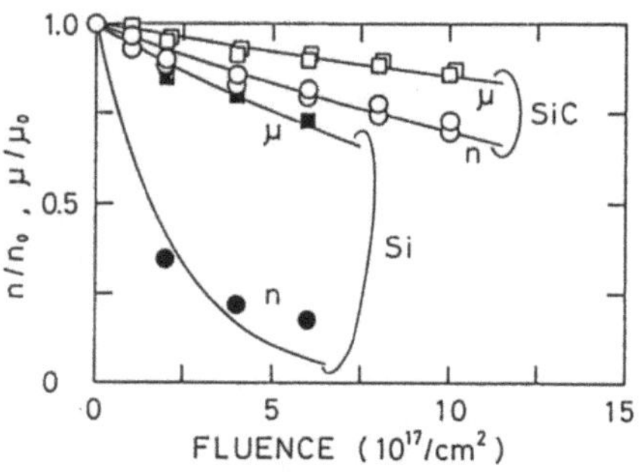

Fig.1 Electron irradiation effects on the carrier density n and the Hall mobility μ at RT for 3C-SiC and Si. n_o and μ_o indicate the carrier density and the mobility before the irradiation, respectively.

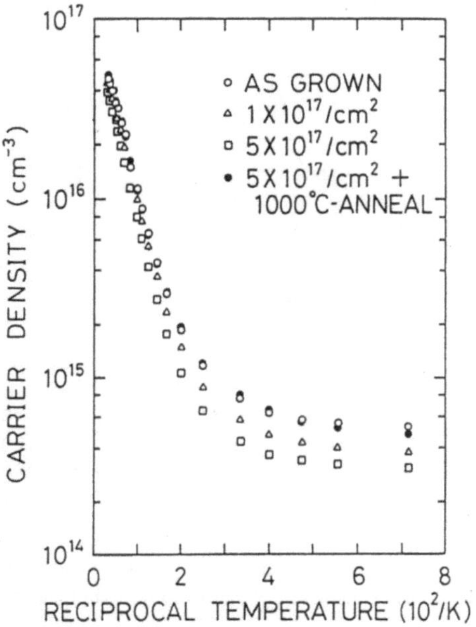

Fig.2 Temperature dependences of the carrier densities for as-grown, electron irradiated, and annealed 3C-SiC epilayers.

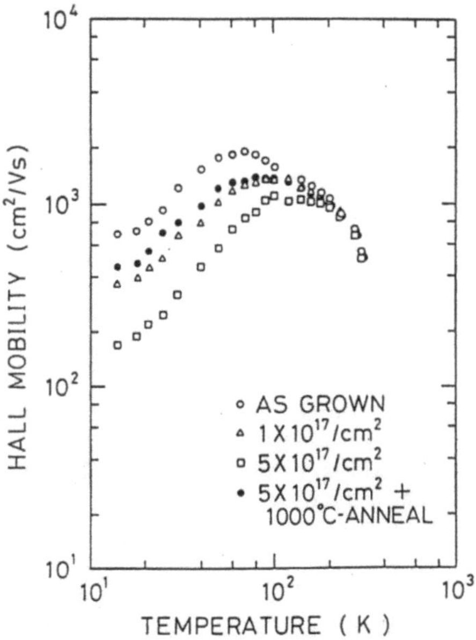

Fig.3 Temperature dependences of the Hall mobilities for as-grown, electron irradiated, and annealed 3C-SiC epilayers.

n-type Si were obtained to be $3.3x10^{-2}cm^{-1}$ and $4.2x10^{-22}Vs$, respectively. Though K_u for 3C-SiC is comparable with that for Si, K_n for 3C-SiC is smaller than that for Si. This result suggests that 3C-SiC epilayers are more tolerant of radiation than Si. The degradation of the electrical properties is due to defects induced by the irradiation. When the electronic levels of defects are much lower than the Fermi level, the defect introduction rate equals the carrier removal rate[6]. The electronic levels of defects induced by the electron irradiation are thought to be deep enough, by analogy with neutron irradiation results[2]. Therefore, the defect introduction rate is estimated to be $0.014cm^{-1}$.

Figure 2 shows the carrier densities as a function of temperature for the epilayers before and after the electron irradiation. The decrease in the carrier density was observed after the irradiation over the temperature range between 14K and RT. The decrease in the carrier densities with the electron fluence gives evidence that electron traps are introduced in 3C-SiC by the irradiation.

The temperature dependences of the Hall mobilities of the epilayers before and after the irradiation are shown in Fig.3. After the irradiation, the mobility decreased significantly at temperatures below 100K. However, the slope of μ vs T curves, which is determined by the major scattering mechanism, was found to be unchanged. The temperature dependence of the mobility of the as-grown 3C-SiC epilayers is dominated by ionized impurity scattering in this temperature region. Therefore, this result means that the amount of charged scattering centers which act like ionized impurities increased by the irradiation.

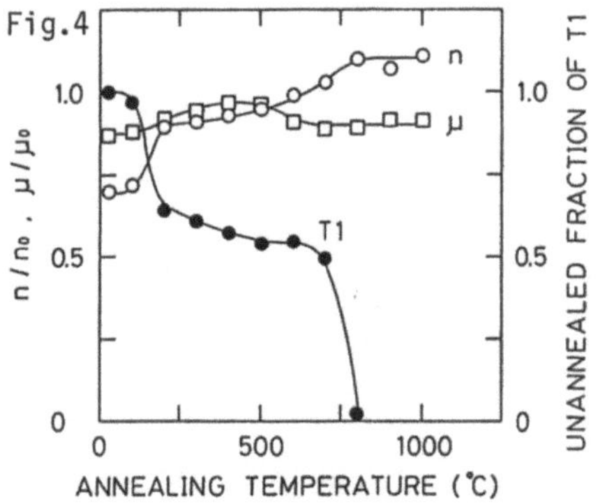

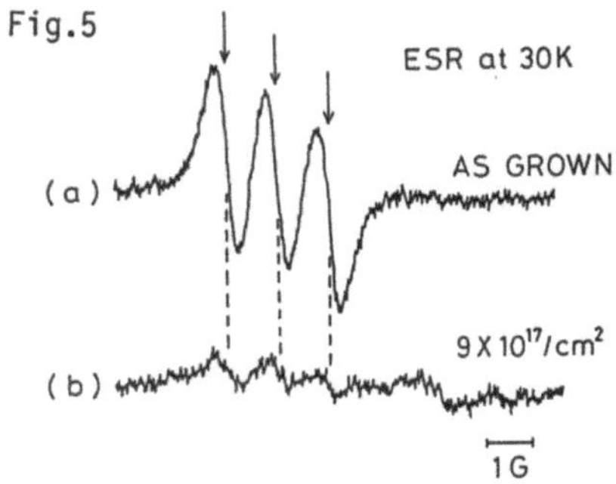

Fig.4 Isochronal annealing behaviors of the carrier density n and the Hall mobility μ at RT for the 3C-SiC irradiated with $1x10^{18}/cm^2$. The annealing behavior of the T1 center is also shown. n_o and μ_o indicate the carrier density and the mobility before the irradiation, respectively.

Fig.5 ESR spectra at 30K for (a) as-grown and (b) the electron irradiated samples with $9x10^{17}/cm^2$. Three-lines hyperfine structure related to N is indicated by the arrows.

The changes in the carrier density and the mobility at RT by the isochronal annealing are shown in Fig.4, as well as that in the ESR intensity of the T1 center at RT. The carrier density recovered fully by the annealing of 800°C. The carrier densities at low temperatures were also fully recovered after the annealing, as shown in Fig.2. The recovery of the carrier density occurred at the same temperatures as the reduction of the T1 intensity. The annealing behavior can be explained by assuming that the T1 centers act as electron traps and the disappearance of the T1 centers results in the increase in the carrier density. On the other hand, the increase in the mobility below 100K was observed after the annealing, as shown in Fig.3. This result is ascribed to the disappearance of the charged scattering centers by the annealing.

Figure 5(a) shows a typical ESR spectrum at 30K for the as-grown epilayers. Three-lines hyperfine structure (hfs) related to N impurities was observed superposed on a single-line structure. The intensity of the hfs decreased by the irradiation, as shown in Fig.5(b). The fluence dependence of the hfs intensity is shown in Fig.6. The hfs intensity also recovered by isochronal annealing as the carrier density did. Therefore, the decrease in the hfs intensity is not caused by the structural change of the defects related with N, but by the shift of the Fermi level. That is, the reduction of the hfs intensity can be explained by the introduction of deep-level traps which capture electrons from the N-related defects. These electron traps act as electron-scattering centers and reduce the mobility. Taking account of the obtained results together, the T1 centers are likely to be dominant electron traps induced by the irradiation.

The result shown in Fig.6 also indicates that the Fermi levels in the epilayers are located close to the electronic level of the N-related defect, considering the Fermi distribution at 30K. It was estimated from

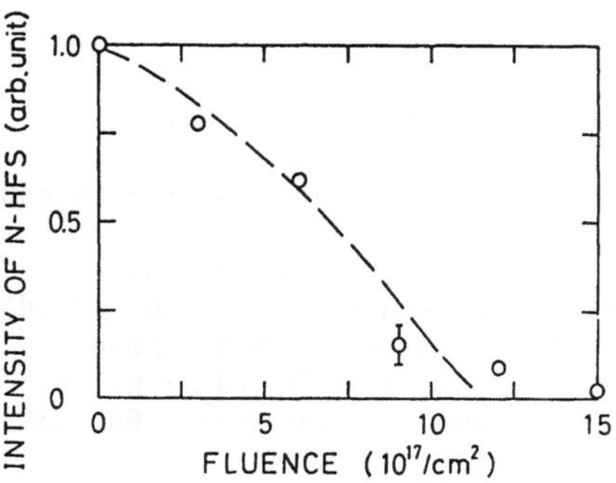

Fig.6 Electron fluence dependence of the ESR intensity of N-related hyperfine structure. The dashed line shows the calculated curve from the model of two donors and a compensation acceptor [ref.7].

the carrier densities at 30K that the Fermi levels in these samples are located at around 20meV below the bottom of the conduction band. Thus, the N-related defect is thought to have the electronic level of about 20meV below the conduction band edge. By using two donors and a compensation acceptor model[7], in which the N-related defects have a donor level of 24meV, the change in the carrier density and the N-related hfs signal can be well interpreted quantitatively, as shown in Fig.6. The defect introduction rate was estimated to be 0.018cm^{-1} from the fluence dependence of the N-related hfs intensity. This value is in good agreement with that obtained from the carrier removal rate.

4. Conclusions

The effects of 1 MeV electron irradiation on electrical properties of CVD-grown 3C-SiC epilayers were studied by Hall and ESR measurements. The carrier density and the Hall mobility of the epilayers decreased with the electron fluence over the range from $1x10^{17}/cm^2$ to $1x10^{18}/cm^2$. The carrier removal rate K_n and the mobility damage constant K_u were obtained to be $1.4x10^{-2}cm^{-1}$ and $3.4x10^{-22}Vs$, respectively. The values K_n and K_u for n-type Si were estimated to be $3.3x10^{-2}cm^{-1}$ and $4.2x10^{-22}Vs$, respectively. From the comparison of these values obtained, 3C-SiC epilayers were found to be more tolerant of radiation than Si. The intensity of the 3-lines hfs related with N impurities was also decreased with the electron fluence. The decrease in the carrier density and the intensity of the N-related hfs is explained by the introduction of electron traps which capture electrons from the conduction band and the N-related defects. Isochronal annealing of the irradiated epilayers indicated that the carrier density recovered at the same temperatures that the T1 centers disappeared. The result is explained by the reduction of the T1 centers which act as electron traps.

References

[1] J.A.Freitas,Jr., S.G.Bishop, J.A.Edmond, J.Ryu, and R.F.Davis, J.Appl.Phys. **61**, 2011 (1987).

[2] V.Nagesh, J.W.Farmer, R.F.Davis, and H.S.Kong, Radiation Effects and Defects in Solids **112**, 77 (1990).

[3] I.Nashiyama, H.Okumura, E.Sakuma, S.Misawa, K.Endo, S.Yoshida, H.Itoh, and N.Hayakawa, in Amorphous and Crystalline Silicon Carbide II, Springer Proceeding in Physics Vol.43, edited by M.M.Rahman, C.Y.-W.Yang, and G.L.Harris (Springer-Verlag, Berlin, 1989) p.100.

[4] H.Itoh, N.Hayakawa, I.Nashiyama, and E.Sakuma, J.Appl.Phys. **66**, 4529 (1989).

[5] G.C.Messenger and M.S.Ash, The effects of radiation on electronic systems (Van Nostrand Reinhold, New York, 1986) p.194.

[6] V.A.J.van Lint, T.M.Flanagan, R.E.Leadon, J.A.Naber, and V.C.Rogers, Mechanisms of Radiation Effects in Electronic Materials, Vol.1 (John Willey & Sons, New York, 1980) p.278.

[7] I.Nashiyama, H.Okumura, E.Sakuma, S.Misawa, K.Endo, and S.Yoshida, in Amorphous and Crystalline Silicon Carbide III, Proc. 3rd International Conference on Amorphous and Crystalline Silicon Carbide and Other Group IV-IV Materials, held at Howard Univ., Washington.D.C., Apr.1990.

Residual Carriers and ESR Centers in Epitaxially Grown 3C-SiC

I. Nashiyama, H. Okumura, E. Sakuma, S. Misawa, K. Endo, and S. Yoshida

Electrotechnical Laboratory, 1-1-4 Umezono, Tsukuba, Ibaraki 305, Japan

Abstract. Both ESR centers and residual carriers in undoped 3C-SiC films epitaxially grown on silicon substrates were studied at temperatures between 4.2 K and 300 K. An ESR signal with broad single-line structure was found at temperatures above 40 K. Its spin density and temperature dependence suggest that the ESR center associated with this signal is a dominant source of residual carriers. As for the complex ESR signal observed below 40 K, we found that the spin density of a ^{14}N hyperfine structure component is much lower than the residual carrier concentration at room temperature. Whereas, the spin density of a single-line component showed strong correlation with the carrier concentration. A two donors compensation model is proposed. The temperature dependences of the carrier concentration and the spin densities are discussed.

1. Introduction

The chemical vapor deposition (CVD) technique of 3C-SiC heteroepitaxial growth on silicon substrates can provide large-area 3C-SiC layers suitable for semiconductor device fabrication. At present, however, a significant amount of electron carriers of the order of 10^{16} cm^{-3} exists even in an undoped 3C-SiC epilayer and concentrations of these residual carriers strongly depend on CVD gas parameters and crystal growth temperature. Reduction of the residual carriers is essential for device applications of this material. However, the origin of these carriers is not completely understood yet[1].

By electron spin resonance (ESR), a complex signal with 3-line structure was observed at temperatures below 40 K for 3C-SiC films[2]. The overall line shape of this complex ESR signal is known to be quite sensitive to the crystal growth conditions[2,3]. In our previous work[4], we showed that this complex signal consists of a ^{14}N hyperfine structure(hfs) component and a single-line component with approximately the same g-value, but with different saturation characteristics and temperature dependences.

Several studies have been made on residual carriers in CVD 3C-SiC films in conjunction with nitrogen donors[5-8]. However, Okumura et al. reported that the nitrogen donor is not a dominant source of the residual carriers[1]. They pointed out a serious disagreement between the fact that the carrier concentration increases with temperatures from 40 K to 300 K by approximately an order of magnitude and the fact that almost all the electrons associated with the complex ESR signal are already excited thermally at 40 K.

In this work, ESR signals are measured for various undoped 3C-SiC films and correlations between spin densities and the carrier concentrations are studied as well as their temperature dependences. The origin of the residual carriers is discussed in terms of ESR spin densities. A model is

Editors: G.L. Harris, M.G. Spencer, and C.Y. Yang

proposed to explain the temperature dependences of spin densities and carrier concentrations simultaneously.

2. Experiments

An undoped 3C-SiC layer was epitaxially grown on a 70x60 mm^2 on-axis Si(001) substrate at 1350°C by using a horizontal CVD system with SiH_4+C_3H_8+H_2 reaction gases. The atomic ratio C/Si of the reaction gas was kept at a fixed value of 2.4 which provided the highest quality 3C-SiC epilayer. 3C-SiC films were prepared for ESR and Hall-effect measurements by dicing the 3C-SiC/Si epitaxial wafer and removing the silicon substrates by chemical etching. The films show n-type conduction with residual carriers from $10^{16} cm^{-3}$ to $10^{17} cm^{-3}$ and thicknesses from 14 to 35 microns depending on the position on the epitaxial wafer. Details of the epitaxial growth of 3C-SiC and their electrical measurements have been described elsewhere[9]. ESR spectra and carrier concentrations were measured at temperatures between 4.2 K and 300 K for various 3C-SiC films prepared from different portions of the epitaxial wafer along the CVD gas flow. Spin densities were determined by using a standardized Mn^+ in MgO.

3. Results

The complex ESR signals were measured at 30 K for various 3C-SiC films and a correlation between the ESR signals and the carrier concentrations was examined. The overall line shape of the ESR signal varied with the position on the epitaxial 3C-SiC/Si wafer along the CVD gas flow as shown in Fig.1. The complex ESR signals were decomposed into hyperfine structure(hfs) components and narrow single-line components by the method stated elsewhere[3], and spin densities of the two components were evaluated separately. The results are compared with carrier densities measured by Hall effect at room temperature in Fig.2. The spin densities of the hfs components are found to be lower than the carrier concentrations by an order

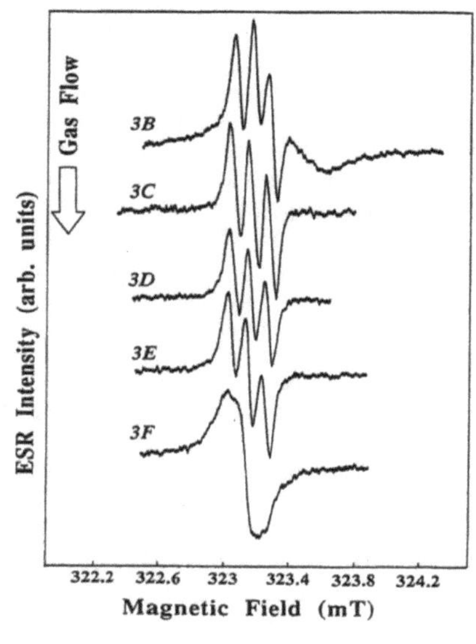

Fig.1 complex ESR signals at 30 K for various 3C-SiC films prepared from different portions of the 3C -SiC/Si epitaxial wafer along the CVD gas flow.

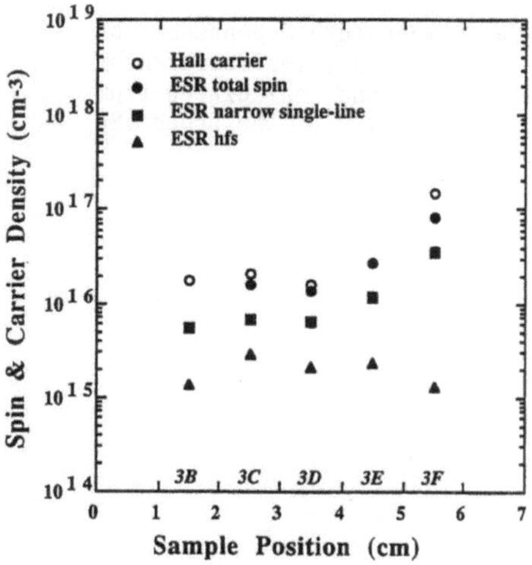

Fig.2 Spin densities measured at 30 K are compared with residual carrier concentrations at room temperature. The total spin densities include the new broad signal and the two components of the complex ESR signal.

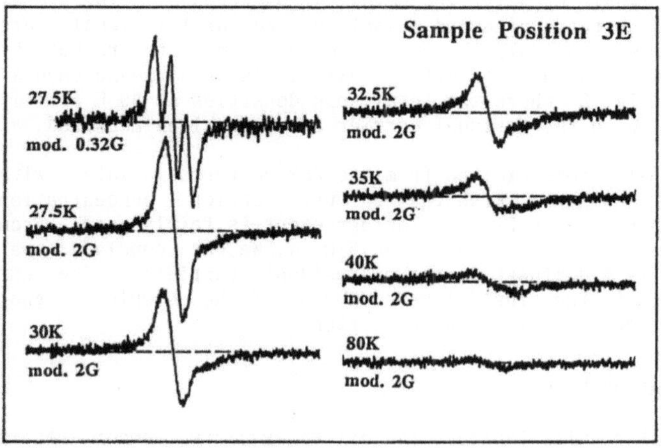

Fig.3 A new broad ESR signal became obvious by increasing the modulation width in the ESR measurements. This signal is superposed on the complex ESR signals below 40 K.

of magnitude and almost independent of the position on the epitaxial wafer. No correlation was observed between them. On the other hand, spin densities of the narrow single-line component indicate a strong correlation with the carrier concentrations. But, the spin densities are not high enough to explain all the residual carriers.

By increasing the modulation width of the ESR measurements, the complex ESR signal was deformed and a new broad single-line signal became obvious as shown in Fig. 3. The complex signal disappeared completely above 40 K, though the broad single-line signal remained. This broad signal is similar in appearance to that reported by Carlos et al.[2] for 3C-SiC samples with higher residual carrier concentrations. The spin density of this broad signal decreased with increasing temperature from 40 K to 150 K. Above

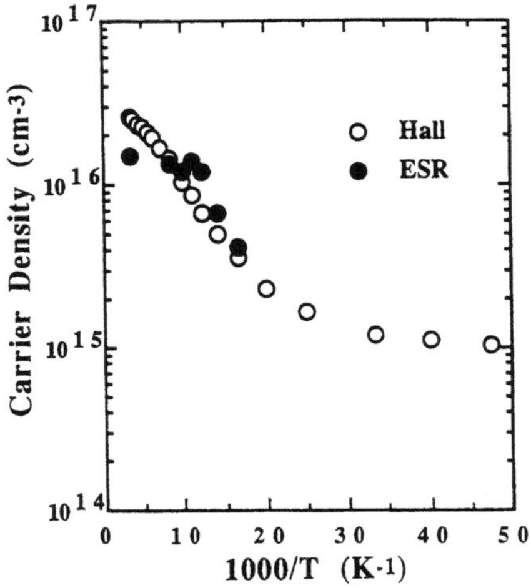

Fig.4 Temperature dependence of the carrier densities measured by Hall effect and calculated from the spin density of the broad ESR signal.

150 K, however, the signal intensity was below the background. The total spin densities including this new broad signal and the complex signal were evaluated at 30 K, where almost all the conduction electrons are trapped by the ESR centers and no saturation effect exists in ESR measurements. Results are shown in Fig. 2, where the total spin densities at 30 K are in excellent agreement with the residual carrier concentrations measured at room temperature.

The density of the electrons excited from the ESR center associated with the new broad signal is compared with the residual carrier concentration as a function of temperature in Fig.4. The agreement is fairly good. From the experimental results shown in Fig. 2 and Fig. 4, we can conclude that the new ESR center is a dominant source of residual carriers. The ESR centers corresponding to the single-line component of the complex signal also contribute to residual carriers to some extent.

4. Two Donors Compensation Model

The above experimental results indicate the existence of two donor levels associated with the complex ESR signal below 40 K and the new broad ESR signal above 40 K. Therefore, we introduce a model consisting of two donor levels and one acceptor level as schematically shown in Fig.5, in order to explain the temperature dependences of the spin densities and carrier concentration. The first donor level with an activation energy E_{d1} corresponds to the new broad signal observed above 40 K, which has a density N_{d1} of an order of $10^{17}cm^{-3}$. The second donor level with an activation energy E_{d2} stands for the narrow single-line and the hfs components observed below 40 K. The density N_{d2} of this level is chosen to be equal to the sum of the both components. Shallow acceptors with an activation energy E_a and a density N_a are assumed to be fully activated at all temperatures, namely, $E_a=0$.

The experimental curve (carrier density vs. reciprocal temperature) obtained by Hall-effect measurements can be well reproduced by the two donors compensation model as shown in Fig 6. The abrupt decrease in the spin density of the complex ESR signal at around 35 K[4] is also well explained by this model. Energy levels and donor concentrations were determined by the best fit of the theoretical curve to the experimental one. Temporary values used in the calculation are given in the figure.

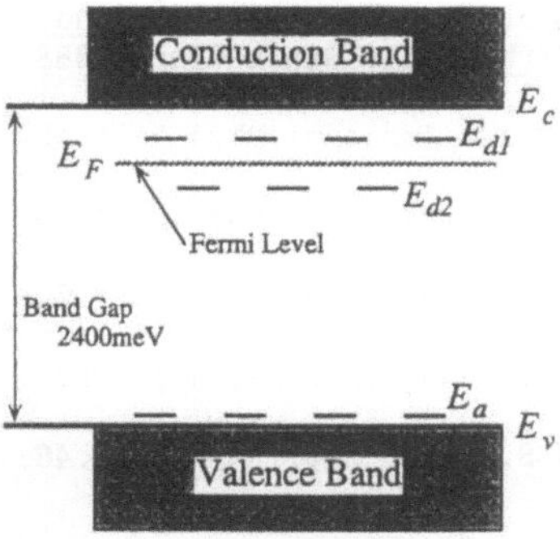

Fig.5 A schematic drawing of the energy levels used for the two donors compensation model.

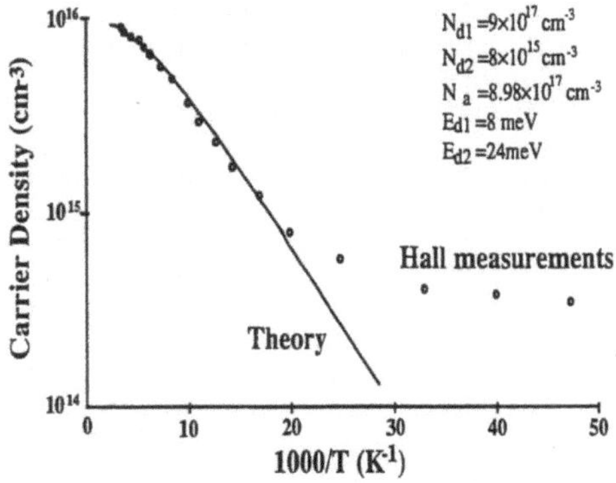

Fig.6 Temperature dependence of the carrier density. The solid line indicates the theoretical values calculated by the two donors compensation model.

5. Summary

Spin densities of the complex ESR signal at 30 K and residual carrier concentration at room temperature were measured for undoped 3C-SiC films with different carrier concentrations. We found that the spin density of the ^{14}N hfs component of the complex signal is lower than the residual carrier concentration by an order of magnitude. No correlation was found between them. Whereas, the spin density of the single-line component showed strong correlation with the carrier concentration. In addition, an ESR signal with a broad single-line structure was found above 40 K. The spin density and its temperature dependence indicate that the ESR center associated with this broad signal is a dominant source of residual carriers. A two donors compensation model can well explain the temperature dependences of the carrier and the spin density above 50 K.

Acknowledgment

The authors would like to thank H. Itoh of Japan Atomic Energy Research Institute for his helpful discussions. The authors are also grateful to I.Sakata of Electrotechnical Laboratory for providing an ESR apparatus and to T. Arai for his assistance in data analyses.

References

1. H.Okumra,M.Shinohara,S.Kuroda,K.Endo, E,Sakuma, S.Misawa, and S.Yoshida, Jpn.J.Appl.Phys.**27**,1712(1988).
2. W.E.Carlos,W.J.Moore,P.G.Siebenmann,J.A.Freitas,Jr.,R.Kaplan,S.G.Bishop, P.E.R. Nordquist,Jr., M. Kong, and R.F. Davis, Proc.Mater.Res.Soc.Symp., Anaheim,**97**,253(1987).
3. S.G.Bishop, J.A.Freitas,Jr., T.A.Kennedy, W.E.Carlos, W.J.Moore, P.E.R. Nordquist,Jr.,and M.L.Gipe, Proc.1st.Int.Conf.Amorphous and Crystalline Silicon Carbide, Washington DC,1987 (Springer,Berlin,1989) p.90.

4. I.Nashiyama,H.Okumura,E.Sakuma,S.Misawa,K.Endo,and S.Yoshida, Proc.2nd. Int.Conf.Amorphous and Crystalline Silicon Carbide II, Santa Clara,1988 (Springer,Berlin,1989) p.100.
5. A.Suzuki,A.Uemoto,M.Shigeta,K.Furukawa,and S.Nakajima,Appl.Phys.Lett.**49**, 450(1986).
6. B.Segall,S.A.Alterovits,E.J.Haugland, and L.G. Matus, Appl.Phys.Lett.**49**, 584(1986).
7. B.Segall,S.A.Alterovits,E.J.Haugland, and L.G. Matus, Appl.Phys.Lett.**50**, 1533(1987).
8. J.A.Freitas,Jr.,S.G.Bishop,P.E.R.Nordquist,Jr., and M.L.Gipe, Appl.Phys. Lett.**52**,1695 (1988).
9. S.Yoshida, K.Sasaki, E.Sakuma, S.Misawa, and S.Gonda, Appl.Phys.Lett.**46**, 766(1985).

Transport Measurements and Shallow Donors in Cubic SiC

W.J. Moore

Naval Research Laboratory, Washington, DC 20375, USA

Abstract. Although there is considerable agreement on the identification of the 54 meV donor seen in photoluminescence in CVD cubic SiC as isolated, substitutional nitrogen, the identification of the shallow donor which dominates the conductivity in this material is less certain. In the traditional view of activation energy for the dominant donors, activation energies take values between 54 meV in the purest materials and approximately 15-20 meV in the more usual CVD cubic SiC grown on Si substrates. This paper presents evidence for reassessing the traditional view and proposes that experiments supporting binding energies greater than 20 meV but less than 54 meV should be reconsidered. Several such experiments are reconsidered and it is shown that reinterpretation to support conduction dominated by binding energies in the 15-20 meV range is reasonable. It is proposed that the 15-20 meV donor which dominates the conductivity is not neutral, isolated, substitutional nitrogen but is some other center. D^- centers or donor complexes are possible origins of this donor.

1. Introduction

Cubic silicon carbide films grown by the CVD process on silicon substrates usually have n-type conductivity at 300 K due to a donor with an apparent binding energy in the 15-20 meV range and which appears to be highly compensated [1-4]. The details of the donor concentration, activation energy, and relative compensation are not substantially changed irrespective of the laboratory growing the material or that making the temperature scanned Hall-effect measurements by which the donor characteristics are determined.

Donor binding energies as small as 15-20 meV are not expected in SiC because the effective mass value for an electron bound to an isolated donor is approximately 50 meV. At high concentrations, the effective donor binding energy can be reduced substantially by several mechanisms [5]. Controlled studies of the effects of high impurity concentration on Si [6] and Ge [7] have been conducted. Extrapolation of these experiments in Si and Ge to SiC is not done easily as the various mechanisms for shifting and broadening the impurity level do not scale the same with effective mass and dielectric constant. However, over the temperature region in which most SiC data are taken ($T > 70$ K), both Si and Ge exhibit a freeze-out characteristic approximating that of isolated impurities with concentrations and compensations which are roughly comparable to those present in SiC samples. Therefore, one would not expect a change in apparent binding energy sufficient to reduce that energy from the approximately 50 meV expected for a donor to the observed 15-20 meV.

Photoluminescence studies of SiC reveal transitions which have been attributed to free-to-bound, donor-acceptor pair, and two electron satellite transitions of a 54 meV donor, usually identified as isolated, substitutional nitrogen [8]. The large difference between the 54 meV binding energy as determined by photoluminescence and the 15-20 meV determined by the thermal equilibrium transport measurement is usually ascribed to relaxation of the lattice in the thermal equilibrium measurement. In a recent experiment [9], the thermal quenching of photoluminescence was used to determine the thermal activation energy of the center providing the 54 meV photoluminescence. In that experiment the thermal activation energy of the donor also was found to be 54 meV. This result suggests that the thermal and optical determinations of activation energy may be much closer than previously thought. The fact that the samples exhibiting a 54 meV binding energy in these optical experiments have a conductivity dominated by a 15-20 meV donor has led to the conclusion that the center providing the 54 meV photoluminescence and the center dominating the conductivity are different [9].

Probably the clearest evidence linking the 54 meV and 15-20 meV binding energies is that of SEGALL et al. [2] in which activation energies from temperature-scanned Hall-effect data for CVD SiC on Si and similar data for vapor grown platelets [10] were plotted as a function of donor concentration. Although the activation energies for the CVD SiC on Si samples fell in the 15-20 meV range, those for the platelets fell in the 20-48 meV range. The combined data set yielded a continuous transition from binding energies near 50 meV for the purest material to 15-20 meV for the material with the highest donor concentrations. The apparent dependence of binding energy on donor concentration resulting from the combined data was in agreement with that observed previously for Si and Ge and suggested that the 54 meV and 15-20 meV centers are the same, differing only in concentration.

In an attempt to resolve these discrepancies and to facilitate improvements in material production, we have reassessed the experimental results which yield donor binding energies in the range from 20 meV to 50 meV in an attempt to determine whether there is clear experimental support for the identification of the 15-20 meV donor as nitrogen in high concentration. If it is possible to establish clearly the existence of donors with these intermediate binding energies then there is persuasive evidence for identifying both the 54 meV and 15-20 meV donors as the same center. If, on the other hand, the experimental evidence for the intermediate values of binding energy is not strong then we may consider models in which the 54 meV and 15-20 meV impurities arise from different centers consistent with the recent photoluminescence result [9].

The data on which these results were based will be reassessed and some representative binding energy determinations by others will be reconsidered. We will suggest that the experimental evidence linking the 54 meV and 15-20 meV centers is not strong and that separate origins for these centers should be considered.

2. Discussion of Experimental Results

Most recent data on the thermal activation of the conduction process in SiC is taken using temperature scanned Hall-effect measurements. Some of these results will be discussed in detail later. We first consider some representative results using other techniques.

An early determination of donor activation energy in CVD SiC on Si was that of KAPLAN et al. [11]. In this work the relative concentration of free carriers was determined by observing the strength of the infrared absorption due to free carrier cyclotron resonance as a function of temperature. The authors explicitly state that they assume the sample to be uncompensated and, therefore, they assume the thermal activation varied as $E/2kT$. We now know that all such samples appear to be highly compensated; the thermal activation varies as E/kT. Therefore, the quoted 38 meV binding energy should have been quoted as 19 meV for compensated material and does not support the existence of binding energies in the 20 to 50 meV region.

A recent study of conductivity in CVD SiC by MOLNAR and KELLNER [12] analyses the thermal activation of the conductivity after the technique of FRITZSCHE [13] and CASEY et al. [14]. Here, binding energies in the 25-32 meV range are calculated from conductivity versus temperature data. However, when Hall-effect data on these same samples are analyzed as carrier density versus temperature, binding energies in the 15-20 meV range were determined. Associating the slope of the conductivity freeze out with impurity activation energy neglects both the temperature dependence of the mobility and that of the conduction band density of states which typically has the form $2M(2\pi m^{*}kT/h^{2})^{3/2}$. Analysis of the Hall-effect data compensates for the assumed temperature dependence of the density of states and analyzes the carrier density with the mobility removed. Conductivity and Hall-effect analyses agree only when these two temperature dependences cancel. That is, only in the special case in which the mobility varies as $T^{-3/2}$. CASEY et al. suggest that the larger activation energies calculated from conductivity analysis of samples with high impurity concentrations are more nearly correct than the energies calculated from carrier density analysis. Although neglecting the dependence of the conduction band density of states and the mobility on temperature tends to yield larger activation energies, these larger energies are not, in general, more accurate than energies from carrier density analysis. If a sample is dominated by impurities in high concentration, conductivity analysis may give an activation energy nearer the correct value for low concentration impurities. On the other hand, if the sample is dominated by shallow impurities in low or intermediate concentration, carrier density analysis gives the correct value and conductivity analysis yields energies that can be significantly too large. Carrier density analysis is always correct as it gives the thermal activation energy for promoting carriers in the conduction process, even though that energy may be reduced by the sample characteristics and the final state of the transition may not be a well-behaved conduction band. The challenge in this case is to determine whether the conduction process is characteristic of isolated impurities thermally promoted to the conduction band or a process modified by other factors such as overlap of impurity states, screening, high compensation, complexing, etc. Activation energies from carrier density analysis, in this case 15-20 meV, are preferred over the higher values derived from conductivity analysis.

In electron spin resonance studies [4] it was originally concluded that the three line nitrogen ESR signal and the 15-20 meV donor dominating the conductivity have temperature dependencies which imply that the ESR signal arises from the neutral 15-20 meV donors. This work has since been reinterpreted [15]. The new result supports identification of the 54 meV center as neutral, isolated, substitutional nitrogen and the 15-20 meV center as another center.

Lastly, we consider the Hall-effect data. It is well-known that Hall-effect data for which the carrier density varies by several orders of magnitude as the temperature is changed can be fitted uniquely with the usual statistical analysis yielding values for impurity density, activation energy, and compensation density which are accurate to substantially better than a factor of two. The principal uncertainty in the fits is the value of the Hall factor which may vary between 1 and 2 and is usually taken to be 1. Carrier densities for donors in SiC typically freeze out with little or no indication of high concentration effects until the temperature is reduced below about 70 K. Below 70 K high concentration effects, including reduction in apparent activation energy, can occur. Often the carrier density changes by less than a order of magnitude over the 300 K to 70 K range. Unique impurity density, activation energy, and compensation results are usually not possible with such a small range. Therefore, there is substantial uncertainty in the values derived from the Hall-effect fits. The following discussion should be viewed as demonstrating that fits to the existing data, using binding energies in the 15-20 meV range, which are superior to the original fits, which used binding energies larger than 20 meV, are possible. Therefore, binding energies in the 20-50 meV range are not necessarily correct.

SEGALL et al.[2] based their discussion of the behavior of activation energy in SiC on data for vapor grown cubic platelets [10] which provide the high purity data and their own data on CVD samples on Si. There is considerable agreement that the data on their CVD samples is correctly fitted and, based on the relatively large range of temperature and the quality of the fit, little room for reinterpretation exists. In addition, their result is in complete agreement with most other careful studies of CVD cubic material conducted more recently.

We have reassessed the data on the platelets to find if that data could be consistent with 15-20 meV donors. Only two of the sets of sample parameters included in the AIVASOVA et al. [10] paper, samples 1 and 6, were accompanied with Hall-effect data which could be associated with them clearly. At least two impurities are present in these samples, one having an activation energy significantly greater than 50 meV and one consistent with the shallow donor which dominates conduction in the CVD SiC on Si. In the following discussion we will consider only the shallow donor.

We have reproduced the data for samples 1 and 2 and have included fits using the conduction band density of states and sample parameters of AIVASOVA et al. in Fig. 1. This fit is not entirely satisfactory (a better fit with these sample parameters can be made if the impurity degeneracy is taken to be 4) and becomes worse if the effective mass is changed from the value of 0.30 m_0 used by AIVASOVA et al. to 0.346m_0 after SEGALL et al. The freeze-out curve predicted for these sample parameters begins at lower temperatures and has more curvature than the data.

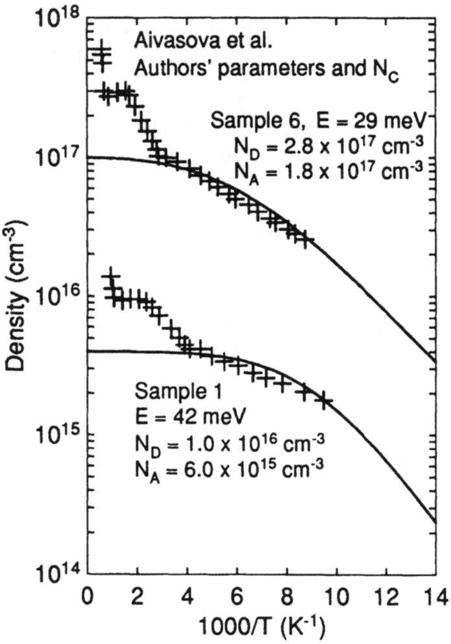

Fig. 1 Carrier density and theory for parameters indicated.

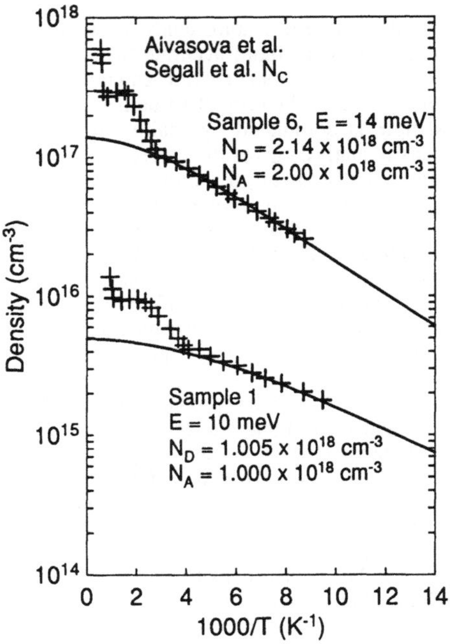

Fig. 2. Best fits to density.

Because these results demonstrate that new fits should be considered, we have increased the compensation (raising the temperature at which freeze out begins) and reduced the binding energy (reducing the freeze-out slope) to get the fits shown in Fig. 2. These are quite good in view of the small temperature range available. However, these fits yield donor activation energies which are 10 and 14 meV, unusually small values. By comparison, 18 meV binding energy fits are given in Fig. 3. These activation energies are about the maximum values compatible with the data and are the fits that we will use. We believe that all the samples included in the AIVASOVA et al. study could be modeled satisfactorily with 15-20 meV donors.

A modified version of the dependence of binding energy on donor concentration is presented in Fig. 4. Here the new data points arising from the reinterpretation of samples 1 and 6 are indicated by arrows extending from the original value, determined by the fits of Fig. 1, to the new values arising from the fits given in Fig 3. We propose that reinterpretation of all Hall data originally plotted by SEGALL et al. would show only activation energies in the 15-20 meV region and no points at higher binding energy.

Other Hall-effect data exists in the literature. We believe that most, if not all, of the fits to such data yielding binding energies greater than 20 meV for the shallow donor dominating the conductivity in undoped samples can be reasonably reinterpreted to yield binding energies in the 15-20 meV range. Therefore, the evidence supporting the existence of binding energies in the 20-50 meV range is not strong.

If there is no strong experimental evidence for binding energies in the 20-50 meV range then it is reasonable to consider that the 15-20 meV center is something other than substitutional, neutral nitrogen. Possible sources for such a donor are complexes with or without nitrogen, D^- centers, perhaps with nitrogen as the donor, or some previously unidentified impurity.

The D^- center is an interesting possibility. Such centers are donors which have weakly attracted a second electron. These centers have been studied extensively in silicon [16] and germanium [17]. In germanium the D^- binding energies range from 1.7 meV to 5 meV, a

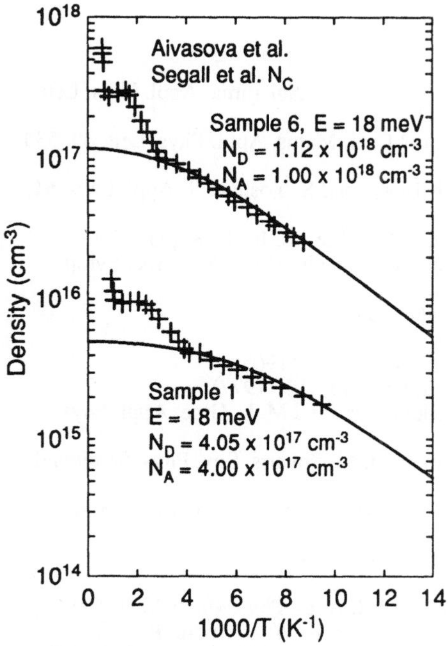

Fig. 3. Compromise fits.

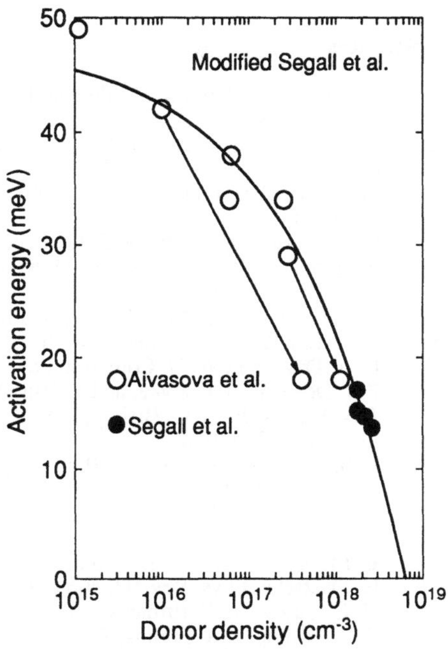

Fig. 4. Modified variation of activation energy with donor density. Arrows indicate the new values from fits of Fig. 3

factor of 2 to 6 smaller than the effective mass donor binding energy, 10 meV. Scaling these values to SiC, in which the effective mass donor binding energy is 50 meV, predicts D^- binding energies in the range from 9 to 25 meV. The unknown donor at 15-20 meV falls in the middle of this range.

In order to have D^- centers there must be excess free electrons. A recent study of inversion domain boundaries (antiphase boundaries) [18] suggests that excess charge could be present in the neighborhood of these boundaries. That charge could provide the required electrons for overcharging donors near these boundaries.

The statistical model for the thermal activation of a D^- center is analogous to that of a compensated normal donor. In the compensated normal donor the role of the compensation is to reduce the number of donors which are able to contribute to the conduction process by thermal ionization of the electron to the conduction band. These compensated donors may, however, still contribute to the recombination process. A weakly excited, 90% compensated donor is one in which 10% of the donors can contribute to the carrier generation process while 90% are recombination centers. This is analogous to a 10% overcharged donor population. In this case also, 10% can contribute to the generation process while the remaining 90% are recombination centers.

This model for the origin of the shallow donor, therefore, has the additional attraction of eliminating the problem of the origin of the large compensation always observed in these samples. It is not necessary to postulate the presence of an unidentified acceptor in concentrations always just a little less than the donor concentration (usually in the range 4-12% less) but never greater than the donor concentration. It is statistically unlikely that an independent acceptor would track the donor concentration so closely. However, it is necessary to postulate that in most samples the population of overcharged donors is in the range 4-12% of the total donor population.

This work was supported, in part, by the Office of Naval Research.

References

1. A. Suzuki, A. Uemoto, M. Shigeta, K. Furukawa, and S. Nakajima, Appl. Phys. Lett. **49**, 450 (1986).
2. B. Segall, S. A. Alterovitz, E. J. Haugland, and L. G. Matus, Appl. Phys. Lett. **49**, 584 (1986).
3. M. Yamanaka, H. Daimon, E. Sakuma, S. Misawa, and S. Yoshida, J. Appl. Phys. **61**, 599 (1987).
4. W. E. Carlos, W. J. Moore, P. G. Siebenmann, J. A. Freitas, Jr., R. Kaplan, S. G. Bishop, P. E. R. Nordquist, Jr., M. Kong, and R. F. Davis, Mater. Res. Soc. Symp. Proc. **97**, 253 (1987).
5. T. F. Lee and T. C. McGill, J. Appl. Phys. **46**, 373 (1975).
6. A. L. Lin, Appl. Phys. Lett. **53**, 776 (1988).
7. E. A. Davis and W. D. Compton, Phys. Rev. **140**, A2183 (1965).
8. P. J. Dean, W. J. Choyke, and L. Patrick, J. Lumin. **15**, 299 (1977).
9. J. A. Freitas, Jr., S. G. Bishop, P. E. R. Nordquist, Jr., and M. L. Gipe, Appl. Phys. Lett. **52**, 1695 (1988).
10. L. S. Aivasova, S. N. Gorin, B. G. Sidyakin, and I. M. Shvarts, Sov. Phys. Semicond. **11**, 1069 (1977).
11. R. Kaplan, R. J. Wagner, H. J. Kim, and R. F. Davis, Solid State Commun. **55**, 67 (1985).
12. B. Molnar and G. Kellner, To be published in Proc. MRS.
13. H. Fritzsche, Phys. Rev. **99**, 406 (1955).
14. H. C. Casey, Jr., F. Ermanis, and K. B. Wolfstirn, J. Appl. Phys. **40**, 2945 (1969).
15. S. G. Bishop, J. A. Freitas, T. A. Kennedy, W. E. Carlos, W. J. Moore, P. E. R. Nordquist, and M. L. Gipe, in: *Amorphous and Crystalline Silicon Carbide*, Vol. 34 of *Proceedings in Physics*, Edited by G. L. Harris and C. Y.-W. Yang (Springer, Berlin 1989).
16. P. Norton, Phys. Rev. Lett. **37**, 164 (1976).
17. S. Narita, Solid State Commun., **53**, 1115 (1985).
18. W. R. L. Lambrecht and B. Segall, Phys. Rev. B **41**, 2948 (1990).

Photoluminescence Properties of 6H-SiC Grown by Step-Controlled Epitaxy

H. Matsunami, H. Nishino, and T. Ueda

Department of Electrical Engineering, Kyoto University, Yoshidahonmachi, Sakyo, Kyoto 606, Japan

Photoluminescence(PL) properties of 6H-SiC single crystals grown by step-controlled epitaxy are studied. The epitaxial growth was carried out at 1500℃ through vapor phase of SiH_4 and C_3H_8 on off-oriented (0001)Si and (000$\bar{1}$)C faces. By the introduction of off-orientation, the growth temperature for high-quality homoepitaxial layers can be reduced by more than 300℃ compared with previous vapor phase growth. Undoped layers show n-type conduction probably due to nitrogen donors incorporated unintentionally through the ambient. Aluminum as a bright luminescent center and acceptor was doped with trimethyl-aluminum. By the temperature dependence of PL spectra and excitation intensity effects on PL spectra, donor(nitrogen)-acceptor(aluminum) pair recombination was confirmed. Doping effects are discussed by the use of PL spectral change and the results of Hall measurements.

1. Introduction

In vapor phase epitaxial(VPE) growth of 6H-SiC, substrate temperatures over 1800℃ had been necessary to obtain good surface morphology on natural(well-oriented) (0001)Si faces[1,2]. Recent efforts in the crystal growth on off-oriented (0001) Si and (000$\bar{1}$)C substrates has brought low temperature (~1500℃) growth of 6H-SiC with mirror-like surfaces[3,4]. By the introduction of off-orientation into substrates, the surface step density is increased, and 6H-SiC is grown laterally from the steps inheriting the stacking order of the substrates, so that 6H-SiC can be grown at lower temperatures by more than 300℃. Hence, the authors named the growth method "step-controlled epitaxy"[3]. So far, for step-controlled epitaxy, the growth mechanism has been elucidated based on the growth experiments, such as effects of off-orientation and surface polarity on crystal structures (polytypes) and surface morphology[5-7].

The polytype of 6H-SiC is a good candidate material for blue light-emitting diodes owing to a wide energy band gap of 2.86eV at room temperature. In addition, 6H-SiC has unequivalent atom sites in the crystal structure, which brings specific photoluminescence.

In this paper, photoluminescence properties of step-controlled epitaxial layers of 6H-SiC are described. Doping effects on PL spectra are also studied.

2. Experimental

Vapor phase epitaxial growth was carried out at 1500℃ in a horizontal quartz tube by using a SiH_4-C_3H_8-H_2 system under atmospheric pressure. Substrates were heated by using RF (radio frequency) induction[6,7]. The growth was carried out after the etching of substrates with HCl gas at 1500℃. The flow rates of SiH_4 and C_3H_8 were 0.15-0.60sccm and 0.1-0.4sccm, respectively in H_2 carrier gas (3slm). Under typical growth condition (SiH_4=0.3sccm and C_3H_8=0.2sccm), the growth rate was 2μm/h.

Undoped layers of 6H-SiC grown by step-controlled epitaxy show n-type conduction. Nitrogen as a donor impurity may be unintentionally incorporated from the ambient. As a bright luminescent center and acceptor, aluminum (Al) was doped during the crystal growth through bubbling of TMA (trimethyl-aluminum) with H_2. The acceptor concentration was changed in the range of TMA flow rate of 0.03-0.18sccm.

As substrates, single crystals of 6H-SiC grown by the Acheson method were used. In this work, both the basal planes of (0001)Si and (000$\bar{1}$)C faces were used. The faces were identified by thermal oxidation utilizing the difference of oxidation rates (oxidation rates on (000$\bar{1}$)C faces is much faster than (0001)Si faces). The off-direction along [11$\bar{2}$0] into the basal planes by 3 to 6 degrees were typically used. The surface of the basal planes was angle-lapped with carborundum, and polished with diamond paste.

The crystal quality of the epitaxial layers was examined using reflection high energy electron diffraction and double-crystal X-ray diffraction methods. Electrical properties were characterized by Hall measurements. Photoluminescence (PL) spectra were measured mainly at 4.2K using 365nm line of a high-pressure Hg lamp as an excitation source.

3. Results and Discussion

3.1 Donor-acceptor pair recombination

In various polytypes of SiC, there exist different atom sites which are crystallographically unequivalent. In 6H-SiC, there are two cubic sites and one hexagonal site[8]. Choyke and Patrick observed three different donor levels attributed to the occupation of the different atom sites by nitrogen (N) impurities[9]. The change of impurity levels owing to different atom sites is named "site effect". Gorban and Krokhmal found a similar site effect for acceptor levels[10]. As to donor-acceptor impurity levels, Ikeda et al. in the authors group studied the PL spectra in detail for single crystals of several polytypes grown from liquid phase[11,12]. They determined donor and acceptor levels by the analysis of PL spectra, although the difference in impurity levels owing to the different kinds of cubic sites was not resolved.

The PL spectra of lightly Al-doped (and unintentionally N-doped) 6H-SiC consist of the following three series.

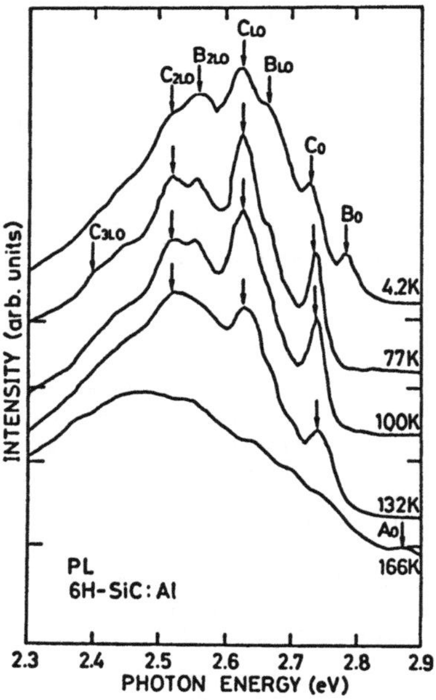

Fig.1 Temperature dependence of photoluminescence spectra for a lightly Al-doped sample. Nitrogen as a donor may be doped through the ambient. A_0: free to acceptor recombination, B_0: zero-phonon line of recombination of electrons at N donors of hexagonal sites and holes at acceptors, C_0: zero-phonon line of recombination of electrons at N donors of cubic sites and holes at acceptors. B_{LO}, B_{2LO}, C_{LO}, C_{2LO}: longitudinal optical phonon replicas of B_0 and C_0, respectively.

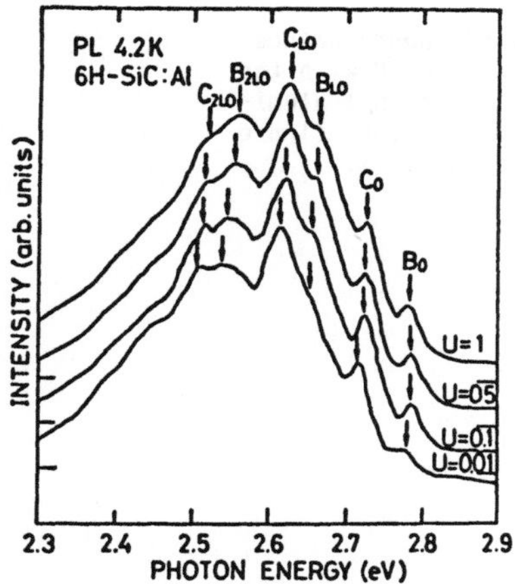

Fig.2 Photoluminescence spectral change for various excitation intensities at 4.2K. Notations in the figure are explained in Fig.1.

(1)A series: free electron in the conduction band to Al acceptors,
(2)B series: donor-acceptor pair recombination by nitrogen at hexagonal sites,
(3)C series: donor-acceptor pair recombination by nitrogen at cubic sites.

Figure 1 shows the temperature dependence of the PL spectra for a typical 6H-SiC doped with aluminum and unintentional nitrogen. In the spectra, zero-phonon peaks showing with zero(0) subscripts and phonon replicas of those zero-phonon peaks are shown. Since electrons bound at N donors of hexagonal sites are transferred to the conduction band with a temperature rise, the peaks of B-series diminish. At higher temperatures, peaks of C-series also diminish by the transfer of electrons bound at N donors of cubic sites to the conduction band. Only the peaks of the A-series remain, which indicates the recombination of free electrons in the conduction band and holes bound at Al acceptors.

In Fig.2, is shown the effect of excitation intensity on PL spectra at 4.2K. Peaks of B- and C-series shift to the higher energy side with the increase of excitation intensity, which indicates the specific feature of the recombination of donor-acceptor pairs. With the decrease in excitation intensity, the peaks of C-series become dominant, which means that the quasi-Fermi levels of electrons under illumination intensity goes down and the occupation number of electrons at N donors of hexagonal sites becomes low. Thus, the recombination of electrons at N donors of hexagonal sites and holes bound at Al acceptors becomes low, indicating the reduction in the intensity of B-series.

3.2 Doping effects

The carrier concentration can be controlled in the range of 10^{17} and $10^{21}cm^{-3}$ by changing the flow rate of TMA between 0.06 and 0.18sccm. The surface was mirror-like and not very different from that of undoped layers. The crystal quality, however, deteriorates when the carrier concentration is above $10^{21}cm^{-3}$, since the full width at half maximum of double crystal X-ray diffraction becomes wider by two times compared with that of substrates.

Typical PL spectra at 4.2K are changed as shown in Fig.3, when the flow rate of TMA is changed. With the increase in TMA flow rate, the peaks of C-series become dominant and those of B-series diminish, which can be explained by the Fermi level going down with higher acceptor doping and the probability of electron occupation at N donors of hexagonal sites becoming smaller than that of cubic sites under the illumination intensity in this study.

When the doping level of aluminum is remarkably increased, the PL spectrum becomes broad and it moves to the lower energy side. The PL intensity both at 4.2K and 77K is decreased when the flow rate of TMA is increased. This is partly due to the deterioration of crystal quality. Nonradiative recombination through Auger processes owing to the large number of holes may also cause the decrease of the PL intensity.

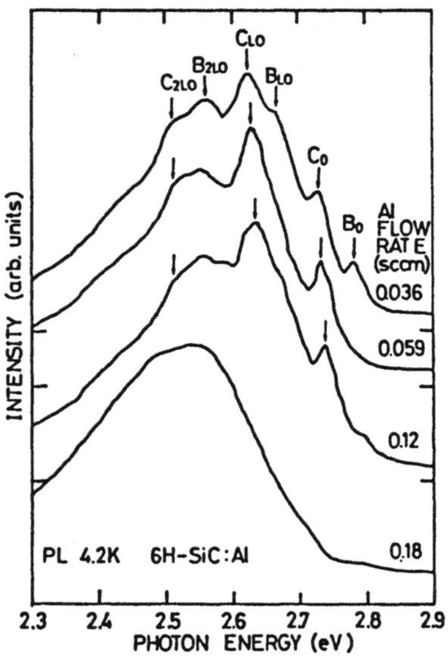

Fig.3 Change of photoluminescence spectra at 4.2K with the flow rate of trimethyl-aluminum. Notations in the figure are explained in Fig.1.

4. Conclusion

Photoluminescence properties were studied for single crystals of 6H-SiC grown by step-controlled epitaxy. The growth was carried out at 1500℃ by using vapor phase epitaxy of SiH_4 and C_3H_8. Donor-acceptor pair recombination was observed in slightly Al-doped samples (unintentionally-doped nitrogen may work as a donor). Temperature dependence of PL spectra and excitation effects on PL spectra indicate the site effects of photoluminescence. Photoluminescence change by Al-doping level is discussed.

Acknowledgment

This work was partially supported by Grant-in-Aid for Scientific Research from the Ministry of Education, Science and Culture, Japan and also by Iketani Science and Technology Foundation.

References

[1] W.v.Muench and I.Pfaffeneder, Thin Solid Films 31 39 (1976).
[2] S.Nishino, H.Matsunami and T.Tanaka, J. Cryst. Growth 45 144 (1978).
[3] N.Kuroda, K.Shibahara, W.Yoo, S.Nishino and H.Matsunami, Extended Abstr. of the 19th Conf. on Solid State Devices and Materials, Tokyo, p.227 (1987).
[4] K.Shibahara, N.Kuroda, S.Nishino and H.Matsunami, Jpn. J. Appl. Phys. 26 L1815 (1987).
[5] H.S.Kong, J.T.Glass and R.F.Davis, J. Appl. Phys. 64 2672 (1988).
[6] T.Ueda, H.Nishino and H.Matsunami, to be published in J. Cryst. Growth.
[7] H.Matsunami, T.Ueda and H.Nishino, to be published in Proc. of 1989 Materials Research Society Fall Meeting.
[8] L.Patrick, Phys. Rev., 127 1878 (1962).
[9] W.J.Choyke and L.Patrick, Phys. Rev., 127 1868 (1962).
[10] L.S.Gorban and A.P.Krokhmal, Sov. Phys. Solid State, 19 733 (1977).
[11] M.Ikeda, H.Matsunami and T.Tanaka, J. Lumin., 20 111 (1979).
[12] M.Ikeda, H.Matsunami and T.Tanaka, Phys. Rev., B22 2842 (1980).

Comparison of Experimental Laue Transmission and Reflection Patterns for 4H, 6H, 15R, and 3C SiC

W.J. Choyke[1], *M. Yoganathan*[1], *and C. Carter, Jr.*[2]

[1]Department of Physics and Astronomy, University of Pittsburgh, Pittsburgh, PA 15260, USA
[2]Cree Research, Inc., 2810 Meridian Parkway, Durham, NC 27713, USA

Abstract. A discussion is given as to how one may readily orient single crystal boule SiC and single crystal epitaxial films of SiC using Laue transmission and reflection x-ray methods. Data are presented for polytypes 4H, 6H, 15R and 3C SiC.

1. Introduction

With the advent of SiC boule growth in Europe, Japan and the US there is a need for quick and reasonably accurate orientation and polytype identification of this material. Despite the fact that the Laue method was applied to SiC platelets well over forty years ago [1] there has been no easily available reference "ATLAS" for such SiC patterns.

We give both transmission and back reflection Laue patterns, with appropriate directions indicated, for (0001)6H, (0001)4H, (0001)15R and (100)3C SiC and a back reflection Laue pattern for (111)3C SiC. We also show the evolution of a 6H SiC back reflection Laue pattern as the boule is rotated away from the (0001) growth axis.

2. Experimental conditions

In our experimental work a Philips x-ray generator is used with usual settings of 40 kV and 15 mA. A standard Polaroid camera back with Polaroid type 57 (ASA 3000) film proves to be entirely satisfactory. In the case of back reflection, the best balance between spatial spot resolution and spot size was found to be for a crystal to film distance of 5.0 cm. Film exposures of 15 minutes were the norm. In the case of Laue transmission patterns, the crystal to film distance was 3.0 cm and two minute exposures were normally adequate. For very thin samples, such as free standing 3C SiC single crystal films, somewhat longer exposures are required. Samples were Lely furnace grown plane parallel platelets with the basal plane perpendicular to the c-axis, Lely grown bulk cubic samples with well defined (100) or (111) facets, modified Lely growth boule slices cut to within a degree of being perpendicular to the c-axis or CVD single crystal films grown on single crystal

Editors: G.L. Harris, M.G. Spencer, and C.Y. Yang

substrates with [100] or [0001] orientations. These samples were mounted on small x-ray goniometers with the basal plane or natural cubic faces perpendicular to the x-ray beam. The angular resolution of the goniometers was about 0.2 degrees. Sample thicknesses varied from a minimum of 10 µm to 2 mm and film exposure times were varied to maximize image contrast.

3. Experimental results

(a) On-axis Laue patterns

Figure 1(a) is a 46% reduction of a Laue transmission photograph of a 6H SiC crystal with the basal plane oriented perpendicular to the x-ray beam. Such a photograph has an easily recognized pattern and can therefore serve as a standard [2,3]. Two important symmetry directions in the basal plane are indicated and in the next section we discuss in detail how one derives this result. In the case of platelets, one frequently has a macroscopic crystalline morphology with well known crystal faces. However, in boule growth or CVD epitaxial single crystal film growth there are no identifiable direction markers and one can use the patterns given in this paper to readily determine desired crystal directions. Figure 1(b) is a back reflection Laue pattern for the same crystal. It is not a photograph but an accurate tracing of a photograph reduced to 46%. In the case of back reflection patterns the image contrast is not nearly as good as for a transmission pattern and thus for clarity we

(a) (b)

Fig. 1(a) and (b). Transmission and back reflection Laue patterns for x-rays hitting 6H SiC along the ĉ axis. (a) Transmission Laue pattern (40 kV, 15 mA, 2 min. exposure, crystal to film distance = 3 cm.). (b) Back reflection Laue pattern (40 kV, 15mA, 15 min. exposure, crystal to film distance = 5 cm.).

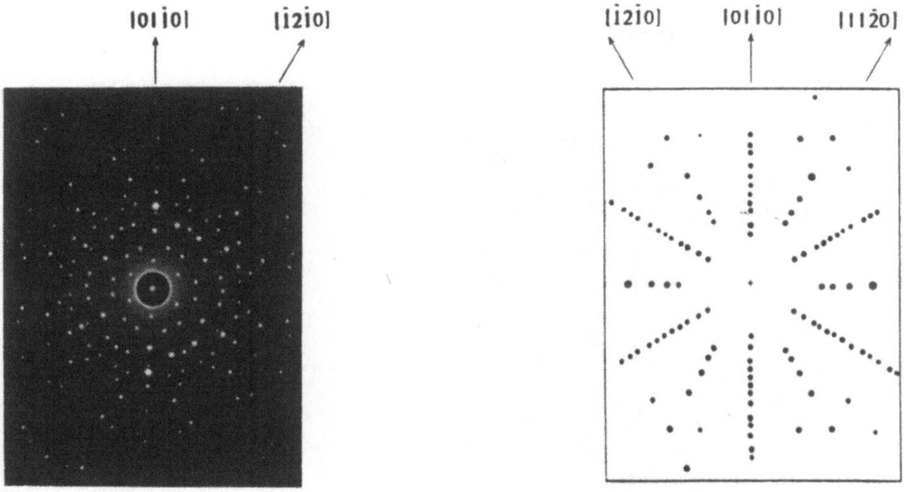

(a) (b)

Fig. 2(a) and (b). Transmission and back reflection Laue Patterns for x-rays hitting 4H SiC along the $\hat{c}$ axis. (a) Transmission Laue pattern (40 kV, 15 mA, 2 min. exposure, crystal to film distance = 3 cm.). (b) Back reflection Laue pattern (40 kV, 15 mA, 10 min. exposure, crystal to film distance = 5 cm.).

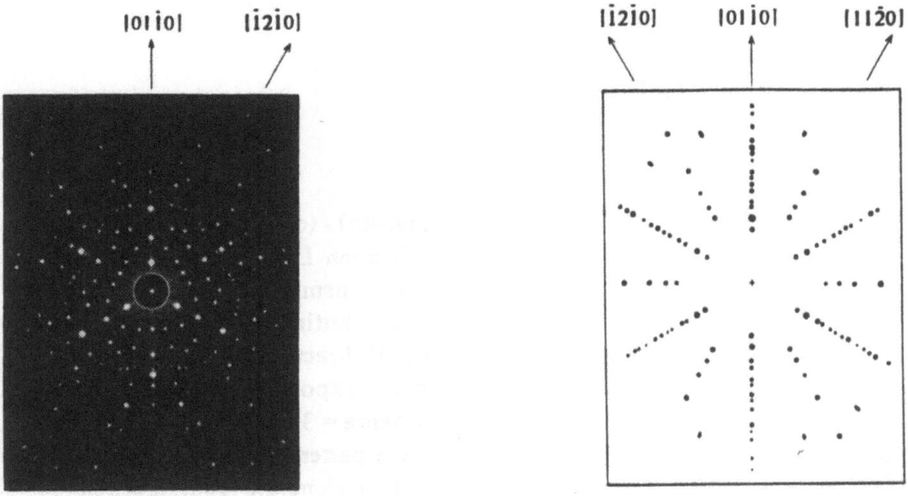

(a) (b)

Fig. 3(a) and (b). Transmission and back reflection Laue patterns for x-rays hitting 15R SiC along the [0001] direction (hexagonal notation). (a) Transmission Laue pattern (40 kV, 15 mA, 2 min. exposure, crystal to film distance = 3 cm.). (b) Back reflection Laue pattern (40 kV, 15 mA, 10 min. exposure, crystal to film distance = 5 cm.).

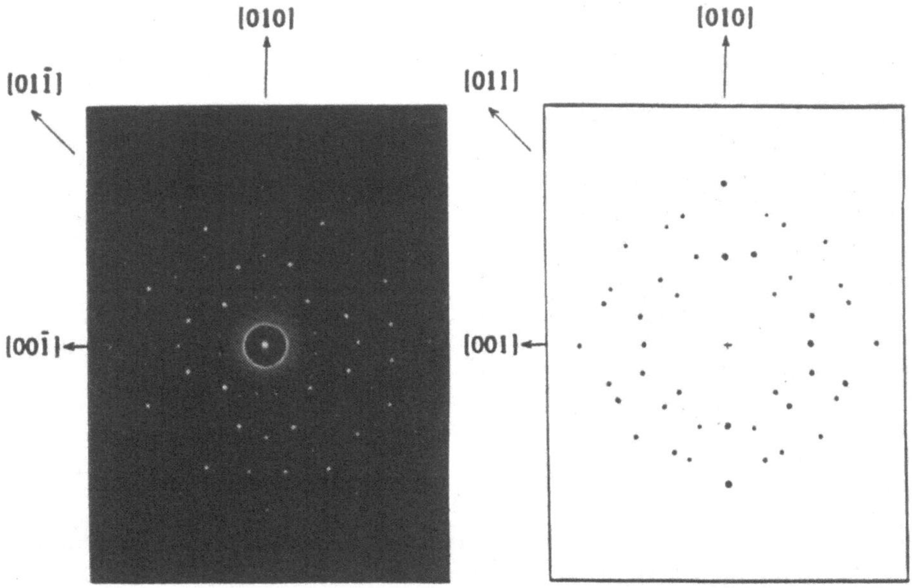

(a) (b)

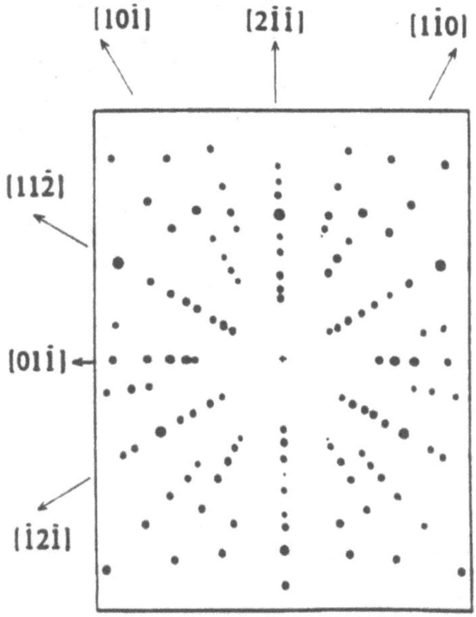

(c)

Fig. 4(a) - (c). Transmission and back reflection Laue pattern of 3C SiC. (a) Transmission Laue pattern for x-rays hitting the sample along the [100] direction (40 kV, 15 mA, 2 min. exposure, crystal to film distance = 3 cm.). (b) Back reflection Laue pattern for x-rays hitting the sample along the [100] direction (30 kV, 15 mA, 15 min. exposure, crystal to film distance = 3 cm.). (c) Back reflection Laue pattern for x-rays hitting the sample along the [111] direction (40 kV 15 mA, 10 min. exposure, crystal to film distance = 5 cm.)

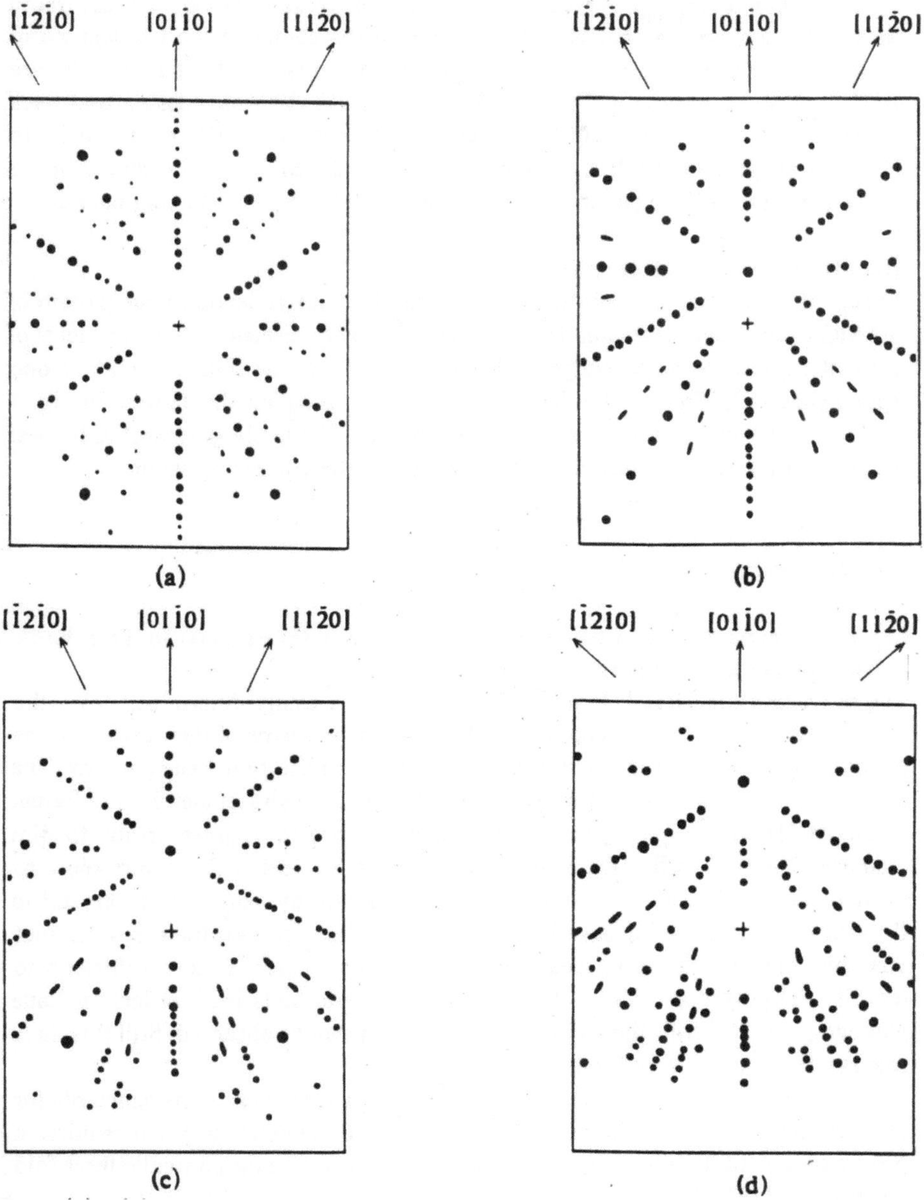

Fig. 5(a) - (d). Evolution of a back reflection Laue pattern of 6H SiC as the boule sample deviates from the [0001] growth direction. (a) On axis (40 kV, 15 mA, 10 min. exposure, crystal to film distance = 5 cm.). (b) 7.5 ° off axis to the incident x-ray beam (40 kV, 15 mA, 10 min. exposure, crystal to film distance = 5 cm.). (c) 11.5 ° off axis to the incident x-ray beam (40 kV, 15 mA, 10 min. exposure, crystal to film distance = 5 cm.). (d) 18.5 ° off axis to the incident x-ray beam (40 kV, 15 mA, 10 min. exposure, crystal to film distance = 5 cm.).

resort to accurate tracings. It should be noted that the cross at the center of the traced back reflection pattern indicates the position of the center of the incident x-ray beam. Similar transmission and back reflection patterns for 4H and 15R SiC are given in Figs. 2 and 3. In Fig. 4(a) and (b) we see the transmission and back reflection patterns for the (100) zincblende modification of SiC(i.e., 3C SiC). In Fig. 4(c) we give the back reflection pattern for a 3C SiC crystal oriented in the [111] direction. All the patterns in Fig. 4 are reduced to 56% of the original size.

(b) Off-axis reflection patterns

In Fig. 5(a) - (d) one can follow the evolution of a back reflection Laue pattern of 6H SiC as the boule is rotated from the [0001] growth direction. The symmetry of the pattern remains quite recognizable even at a tilt of 7.5° (Fig. 5(b)) but as one approaches 20°, (Fig. 5(d)), it becomes difficult to recognize the pattern. In Fig. 5 we used a reduction of 51% from the original size. If a boule grows off axis these patterns will provide a quick method for estimating the off-axis deviation.

4. Discussion

(a) Determination of polytypes of SiC from transmission and back reflection Laue patterns

The polytypes 6H, 4H, 15R and 3C of SiC can be easily determined from the transmission Laue photographs, as they have unique patterns of their own. But the polytype determination is not so easy from the back reflection Laue patterns. The back reflection Laue patterns for 6H, 4H and 15R SiC are almost identical. Different planes of 6H and 4H SiC give the same spot distance for a given crystal to film distance. The back reflection Laue spot distances for 15R SiC are very close to those of 6H or 4H SiC. We could solve this problem by using a large crystal to film distance. But the increase in crystal to film distance also increases the spot sizes and the uncertainty in measurements of spot distances makes it difficult to identify the polytype of SiC. If we take a perfect on axis back reflection Laue photograph of 15R SiC, then a three fold symmetry can be observed. Still this three fold symmetry is not very apparent.

In the case of 3C SiC, the back reflection Laue pattern is easily recognizable for (100)3C SiC, as it has a unique pattern of four fold symmetry, but it requires a precise on axis back reflection Laue photograph to recognize the particular three fold pattern of (111)3C SiC.

(b) Method of identification of Laue spots for (0001)4H, (0001)6H, (0001)15R, (100)3C, and (111)3C SiC.

We will first discuss the method of identifying $(10\bar{1}l)$ x-ray scattering from (0001)4H and (0001)6H SiC. Fig. 6(a) and (b) show the Bragg reflections of an x-ray beam directed along the [0001] direction of a SiC crystal, reflecting from two

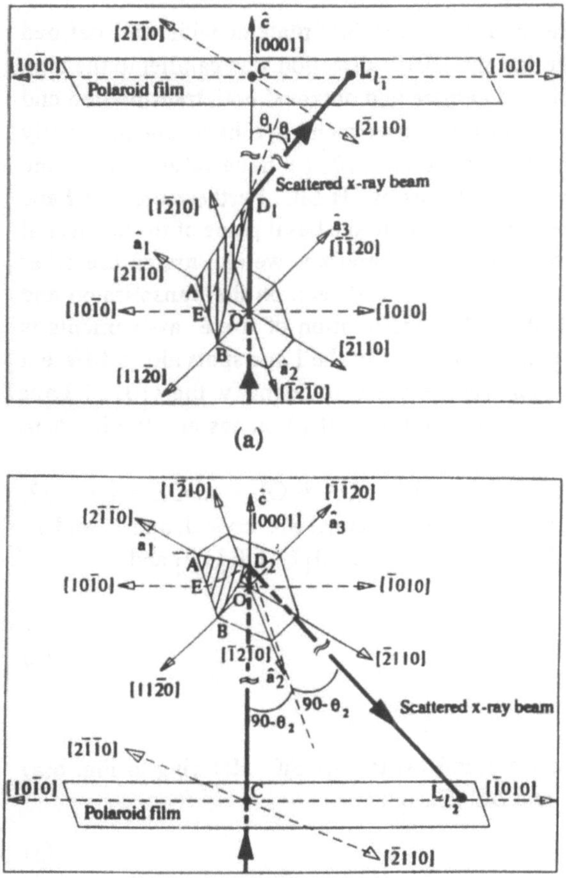

Fig. 6(a) and (b). Bragg reflection from the $(10\bar{1}l)$ plane of a hexagonal crystal, giving: (a) A transmission Laue spot, and (b) A back reflection Laue spot, for x-rays hitting the sample along the [0001] direction.

different $(10\bar{1}l)$ planes, resulting in transmission and back reflection Laue spots, respectively. The $(10\bar{1}l)$ planes ABD_1 and ABD_2 in Fig. 6(a) and (b) make angles θ_1 and θ_2 with the incident x-ray beam (i.e., $\angle OD_1E = \theta_1$ and $\angle OD_2E = \theta_2$). Intensity maxima are observed as spots ($L_{\ell 1}$ and $L_{\ell 2}$), on the polaroid films, when the angles θ_1 and θ_2 satisfy the Bragg condition, $2d_1 \sin\theta_1 = n_1 \lambda_1$ and $2d_2 \sin\theta_2 = n_2 \lambda_2$. Here d_1 and d_2 are the lattice spacings between the $(10\bar{1}l)$ planes shown in Fig. 6(a) and (b) respectively, n_1, n_2 are integers, and λ_1, λ_2 are two wavelengths in the 40 kV x-ray continuum. The spot C in Fig. 6(a) and (b) is due to the undeviated x-ray beam.

From the geometry of Fig. 6(a) and (b), EO ($[10\bar{1}0]$) is parallel to $CL_{\ell 1}$ and $CL_{\ell 2}$. In other words, if we know the $[10\bar{1}0]$ direction of the hexagonal SiC crystal, then we know the $(10\bar{1}l)$ Laue reflections on the polaroid film, as these spots are aligned in a direction parallel to the direction $[10\bar{1}0]$. In order to experimentally verify basal plane directions, we mounted a 6H SiC platelet with well defined hexagonal facets in such a manner that the $[10\bar{1}0]$ direction was parallel to the long axis of the film cassette of the polaroid camera and obtained both transmission and back reflection Laue patterns. From this experiment, we have unequivocally determined the $[10\bar{1}0]$ direction, and hence the $(10\bar{1}\,l)$ Laue reflections in the transmission and back reflection Laue patterns of 6H SiC. Furthermore, the Laue pattern, taken with x-rays incident perpendicular to the basal plane of the hexagonal SiC crystal mirrors that of the basal planes. Therefore, we can assign the other symmetry directions such as $[01\bar{1}0]$, $[11\bar{2}0]$, $[\bar{1}2\bar{1}0]$, etc. on the transmission and back reflection Laue patterns of 6H SiC. Confirmation of these assignments is obtained by constructing scales which will identify the Laue spots along different symmetry directions. One can construct a scale to identify the $(10\bar{1}\,l)$ Laue reflections, if one knows the angles between the $(10\bar{1}\,l)$ planes and the incident x-ray beam.

In Fig. 6(a) and (b), $OD_1 = c/\ell_1$, $OD_2 = c/\ell_2$, $OA = OB = a$, $OE = a\sqrt{3}/2$, where a and c are the lattice parameters of the hexagonal SiC crystal, and ℓ_1 and ℓ_2 are positive real numbers. In ΔOED_1 and ΔOED_2, $\tan(\theta_1) = OE/OD_1$ and $\tan(\theta_2) = OE/OD_2$.

$$\tan(\theta_1) = (a\sqrt{3}/2)/(c/\ell_1). \qquad (1)$$

$$\tan(\theta_2) = (a\sqrt{3}/2)/(c/\ell_2). \qquad (2)$$

$c/a = 4.907$ [4] for 6H SiC and $c/a = 3.266$ [4] for 4H SiC. Refering to Fig. 6(a) and (b), we see that in $\Delta D_1CL_{\ell 1}$ and $\Delta D_2CL_{\ell 2}$,

$$CL_{\ell 1} = CD_1 \tan(2\theta_1). \qquad (3)$$

$$CL_{\ell 2} = CD_2 \tan(180-2\theta_2). \qquad (4)$$

For our experimental Laue patterns, we used a crystal to film distance 3 cm(CD_1 = 3 cm) for all transmission Laue photographs and 5 cm(CD_2 = 5 cm) for all back reflection Laue photographs, the one exception being 3 cm for (100)3C SiC.

In the case of $(10\bar{1}\,l)$ reflections, a transmission Laue spot distance $CL_{\ell 1}$ and a back reflection Laue spot distance $CL_{\ell 2}$ can be calculated for different ℓ_1, ℓ_2 values from Eqs. (1), (2), (3), and (4). These calculations of spot distances versus ℓ- values for $(10\bar{1}\,l)$ reflections for 6H SiC are given in the form of scales (A) and (B) of Fig. 8 and further numerical details on Table I. Following a similar procedure, we have also constructed scales to identify $(11\bar{2}\,l)$ transmission and back reflection Laue spots of 6H SiC. These scales are shown in Fig. 8(C) and (D) and further numerical details are given in Table I.

In the case of 4H SiC, we have also verified the $[10\bar{1}0]$ direction, by matching the experimental (transmission and back reflection)Laue spots to the constructed $(10\bar{1}\,l)$ scales. The constructed scales for 4H SiC for $(10\bar{1}\,l)$ and $(11\bar{2}\,l)$Laue reflections are given in Fig. 9 and further numerical details are given in Table II.

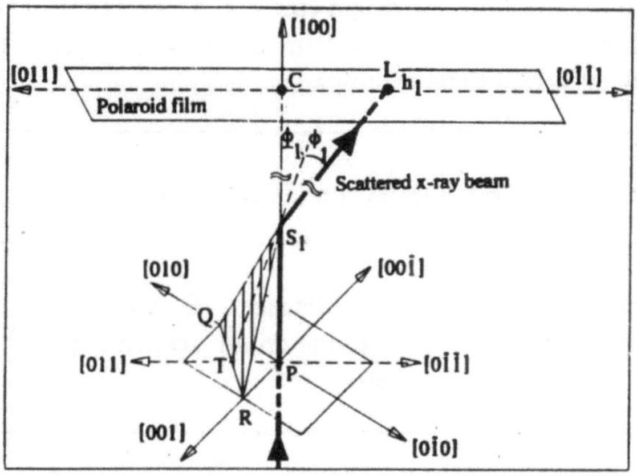

(a)

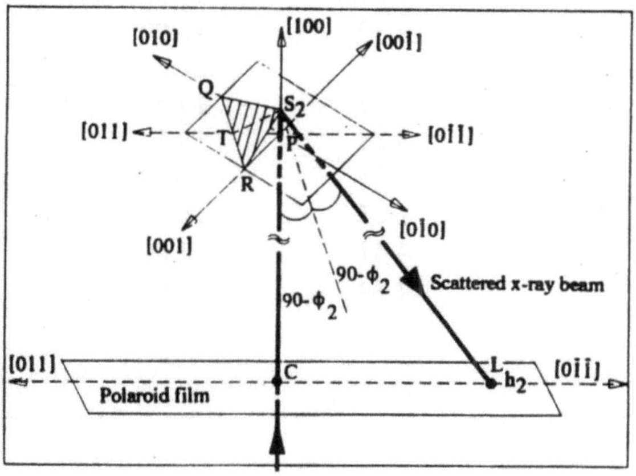

(b)

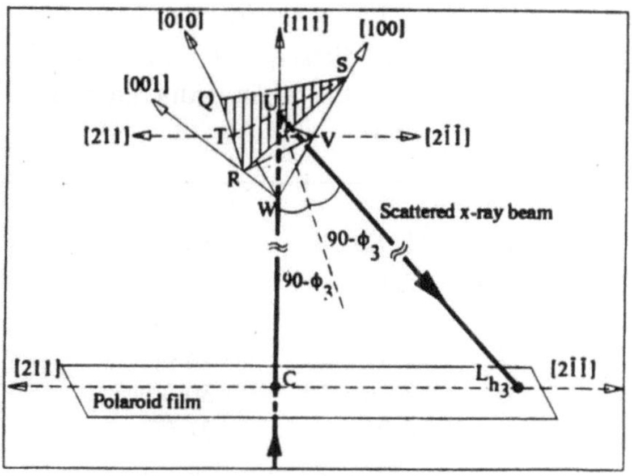

(c)

Fig. 7(a) - (c). Bragg reflection from the (h11) plane of a cubic crystal, giving: (a) A transmission Laue spot, for x-rays hitting the sample along the [100] direction,
(b) A back reflection Laue spot, for x-rays hitting the sample along the [100] direction, and (c) A back reflection Laue spot, for x-rays hitting the sample along the [111] direction.

Table I. Calculated Laue spot distances from the origin (see Fig. 8) for $(10\bar{1}\ell)$ and $(11\bar{2}\ell)$ reflections for 6H SiC. The crystal to film distance is 3 cm for transmission and 5 cm for back reflection Laue photographs.

Transmission $(10\bar{1}\ell)$ - Scale A			Back reflection $(10\bar{1}\ell)$ - Scale B		
Spot	Spot distance from the origin (cm)	ℓ - value for scattering plane	Spot	Spot distance from the origin (cm)	ℓ - value for scattering plane
a	1.09	1.00	a	1.48	39.00
b	1.71	1.50	b	1.83	32.00
c	2.04	1.75	c	2.39	25.00
d	2.41	2.00	d	2.76	22.00
e	2.83	2.25	e	3.17	19.50
f	3.28	2.50	f	3.38	18.50
g	3.80	2.75	g	3.75	17.00
h	5.10	3.25	h	4.05	16.00
			i	4.41	15.00
			j	4.84	14.00
			k	5.54	12.75

Transmission $(11\bar{2}\ell)$ - Scale C			Back reflection $(11\bar{2}\ell)$ - Scale D		
Spot	Spot distance from the origin (cm)	ℓ - value for scattering plane	Spot	Spot distance from the origin (cm)	ℓ - value for scattering plane
a	1.28	2.00	a	1.92	53.00
b	2.02	3.00	b	2.13	48.00
c	2.93	4.00	c	2.47	42.00
			d	2.95	36.00
			e	3.66	30.00
			f	4.91	24.00

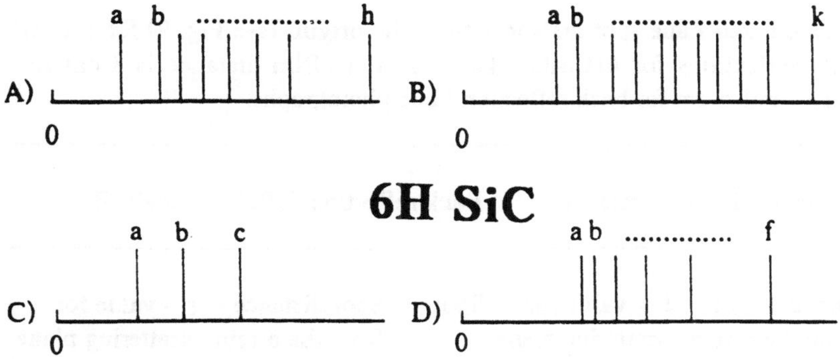

Fig. 8. Transmission and back reflection Laue scales for 40 kV x-rays hitting 6H SiC along the $\hat{c}$ axis. The (A) scale is for transmission Laue spots scattered by $(10\bar{1}l)$ planes. The (B) scale is for back reflection Laue spots scattered by $(10\bar{1}l)$ planes. The (C) scale is for transmission Laue spots scattered by $(11\bar{2}l)$ planes. The (D) scale is for back reflection Laue spots scattered by $(11\bar{2}l)$ planes. See Table I for spot identification and spot distances from the origin.

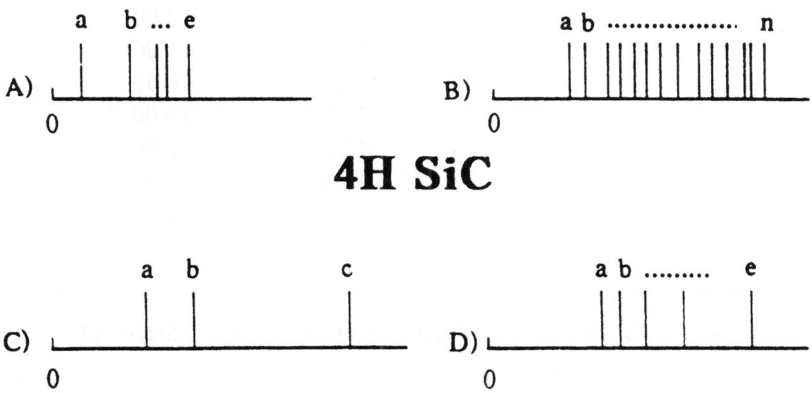

Fig. 9. Transmission and back reflection Laue scales for 40 kV x-rays hitting 4H SiC along the $\hat{c}$ axis. The (A) scale is for transmission Laue spots scattered by $(10\bar{1}l)$ planes. The (B) scale is for back reflection Laue spots scattered by $(10\bar{1}l)$ planes. The (C) scale is for transmission Laue spots scattered by $(11\bar{2}l)$ planes. The (D) scale is for back reflection Laue spots scattered by $(11\bar{2}l)$ planes. See Table II for spot identification and spot distances from the origin.

Table II. Calculated Laue spot distances from the origin (see Fig. 9) for $(10\bar{1}\ell)$ and $(11\bar{2}\ell)$ reflections for 4H SiC. The crystal to film distance is 3 cm for transmission and 5 cm for back reflection Laue photographs.

Transmission $(10\bar{1}\ell)$ - Scale A			Back reflection $(10\bar{1}\ell)$ - Scale B		
Spot	Spot distance from the origin (cm)	ℓ - value for scattering plane	Spot	Spot distance from the origin (cm)	ℓ - value for scattering plane
a	0.81	0.50	a	1.48	26.00
b	1.71	1.00	b	1.77	22.00
c	2.23	1.25	c	2.19	18.00
d	2.41	1.33	d	2.42	16.50
e	2.83	1.50	e	2.69	15.00
			f	2.91	14.00
			g	3.17	13.00
			h	3.49	12.00
			i	3.89	11.00
			j	4.13	10.50
			k	4.41	10.00
			l	4.72	09.50
			m	4.84	09.33
			n	5.09	09.00

Transmission $(11\bar{2}\ell)$ - Scale C			Back reflection $(11\bar{2}\ell)$ - Scale D		
Spot	Spot distance from the origin (cm)	ℓ - value for scattering plane	Spot	Spot distance from the origin (cm)	ℓ - value for scattering plane
a	2.02	2.00	a	2.13	32.00
b	2.92	2.66	b	2.47	28.00
c	5.85	4.00	c	2.95	24.00
			d	3.66	20.00
			e	4.91	16.00

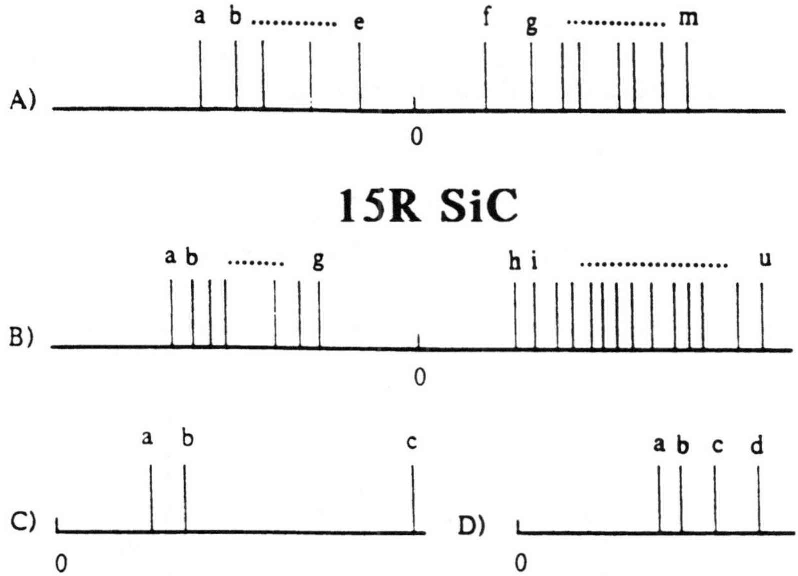

Fig. 10. Transmission and back reflection Laue scales for 40 kV x-rays hitting 15R SiC along the $\hat{c}$ axis. The (A) scale is for transmission Laue spots scattered by $(10\bar{1}\ell)$ planes. The (B) scale is for back reflection Laue spots scattered by $(10\bar{1}\ell)$ planes. The (C) scale is for transmission Laue spots scattered by $(11\bar{2}\ell)$ planes. The (D) scale is for back reflection Laue spots scattered by $(11\bar{2}\ell)$ planes. See Table III for spot identification and spot distances from the origin.

Turning to the identification of the $(10\bar{1}\ell)$ Laue reflections of (0001)15R SiC, we see that we can follow the same scheme (c/a = 12.272 [4]) as for (0001)4H SiC or (0001)6H SiC, except that certain $(10\bar{1}\ell)$ reflections of 15R SiC are missing [1] for the condition $\ell - 1 \neq 3m$, where m is an integer. For example, the $(10\bar{1}\ell)$ Laue reflections with $\ell = +2, +3, +5, +6, +8, +9, ..$ and $\ell = -1, -3, -4, -6, -7, -9, ...$ will be missing. The rhombohedral crystals, unlike the hexagonal crystals, have a three fold symmetry with respect to the [0001] direction. In other words, the Laue spots for (0001)15R SiC, scattered from the $(10\bar{1}\ell)$ and the $(10\bar{1}\bar{\ell})$ planes, unlike those of the hexagonal SiC crystals, will not be observed at equal distances on either side of the origin. The $(10\bar{1}\ell)$ and $(11\bar{2}\ell)$ Laue scales, both for transmission and back reflection, are given in Fig. 10 and further numerical details are given in Table III.

Next we consider the zincblende modification of SiC and the method of identifying the (h11) x-ray scattering from (100)3C SiC. Fig. 7(a) and (b) show the Bragg reflections of an x-ray beam directed along the [100] direction of a 3C SiC crystal, reflecting from two different (h11) planes, resulting in transmission and back reflection Laue spots (L_{h1} and L_{h2}) respectively. The (h11) planes QRS_1 and QRS_2 in Fig. 7(a) and (b), make angles $\O_1$ and $\O_2$ with the incident x-ray beam

Table III. Calculated Laue spot distances from the origin (see Fig. 10) for $(10\bar{1}\ell)$ and $(11\bar{2}\ell)$ reflections for 15R SiC. The crystal to film distance is 3 cm for transmission and 5 cm for back reflection Laue photographs.

Transmission $(10\bar{1}\ell)$ - Scale A			Back reflection $(10\bar{1}\ell)$ - Scale B		
Spot	Spot distance from the origin (cm)	ℓ - value for scattering plane	Spot	Spot distance from the origin (cm)	ℓ - value for scattering plane
a	3.41	- 6.40	a	3.92	- 41.00
b	2.85	- 5.66	b	3.59	- 44.00
c	2.42	- 5.00	c	3.32	- 47.00
d	1.66	- 3.66	d	3.08	- 50.00
e	0.86	- 2.00	e	2.29	- 65.00
f	1.14	+2.60	f	1.90	- 77.00
g	1.84	+4.00	g	1.58	- 92.00
h	2.36	+4.90	h	1.54	+94.00
i	2.63	+5.33	i	1.85	+79.00
j	3.25	+6.20	j	2.21	+67.00
k	3.49	+6.50	k	2.46	+61.00
l	3.92	+7.00	l	2.76	+55.00
m	4.31	+7.40	m	2.94	+52.00
			n	3.16	+49.00
			o	3.40	+46.00
			p	3.70	+43.00
			q	4.05	+40.00
			r	4.28	+38.33
			s	4.49	+37.00
			t	5.04	+34.00
			u	5.42	+32.33

Transmission $(11\bar{2}\ell)$ - Scale C			Back reflection $(11\bar{2}\ell)$ - Scale D		
Spot	Spot distance from the origin (cm)	ℓ - value for scattering plane	Spot	Spot distance from the origin (cm)	ℓ - value for scattering plane
a	1.49	5.75	a	2.24	115.00
b	2.02	7.50	b	2.58	101.00
c	5.65	14.75	c	3.11	86.00
			d	3.79	73.00

(i.e., $\angle TS_1P = Ø_1$ and $\angle TS_2P = Ø_2$). In Fig. 7(a) and (b), PR = PQ = a, $PS_1=a/h_1$, $PS_2 = a/h_2$ and $TP = a/\sqrt{2}$, where a is the lattice parameter of 3C SiC and h_1 and h_2 are positive real numbers. From ΔTS_1P and ΔTS_2P,

$$\tan Ø_1 = (a/\sqrt{2})/(a/h_1) = h_1/\sqrt{2}. \quad (5)$$

$$\tan Ø_2 = (a/\sqrt{2})/(a/h_2) = h_2/\sqrt{2}. \quad (6)$$

From ΔCS_1L_{h1} and ΔCS_2L_{h2},

$$CL_{h1} = CS_1 \tan (2Ø_1). \quad (7)$$

$$CL_{h2} = CS_2 \tan (180\text{-}2Ø_2). \quad (8)$$

From the geometry of the reflections, we see that the (h11) reflections are observed on the polaroid film in a direction parallel to the [011] direction. Following a similar procedure as we have applied previously for 4H or 6H SiC, we can construct scales to identify the (h11) transmission and back reflection Laue spots for (100)3C SiC using Eqs. (5), (6), (7) and (8). Scales to identify the (h11) and (h10) Laue reflections, both transmission and back reflection , for (100)3C SiC are given in Fig. 11 and further numerical details are given in Table IV.

Fig. 7(c) shows the Bragg reflection of an x-ray beam directed along the [111] direction of a 3C SiC crystal, reflecting from a (h11) plane, resulting in a back reflection Laue spot L_{h3}. The (h11) plane QSR_3 in Fig. 7(c) makes an angle $Ø_3$ with the incident x-ray beam (i.e., $\angle TUW = Ø_3$). In Fig. 7(c), $SW = a/h_3$ (from the definition of Miller indices (hk) of a plane of a cubic crystal ($h = h_3$, k= 1, ℓ = 1)), RW = QW = VW = a and $\angle UTV = 90 - Ø_3$, where a is the lattice parameter of 3C SiC, and h_3 is a positive real number. The angle Ø between the planes (hkℓ) and $(h'k'\ell')$ of a cubic crystal is given by Eq. (9) [5].

$$\cos\phi = \frac{(h\times h' + k\times k' + \ell\times \ell')}{\sqrt{(h^2+k^2+\ell^2)(h'^2+k'^2+\ell'^2)}} \quad (9)$$

Therefore, the angle between the planes (h11) and (111), that is, the angle between the planes QRS and QRV, $\angle UTV$ is given by Eq. (10), if we substitute $h = h_3$, $k = 1$, $\ell = 1$, $h' = 1$, $k' = 1$ and $\ell' = 1$ in Eq. (9).

$$\cos(90-\phi_3) = \frac{(h_3+2)}{\sqrt{3h_3^2+6}} \quad (10)$$

In ΔUCL_{h3}

$$CL_{h3} = CU \tan (180\text{-}2Ø_3). \quad (11)$$

Table IV. Calculated Laue spot distances from the origin (see Fig. 11) for (h10) and (h11) reflections for (100)3C SiC and (h11) reflections for (111)3C SiC. The crystal to film distance is 3 cm for Laue Photographs taken with (100)3C SiC and 5 cm for the back reflection Laue photograph taken with (111)3C SiC.

Transmission (h10) - Scale A (100)3C SiC			Back reflection (h10) - Scale B (100)3C SiC		
Spot	Spot distance from the origin (cm)	h - value for scattering plane	Spot	Spot distance from the origin (cm)	h - value for scattering plane
a	2.25	0.33	a	2.25	3.00
b	4.00	0.50	b	4.00	2.00

Transmission (h11) - Scale C (100)3C SiC			Back reflection (h11) - Scale D (100)3C SiC		
Spot	Spot distance from the origin (cm)	h - value for scattering plane	Spot	Spot distance from the origin (cm)	h - value for scattering plane
a	1.50	0.33	a	1.84	5.00
b	2.42	0.50	b	2.42	4.00
c	3.10	0.60	c	3.64	3.00
d	3.64	0.66			

Back reflection(h11) - Scale E (111)3C SiC			Back reflection (h11) - Scale E (111)3C SiC		
Spot	Spot distance from the origin (cm)	h - value for scattering plane	Spot	Spot distance from the orgin (cm)	h - value for scattering plane
a	4.83	2.20	i	1.43	1.33
b	4.04	2.00	j	1.55	1.36
c	3.50	1.86	k	1.87	0.66
d	3.27	1.80	l	2.28	0.60
e	2.73	1.66	m	2.58	0.56
f	2.49	1.60	n	3.10	0.50
g	2.11	1.50	o	3.73	0.43
h	1.71	1.40	p	4.04	0.40

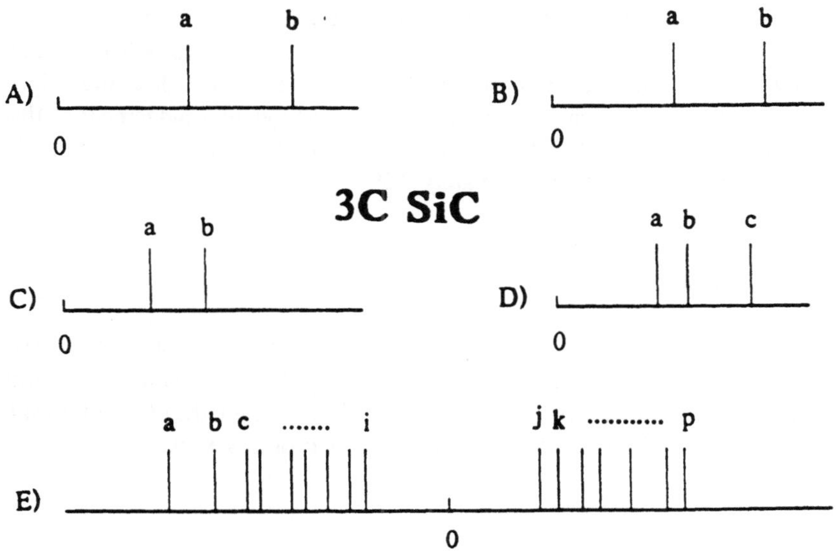

Fig. 11. Transmission and back reflection Laue scales for x-rays hitting 3C SiC along [100] and [111] directions. The (A) scale is for transmission Laue spots for 40kV x-rays hitting the sample along the [100] direction, scattered by (h10) planes. The (B) scale is for back reflection Laue spots for 30 kV x-rays hitting the sample along the [100] direction, scattered by (h10) planes. The (C) scale is for transmission Laue spots for 40 kV x-rays hitting the sample along the [100] direction, scattered by (h11) planes. The (D) scale is for back reflection Laue spots for 30 kV x-rays hitting the sample along the [100] direction, scattered by (h11) planes. The (E) scale is for the back reflection Laue spots for 40 kV x-rays hitting the sample along the [111] direction, scattered by (h11) planes. See Table IV for spot identification and spot distances from the origin.

From the geometry of the (h11) reflections, we can see that CL_{h3} is parallel to TV([211]). We can construct a scale to identify the (h11) back reflection Laue spots of (111)3C SiC, using equations (10) and (11). The (h11) back reflection Laue scale for (111)3C SiC is given in Fig. 11(C). Once we confirm the (h11) Laue reflections, we know the [211] direction of the crystal, and therefore we can assign the other symmetry directions, from the symmetry of the (111) face of the 3C SiC crystal.

5. Conclusions

Transmission Laue patterns may be used as an easy means to identify polytypes and to determine the azimuthal directions of boule grown SiC. However, slices thinner than 2 mm are required.

Back reflection Laue patterns may be used for c-axis determination of boule samples and for determination of azimuthal directions. Distinguishing 6H, 4H and 15R SiC by this method is not impossible but it is not easy enough to make it a technique of choice. Reflection patterns can very quickly yield a judgement of the boule alignment with respect to the c-axis. Accuracy of aligning a crystal with respect to the principal axis is approximately 0.5°.

6. Acknowledgements

We wish to thank N.J. Doyle for helpful discussions with respect to x-ray techniques, Wayne Bayer for much help in obtaining the Laue photographs and M.K. Jayananda for help with computer graphics. We also thank the office of Naval Research (Grant# N00014-89 J-1517) for partial support of this work.

References

1. Mitchell, R.S., The American Minerologist Vol.38, 60 (1953).
2. H. Posen, in Silicon Carbide - 1973, edited by R.C. Marshall, J.W. Faust, Jr. and C. E. Ryan, University of South Carolina Press, Columbia, South Carolina, pp.246-253 (1974).
3. Y. Tung and J.W. Faust, Jr., in Silicon Carbide - 1973, edited by R.C. Marshall, J.W. Faust, Jr. and C.E. Ryan, University of South Carolina Press, Columbia, South Carolina, pp.254-258 (1974).
4. Gmelin Handbook of Inorganic Chemistry, 8th ed., Silicon-supplement, vol.B2, Springer-Verlag (1984).
5. Cullity, B.D.: Elements of X-ray Diffraction, Addison-Wesley Publishing Company, Inc., pp.502 (1978).

Ohmic Contacts to Silicon Carbide Devices

M.M. Anikin, M.G. Rastegaeva, A.L. Syrkin, and I.V. Chuiko

A.F. Ioffe Physical-Technical Institute, Academy of Sciences, 194021 St. Petersburg, Russia

One of the bottlenecks in silicon carbide device technology [1,2] is the ohmic contact. Common materials for ohmic contacts to SiC-6H devices are aluminium for p-type regions [1,2,3] and nickel or chromium [1,4,5] for n-type regions. The aluminium contact imposes a limit of 500°C on the operating temperature of SiC devices. The specific contact resistance between nickel or chromium and heavily doped SiC is as high as r_c-3x10^{-4}Ohm cm^2. In this paper we investigate the fabrication process for W-or W/Au- based ohmic contacts to n- and p-type SiC-6H, and specifically addresses the question as to how the contact performance depends on such factors as substrate orientation, the method of preparation of the surface, the changes in the properties of epitaxially grown films or Lely grown wafers after placement of the contact, the height of the burn-in temperature, and the level of uncompensated impurity concentration.

Experimental procedure

Deposition of the contacts was carried out in vacuum by evaporation of W and Au in a single process. In the case of p-typeSiC-6H, the W and Au were deposited on an 0.01-0.05 μm thick Al layer previously grown by magnetron sputtering. The burn-in was done using electron beam heating in vacuum (10^{-6}Torr); the duration of the burn-in step was 120 sec. Three methods for initial wafer preparation were used: etching in KOH melts; the rough polishing treatment with 5 μm diamond paste, followed by ion cleaning in argon plasma (applied in the case of (0001)C- and (0001)Si-faced wafers); high-precision - sublimation etching (giving a polishing effect to (0001)Si and

(0001)C surfaces). The epitaxially grown layers had the orientation of (0001)C and (0001)Si. Their initial preparation consisted of successive rinses in organic solvents and deionized water. Following surface preparation, the wafers were overlaid with the contact material on both sides, and circular contact areas 300μm across and 1 mm apart were formed on one of the surfaces. Contact resistance was determined by four-point method or from the total resistance of the structures, corrected for the spreading resistance of the sample. The morphology of the W-SiC interface (which was exposed by stripping off the contact material overlayers) was examined by raster electron microscopy. The cross sectional composition of the material under the contact was determined by Auger spectroscopy in conjunction with layer- to-layer etching.

Experimental results

Electrical characteristics of the contacts.

At the levels of uncompensated donor concentration from $3x10^{18}$ to $1x10^{19}$ cm^{-3}, Lely grown SiC (0001)C-faced wafers exhibited a specific range of burn-in temperatures for nonrectifying contacts that varied according to the method of prior surface preparation. For the wafers etched in KOH, this temperature range, which lies around 1600 C, is narrow, while the mechanical roughening is associated with its broadening down to 1200°C. The sublimation grown (0001)C-faced layers with the concentration of uncompensated donors of $5x10^{17}$ to $1x10^{18}$ cm^{-3} had the optimum burn-in temperature range lying between 1200° and 1400°C. The specific contact resistance on the Lely grown (0001)C- faced substrates was found to decrease with increasing impurity concentration, Fig.1, from $1x10^{-2}$ Ohm cm^2 to $5x10^{-4}$ Ohm cm^2.

In the case of (0001)Si-faced single crystals with the concentration of uncompensated donors $3x10^{18}$ $-1x10^{19}$ cm^{-3}, the temperature range for ohmic contact formation was 800-1400°C for samples grown by the Lely method on KOH etched and roughened substrates, and it was 1200 -1800°C for sublimation

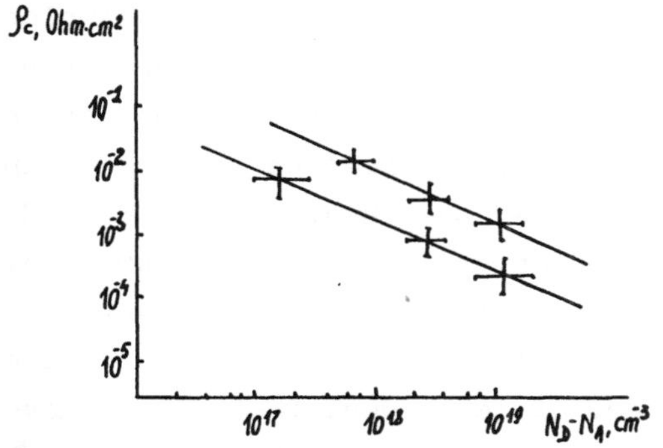

Fig.1.Specific contact resistance of the W- n- type SiC-6H contact as a function of doping level for (A) (0001)Si and (B) (0001)C faces

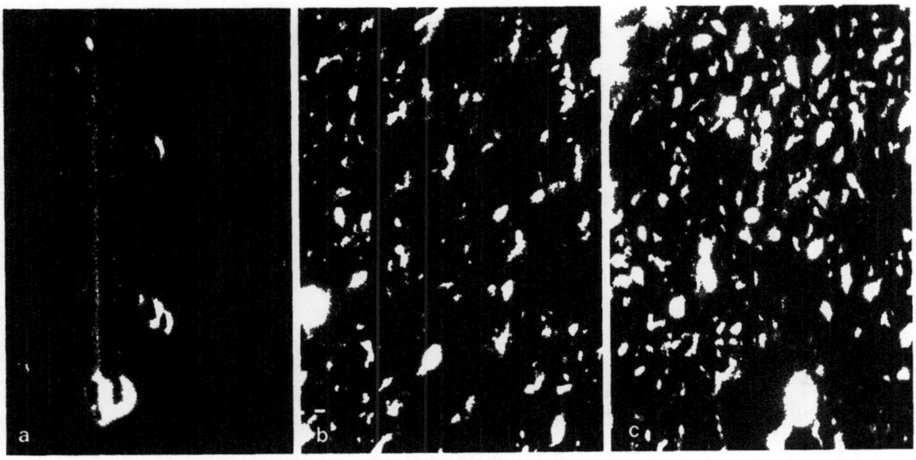

Fig.2.Micrographs (×500) showing the SiC-6H surface after removal of W contacts that had been burned-in at 1600°C (a) the 10^{-4} Ohm cm^2 specific contact resistance; (b) the $3x10^{-2}$ Ohm cm^2 specific contact resistance; (c) the contact with a nonlinear I/V characteristic

etched samples. For epitaxial layers with impurity concentration level of $1x10^{17}$ cm^{-3}, this temperature range ranges between 1000° C and 1400° C. The specific contact resistance for the (0001)Si orientation ranged from $1x10^{-4}$ to $5x10^{-3}$ Ohm cm^2, depending on the impurity concentration level, Fig.1. The morphological examination of the W-SiC interface revealed the presence of local melted regions in the samples where contact formation was done at high burn-in temperatures. These regions range in size from a few parts of a micron to a few microns. The highest density of such regions occurred in samples of both orientations that had been heated to 1800° C. The micrographs of KOH etched monocrystalline (0001)C surfaces, taken after the removal of the tungsten overlayer are presented in Fig.2. Although the melted regions that had formed at about 1600°C are quite large, it is evident that they have no reducing effect on the contact resistance: the linear I/V characteristic of the contact is clearly not due to the presence of melted regions. The contacts to p-type SiC-6H samples were produced by burning- in the Al-W- AuW-W overlayer at 1800°C for 120sec. The typical specific contact resistance values achieved were of the order of (2- 5)x10^{-4} Ohm cm^2 .

The structure of the W/SiC contact:

Wafer composition within the contact region was determined by Auger electron spectroscopy in conjunction with layer-to-layer etching. Fig.3 shows the intensities of Auger peaks:

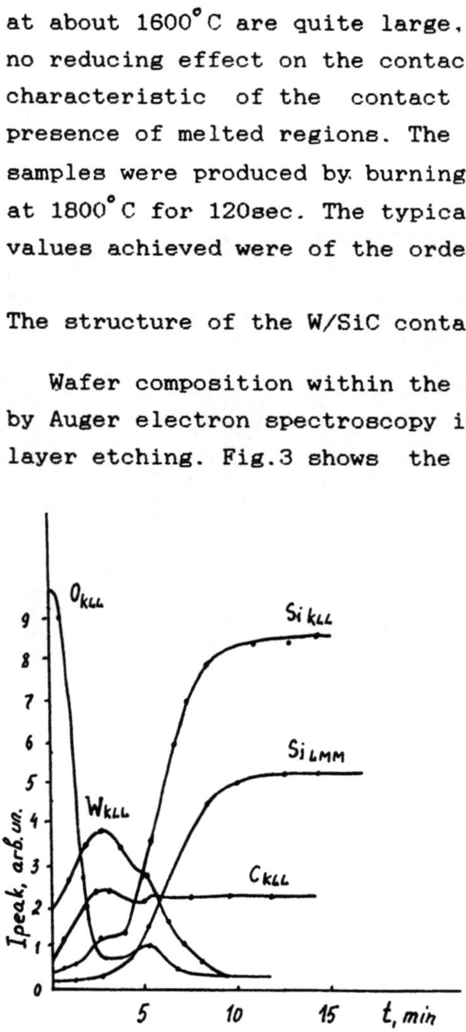

Fig.3.Intensities of Auger peaks of W,Si,C,and O as a function of etching time;the contact to 6H-SiC (0001)C-faced,Lely grown wafer was subjected to layer-to-layer etching

Si_{lmm}, Si_{kll} , O_{kll}, C_{kll} and W_{kll} as a function of etching time for tungsten contact to the KOH etched SiC(0001)C face, which has been fabricated at 1600° C.The carbon, whose presence is evident in the deposited tungsten, must have been captured from the atmosphere during deposition, since its peak intensity shows no temporal dependence on the etching time. The etching time dependence observed in the cause plot peak intensity due to silicon is attributable to diffusion of this material from SiC to tungsten. As can be seen, heating results in a relatively large amount of silicon present in the tungsten contact, and a corresponding broadening of the Si-depleted SiC region adjacent to the contact area. The presence of the resulting WSi_2 inter-layer has also been reported for β-SiC [7], which lends support to our tentative identification of this layer in SiC-6H. The intensity of the oxygen peak has also been found to increase at the W-SiC interface-due to the presence of oxides. The oxygen-enriched region does not disappear with heating. This can probably account for the fact that contact resistance was found to vary from sample to sample and even within a sample.

From the above, the structure of W-SiC ohmic contact emerges as consisting of the following layers:W-WSi- Si-depleted SiC-SiC of stoichiometric composition. In addition, the metallurgical boundary is interspersed with oxide traces. The disordered structure of SiC in the contact area provides conditions for carrier tunneling.

Discussion

The differences in the conditions under which ohmic contacts form on sublimation etched (0001)Si-faced substrates and KOH etched (0001)C-faced substrates are due to the difference in the properties of oxides present on these faces and the different conditions for dissociation: the (0001)C face undergoes oxidation much more readily than the (0001)Si face, while self- diffusion of silicon occurs much more rapidly from (0001)C face in 6H-SiC. The conditions for contact formation on epitaxial layers as opposed to etch-polished substrates are different because of the difference in the respective values of intrinsic point defect concentration: the bias in the stoichiometric composition toward silicon is much larger in epitaxial layers than in substrates. Also, because silicon diffuses much faster from epitaxial layers than from substrates, contacts to the epitaxial layer can form at lower burn-in temperatures. That KOH etching facilitates contact formation on the (0001)C faces but not on (0001)Si faces is explained by the difference in the KOH effect for the two orientations, so that while the (0001)C face becomes highly polished, the roughness of the (0001)Si face remains as severe as is normal for a mechanically roughened surface. The significantly disordered state and the related low temperature dissociation represent one of the three features that make the conditions for contact formation identical on KOH etched (0001)Si face and also roughened surfaces of both orientations. The other two are the presence of inhomogeneities in the space charge region and a significant increase in the effective contact area. These unifying features are so prominent that the differences due to orientation virtually disappear. The burn-in temperature of 1800°C makes the contact rectifying. This is probably because the region of Si-depleted and disordered SiC deepens, accompanied by spoiling of the near-surface regions.

References

1. M.M.Anikin, I.V.Popov, V.E.Sevast´yanov, A.L.Syrkin, A.V.Suvorov, V.E.Chelnokov and G.P.Shpynev: Sov.Tech.Phys.Lett., 10, (1984), p.444
2. M.M.Anikin, A.A.Lebedev, I.V.Popov, A.M.Strelchuk, A.V.Suvorov, A.L.Syrkin and V.E.Chelnokov: Fiz.Tekh.Poluprovodn., 20, (1986), p.844 (in Russian).
3. S.N.Vainshtein, V.A.Dmitriev, A.L.Syrkin and V.E.Chelnokov: Sov. Tech.Phys.Lett., 13, (1987), p.413
4. I.L.Vasiliev and T.G.Boeva: Electronic Technology Reviews, 10(212), (1970), (in Russian)
5. A.V.Naumov, S.V.Nikitin, A.T.Ostroumov and Yu.A.Vodokov: Fiz. Tekh.Poluprovodn., 21, (1987), p.377(in Russian).
6. M.Shur: "GaAs Devices and Circuits" (Plenum Press, New York, London, 1987), p.148-151.

7.K.M.Geib,J.E.Mahan and C.W.Wilmsen: W/SiC contact resistance at elevated temperatures: Proceedings of the 2nd Inter.Conf. on Amorphous and Crystalline SiC, Santa Clara, CA, December, (1988),p.15-16, Springer-Verlag,Vol.34.

Surface Micromachining of n-Type β-SiC Using Laser Assisted Photoelectrochemical Etching

J.S. Shor, X.G. Zhang, M.N. Ruberto, and R.M. Osgood*

Columbia University Microelectronics Sciences Laboratories,
New York, NY 10025, USA
*Also: Kulite Semiconductor Products, Leonia, NJ 07605, USA

In this paper, we report a new, nonthermal wet etching technique for ß-SiC using laser assisted photoelectrochemistry. In this process, UV laser excitation causes holes to be photogenerated in the semiconductor. An external anodic bias forces these holes up to the surface, where they cause dissolution to occur in an aqueous HF solution. Etch rates as high as 100 μm/min have been obtained by this method. The etching conditions, trench morphologies, and electrochemical considerations of this process are discussed.

1. Introduction

SiC has long been considered a potentially useful semiconductor for high temperature applications. However, difficulties in growing single crystals have prevented the emergence of silicon carbide technology in the past. Recent breakthroughs in crystal growth [1] have sparked interest in SiC devices and processing techniques. Among the technical challenges that remain is the selective etching of SiC. Currently, the most effective etching technique for SiC is reactive ion etching (RIE), with reported etch rates as high as 0.2 μm/min [2]. In this paper, we report a new, low temperature, laser assisted photoelectrochemical etch of ß-SiC, resulting in well controlled etch rates between 1-100 μm/min. This technique has been demonstrated successfully for micro-scale etching of other semiconductors [3,4].

Limited attempts have been made to etch SiC using photoelectrochemical methods in the past. M.M. Carrabba et. al. [5] reported etching n-type SiC diffraction gratings under anodic potentials using a uniform light source. The electrochemical studies indicated that the presence of HF in aqueous solutions is important to electrochemically etch SiC [6,7]. Glerria and Memming [6] reported that SiC dissolves in aqueous H_2SO_4 solutions at anodic potentials through the formation of a passivating layer, which was suggested to be SiO_2, for it is easily etched away by HF. Morisaki et. al. reported a similar result [7]. Harris et. al. [8] reported using aqueous HF solutions to delineate SiC p-n junctions. Although SiC has been shown to be unstable in aqueous solutions at anodic potentials and etchable in HF solutions, no etch rate data has been reported. Also, electrochemical measurements such as I-V curves and flatband potentials were reported for α-SiC, but there is no similar data for ß-SiC.

2. Experimental

Epitaxial layers (10 μm thick) of unintentionally doped n-type ($n=10^{17}/cm^3$) ß-SiC were grown on p-Si by CVD at the NASA Lewis Research Center. Aluminum was evaporated onto a small region of the SiC epilayer for ohmic contact. Electrodes were formed by encapsulating the SiC samples with polyimide tape, exposing only the bare SiC. The electrochemical cell was made of Teflon, and a Pt wire was used as a counter electrode along with a saturated calomel reference electrode (sce).

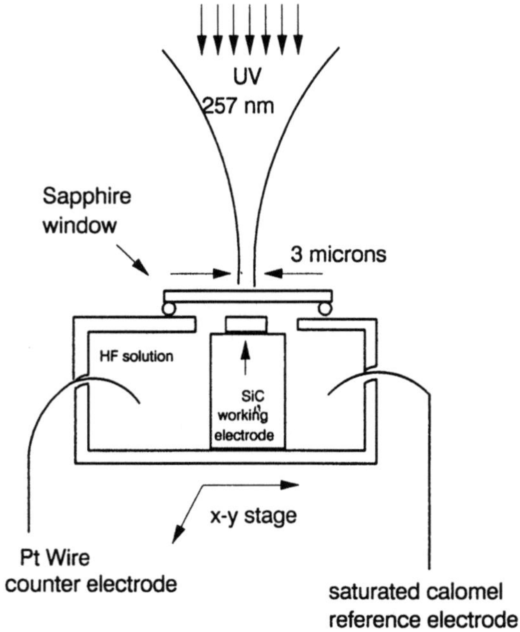

Fig. 1: Photoelectrochemical cell for laser etching. The SiC is held at a fixed potential with respect to the saturated calomel reference electrode (sce) while current flows between the SiC sample and a platinum wire counter electrode. The incident laser beam is a frequency doubled Ar^+ laser at 257 nm. Microstructures are fabricated by moving the sample on an xy stage.

The potential of SiC electrode was controlled with a potentiostat, and photocurrent was plotted on an X-Y recorder. To obtain a Mott-Schottky plot, a function generator was used to produce an AC potential which was fed into the potentiostat. AC currents were measured using a lock in amplifier. The electrolytes used were aqueous HF solutions. In some experiments small amounts of hydrogen peroxide were added to see the effect on the morphology of the etched features.

Before the laser etching began, the sample was photoelectrochemically etched in the etching solution under room light for 1 hour at 1 V_{SCE}. The photocurrent density during this uniform etching was about 3 $\mu A/cm^2$. The laser source was a frequency doubled Ar^+ laser at 257 nm, which was focused to approximately a 3 μm spot with a microscope objective. While a constant anodic potential in the range 0.6-2.0 V was applied, the laser beam was scanned across the sample at a controlled rate to produce microtrenches 1-4 μm deep. The morphology and depth of these structures were examined under a SEM. The dwell time of the laser was calculated by dividing the beam spot by the scan speed. Etch rates were determined by dividing the trench depth by the dwell time. The electrochemical setup is shown in figure 1.

3. Results

3.1 ETCHING: Ultra fast etch rates of 10-100 μm/min were achieved using this etch method. In figure 2a, the cross section of a typical etched trench is exhibited. This trench was etched at an anodic potential of 1 V_{sce} and a laser power density of 450 W/cm^2 in H_2O:HF:H_2O_2 (205:5:1). An etch rate of 45 μm/min was obtained under these conditions.

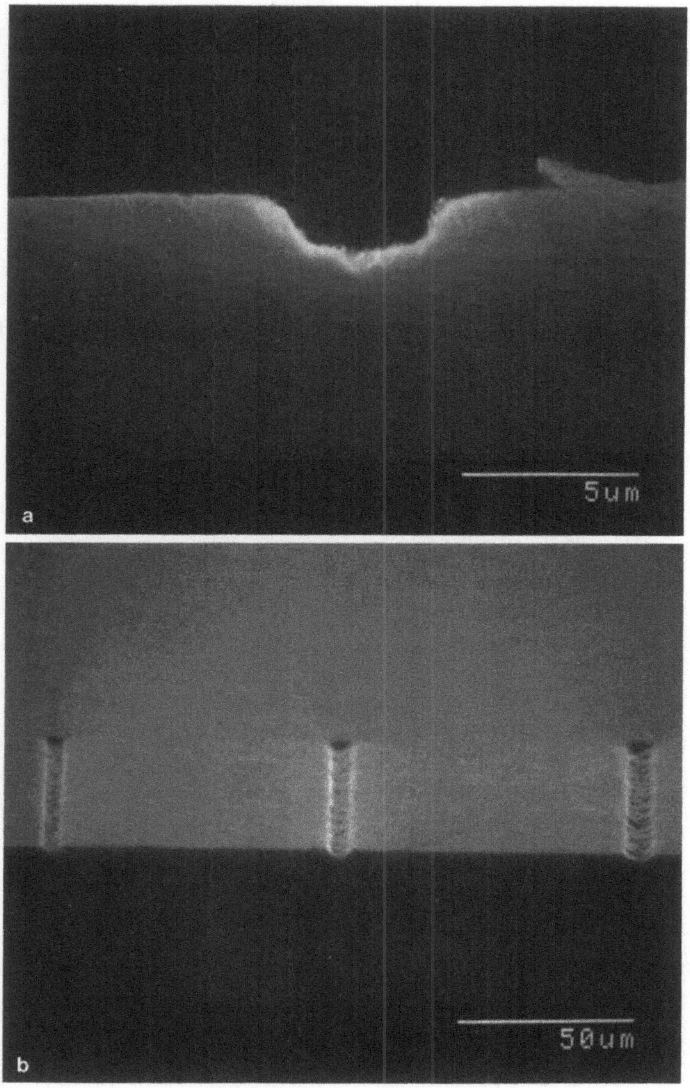

Fig. 2a, b: Laser photochemically etched trenches in n-type ß-SiC. These were etched at 257 nm. in a H_2O:HF:H_2O_2 (205:5:1) solution at 1 V_{sce} at power densities ranging from 450-900 W/cm^2.

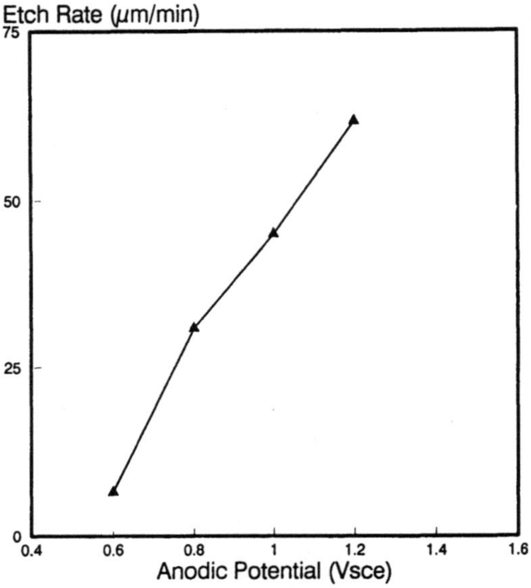

Fig. 3: Etch Rate vs. Anodic Potential for n-type ß-SiC at a laser intensity of 850 W/cm^2 at 257 nm. in H_2O:HF:H_2O_2 (205:5:1).

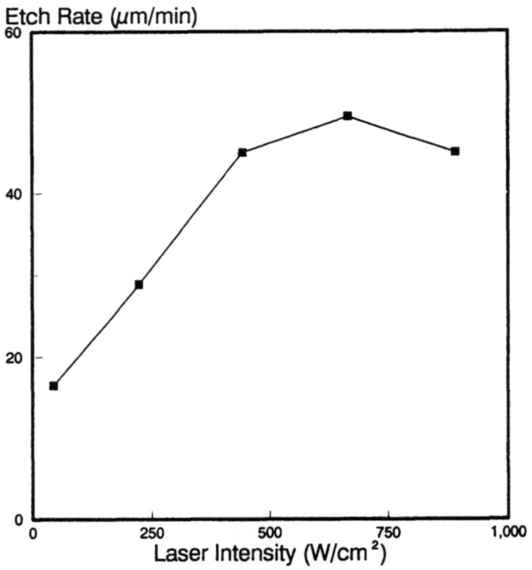

Fig. 4: Etch Rate vs. Laser Intensity for n-type ß-SiC under illumination at 257 nm. and an applied potential of 1 V_{sce} in H_2O:HF:H_2O_2 (205:5:1).

In order to determine the optimal conditions for etching, the effects of laser intensity, potential, and electrolyte composition on the etch rate and morphology were studied. Figure 3 shows the dependence of etch rate on the applied anodic potential for a power density of ≈ 850 W/cm^2 in a solution of H_2O:HF:H_2O_2 (205:5:1). The onset of etching occurs at about 0.5 V_{SCE} and increases with potential up to 1.2 V_{SCE}, and saturates at higher potentials. Figure 4 is a plot of the variation of etch rate with light intensity at a constant potential of 1 V_{SCE} in solution of H_2O:HF:H_2O_2 (205:5:1). At low intensities, the etch rate increases with increasing light intensity. There does not seem to be a minimum power density necessary for etching. At a light intensity of 500 W/cm^2, the etch rate begins to saturate.

Rapid etching occurred only in illuminated areas. The shape of the trenches at low laser power densities followed the gaussian intensity distribution of the laser beam. Also, as light intensity increases, trench widening takes place. As shown in Fig. 2b, the trench at the left was etched at 450 W/cm^2 and is 4 μm wide, while the two next to it were etched at 650 and 891 W/cm^2 and are 6 and 8 μm wide respectively.

The morphology of etched features obtained thus far are rough on the submicron scale. Addition of small amounts of H_2O_2 (½%) to the aqueous HF solutions seems to produce somewhat smoother features. However, the existence of H_2O_2 in the HF solution causes the etch rate to decrease. For example, with the addition of ½% H_2O_2 at 1 V_{sce} and 275 W/cm^2, the etch rate drops from 52 μm/min to 35 μm/min. At higher H_2O_2 concentrations (2.5%), this etch rate decreases to 3.5 μm/min.

3.2 ELECTROCHEMICAL MEASUREMENTS: I-V curves of the SiC electrode in H_2O:HF:H_2O_2 (205:5:1) are plotted for various laser intensities in figure 5. These curves were taken while the laser was scanning across the sample in order to prevent the beam from etching strait through to the silicon substrate. There was a very small dark current which increased linearly with the potential up to 4 μm/cm^2 at 2 V_{sce}. This current density represents a dark etch rate of ≈ 0.4 Å/min, assuming 8 equivalents/mole for the dissolution of SiC. The onset of high photocurrents occurs at about 0.4 V_{sce}. These currents increase with potential and eventually saturate at higher potentials. The saturation current increases with light

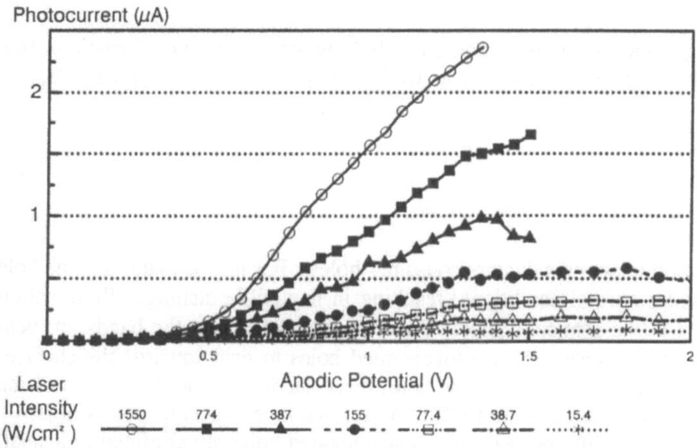

Fig. 5: i-V curve for n-type ß-SiC illuminated at various intensities at a wavelength of 257 nm. in an H_2O:HF:H_2O_2 (205:5:1) solution. The onset of photocurrents, and hence etching occurs at ≈ 0.5 V_{sce}.

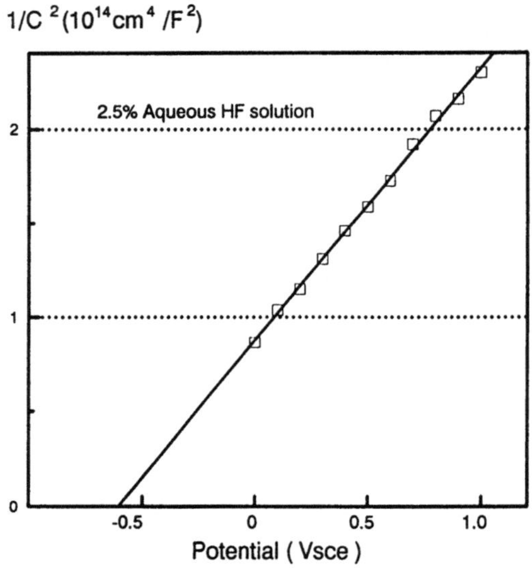

Fig. 6: Mott-Schottky plot for n-type ß-SiC in a 5% HF aqueous solution. The flat band potential exhibited is -0.55 V_{sce} while the carrier concentration is 10^{17} cm^{-3}.

intensity, as seen in Figure 5. Although the photocurrents are in the μA range, the current densities can be as high as 30 A/cm^2, since the laser beam is focused to such a small spot (3μm). The cathodic I-V measurement indicates that the onset of cathodic current is about -0.5 V_{sce}. The cathodic reaction is attributed to the hydrogen evolution.

Figure 6 is a Mott-Schottky plot which shows the capacitance-voltage dependance according to the Mott-Schottky equation:

$$1/C^2 = (2/qN_{sc}\epsilon A^2)(V - V_{fb} - kT/q) \quad .$$

The slope of the plot gives a carrier concentration of 10^{17}, which has been confirmed by Hall measurements. The extrapolation of the plot to $1/C^2 = 0$, yields a flatband voltage V_{fb} = -0.55V_{sce}.

4. Discussion

In general, the dissolution of semiconductors consumes holes. For n-type material, the hole density is very low under normal conditions, resulting in negligible etching. Illumination generates photocarriers in the semiconductor. Under anodic potentials, the bands are bent upwards in n-type material, causing the photogenerated holes to drift towards the surface, while electrons are driven into the bulk. These holes react, leading to the dissolution of the material. Rapid etch rates can be observed when the surface hole concentration is high.

Because the laser beam is highly focused, the photogenerated holes are confined to a small area, as is the region of material dissolution. This localized dissolution leads to the formation of well defined etched features. The shape and width of a laser etched trench is dependent on the material's minority carrier diffusion length and the laser intensity [9].

From the I-V curves and the flatband potential, it is noted that a large band bending, about 1 V, is needed for the etching to occur. The low anodic photocurrents at potentials lower than 0.5 V_{sce} are probably due to recombination in the space charge layer or surface recombination. According to previous analysis, the potential for the onset of photocurrents can be considerably higher than the flatband potential for an n-type semiconductor when there is a high density of surface states [10]. ß-SiC grown on silicon has a very large defect density which can cause this recombination to occur. Thus, etching can only occur when a large enough potential is applied to allow holes to reach the surface without recombining. At very high potentials, most photogenerated holes are driven to the surface. Beyond these potentials, the photocurrents, and hence the etch rates, are no longer dependent on further band bending.

The thickness of the space charge layer can be calculated based on the flatband potential. At an applied potential of 1 V_{sce} the space charge layer is ≈ 1300 Å, assuming no contributions from defects. At the wavelength used (257 nm), the absorption depth of ß-SiC is ≈ 1000 Å [11]. Thus, the majority of photocarriers are generated in the high field region. The optical absorption of ß-SiC exhibits a large wavelength dependence. For example, at 380 nm, the absorption depth is 10 μm, allowing only 1% of photocarriers to be generated in the depletion layer. This suggests that photoelectrochemical etching of SiC will vary highly with wavelength. This has been demonstrated by CARRABBA et. al. [5] who show a large dependence of photocurrent on wavelength over the range 380-530 nm.

There are several explanations for the roughness of the etched features. For example, surface analysis experiments [12] have indicated that a carbon rich layer exists on the surface when n-type ß-SiC is electrochemically etched in HF. This carbon may be in the form of clusters, and may cause self-masking to occur. Addition of an oxidizer, such as H_2O_2, may help dissolve the carbon, resulting in a higher quality etch. The rough morphology may also be due to anodic photo-oxidation of carbon resulting in the formation of micro CO_2 bubbles which can redirect the incident light and prevent reactions to occur inside the bubbles. Also, the large defect density of SiC/Si can cause preferential etching to occur on defect sites, resulting in a nonuniform etch. These phenomena are currently being investigated in order to optimize the process.

5. Conclusion

Laser assisted photoelectrochemical techniques have been used to etch n-type ß-SiC. Etch rates of 1 - 100 μm/min have been achieved and can be controlled easily by varying the laser intensity and the anodic potential. Electrochemical measurements indicate that the flat band potential is -0.55 V_{sce} and that a band bending of 1 V. is needed for the dissolution reactions to occur. The rapid etch rates and relatively low light intensity requirements allow for the use of focused direct writing as well as broad illumination etching. Finally, the etching can be monitored in situ by simply measuring the anodic photocurrents.

6. Acknowledgements

We would like to thank Tony Powell of NASA Lewis for his useful comments and for generously supplying the ß-SiC samples. We would also like to thank D. Goldstein and A.D. Kurtz of Kulite Semiconductor and D. Podlesnik of Columbia University for helpful discussions, as well as S. Goodman and A. Boseman for their technical assistance.

References

1. J.A. Powell, L.G. Matus and M.A. Kuczmarski: J. Electrochem. Soc. **134**, 1558 (1987)
2. J.W. Palmour, R.F. Davis, P. Astell-Burt, P. Blackborow: *Science and Technology of Microfabrication*, ed. by R.E. Howard, E.L. Hu, S. Namba and S.W. Pang (Mat. Res. Soc., Pittsburgh, 1987) p. 185
3. D.V. Podlesnik, H.H. Gilgen: *Laser Chemical Processing for Microelectronics*, (Cambridge University Press, 1989) p. 109
4. D.V. Podlesnik, H.H. Gilgen, R.M. Osgood: Appl. Phys. Lett. **48**, 496 (1986)
5. M.M. Carrabba, J. Li, J.P. Hachey, R.D. Rauh and Y. Wang: Electrochem. Soc. Confr. Oct. 15-20, 1989 Hollywood, Florida
6. M. Gleria and R.Memming: J. Electroanal. Chem. **65**, 163 (1975)
7. H. Morisaki, H. Ono and K. Yazawa: J. Electrochem. Soc. **131**, 2081 (1984)
8. G.L. Harris and K. Fekade: J. Electrochem. Soc. **135**, 405 (1988)
9. M.N. Ruberto, A.E. Willner, X. Zhang, D.V. Podlesnik, R.M. Osgood: to be submitted to J. Electrochem. Soc.
10. J. Reichman: Appl. Phys. Lett. **36**, 57 (1980)
11. H.R. Philipp and E.A. Taft: *Silicon Carbide* (Pergamon Press, 1960) ed. by J.R. Connor and J. Smiltens) p. 366
12. To be published

Microplasmas and Current Fluctuations in Silicon Carbide p-n Structures

B.S. Kondratjev and I.V. Popov

A.F. Ioffe Physical-Technical Institute, Academy of Sciences, 194021 St. Petersburg, Russia

Carrier transport via microplasmas (MP) in silicon carbide p-n structures has been studied experimentally in a wide temperature range, 300-700 K. A new approach to data analysis was employed which relies on consideration of fluctuations of the current pulsations using results of simultaneous measurements of major parameters determining the MP current (amplitudes, durations and pauses). It has been ascertained that a model presently in use which assumes the signal to be digital is inapplicable in view of amplitude fluctuations, nonexponential distributions of temporal characteristics and the existence of statistical correlation between the latter. Major dynamic characteristics of the MP process have been measured and their characteristic properties determined, in particular, a sign-alternating behaviour of the temperature coefficient of the breakdown voltage (TCBV). The sign of TCBV was shown to be positive for T>360 K.

Microplasmas and related phenomena are the classical subject of research into breakdown processes in real p-n structures. The specific character of impact ionization processes and of carrier multiplication in small volumes inside the space charge region of a p-n junction leads to fluctuations of current in time [1]. These fluctuations were shown to have their origin in some stochastic pulsed processes $I_v(t)$, whose statistical characteristics are strongly affected by external factors (voltage and temperature). A range of voltages where current pulsations are observed is usually characterized by an instability interval ΔV. It is known [2] that a local increase of voltage V within ΔV for an individual microplasma (MP) makes the average of the MP pulses longer; thus an MP serves to sustain the avalanche current I_1 for a longer time, which is referred to as an increase of the probability of a MP to turn on. A number of major dependences observed experimentally are usually described by fitting parameters of the McIntyre-Heitz model [2]. These parameters are the MP breakdown voltage V_b, MP series resistance R_s and probability functions p_{01}, p_{10} which are, respectively, MP turn-on and turn-off probabilities in a unit time. A relation between experimentally observed pulse durations τ_1 and pauses τ_0 in a random sequence of pulses, on the one hand, and the above probabilities on the other has the form $p_{01} = 1/\langle\tau_0\rangle$, $p_{10} = 1/\langle\tau_1\rangle$. These relations were derived from a mathematical analysis of the structure of a random pulsed process [3,4].

Temperature variations are known to affect parameter values in a predictable way, which is used for diagnosis of breakdown phenomena. In experiments on breakdown in conditions when MP is a predominant factor it is possible, from measured dependences of the breakdown voltage on temperature $V_b(t)$, to determine in particular for individual MP's, a value and sign of the temperature coefficient of the breakdown voltage (TCBV) $\beta = (1/V)(dV/dT)$ [1]. Detailed studies of MP fluctuations in silicon carbide structures are lacking, with the sole exception of [5]. Here we present a study of MP parameters and MP fluctuations in silicon carbide p-n structures (on SiC-6H p^+-n structures, orientation C//E, with an abrupt junction). The growth technique is described in [6,7]. A study of individual MP's was performed on a number of mesas 300 μm in diameter etched on the same substrate. Concentration N_d-N_a in the epitaxial layer was approximately 3×10^{18} cm^{-3}.

The breakdown was found to develop in the bulk via single MP's. Most of the research was carried out using an installation described in [4,8]. The installation contained a measuring and data processing assembly, utilized for statistical and temporal analysis of every single realization of the MP process [4], in particular, for determination of average values in a stochastic process $I_v(t)$, of divergences, and for calculating and plotting distribution histograms and

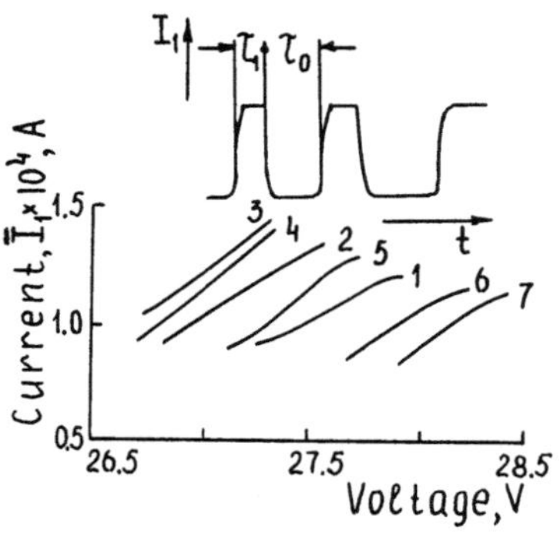

Fig.1. The dependence of an average amplitude of pulsations I_1 on voltage V (in a region where fluctuations of amplitudes of pulses in a series are hardly observable) at different temperatures. In the inset, an oscillogram of current pulses is shown.

(auto-, cross-) correlational functions. Resolution of time intervals was better than 1.5 μs. The maximum number of pulses in a set for analysis, N = 8×10^3, was sufficient for obtaining the statistical accuracy required.

The present work is, as far as we know, the first time that an MP fluctuation value has been determined from a great number of current pulses registered continuously (without interruption). Measurements on SiC p-n structures were performed so that every individual MP was registered and its parameters determined. Data given in figures 1 to 4 refer to the first MP to turn on in one of the mesas. From data obtained on the whole set of structures it was found that relatively low-voltage MP's (V_b=25-50 V) had the following parameters: $R_s=(1.1-2) \times 10^4$ Ohm and maximum amplitude of current in MP I_1=80-120 μA. The pulse shapes (see inset in Fig.1) and the form of $I_1(V)$ curves at T,K: 1 - 293, 2 - 318, 3 - 368, 4 - 448, 5 - 543, 6 - 643, 7 - 688 demonstrate an agreement with results of research on low-voltage MP's in other materials, for example, Si. In addition, the pulse shapes and the analogy with Si warrant a conclusion that the thermal component of the series resistance is of small importance and only the spreading resistance and the space-charge resistance should be taken into account [1].

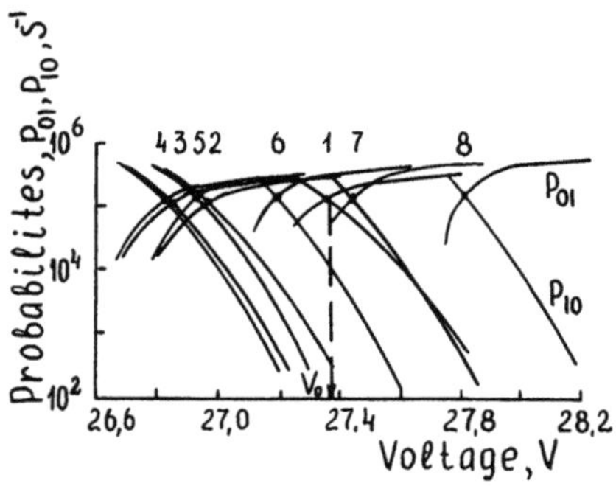

Fig.2.Paired dependences of the turn-on and turn-off probabilities on voltage at different temperatures T,K. The arrow indicates voltage V for curve I (T=293).

Consider the typical dependences $p_{01}(V)$ and $p_{10}(V)$ given in Fig.2. Numbers 1 to 8 correspond to temperatures T,K: 1 - 293, 2 - 330, 3 - 355, 4 -395, 5 - 455, 6 - 530, 7 - 595, 8 - 680. An analysis of the functional form and numerical parameters of the dependences $p_{01}(V)$ and $p_{10}(V)$ in the temperature range studied (see, for example, curve 1 in Fig.2) evidences an agreement with the general behavior of probabilities for turn-on and turn-off of current MP's, investigated most thoroughly for Si [2,4]. The turn-on probability is found to tend to some limit and the turn-off probability to decrease as V (or I_1) is increased, which is seen by comparing data in Figs. 1 and 2. Similar studies are available for Si [4,8], undertaken in a search for a model of the dynamics of MP pulsations. Investigations of the pulsations model [3,4] gave evidence that an important role is played by one- dimensional distribution functions of densities of pauses, pulse durations and their sums $W(\tau_1)$, $W(\tau_0)$ and $W(\tau_2=\tau_1+\tau_0)$, respectively. These dependences measured at a fixed temperature and at voltage $V \geqslant V_0$ (Fig.3) are nonexponential. In a characteristic region $V \cong V_0$, where $p_{01}= p_{10}$, an exponential distribution of pause and pulse durations $W(\tau_j)\sim\exp(-c.\tau_j)$; j=0,1 measured in the continuous mode could be expected a priori (as was usually done [3,9]), which corresponds to the Markovian (or semi-Markovian) character of the MP fluctuation process. In addition, statistical correlations between parameters τ_0, τ_1

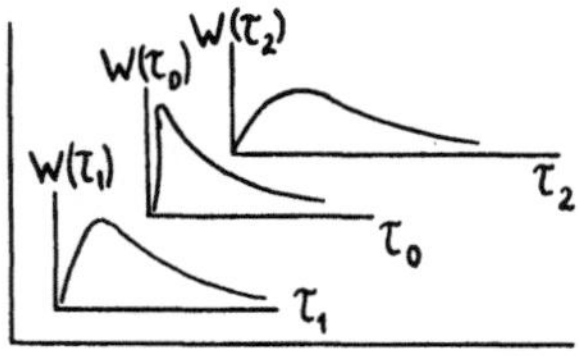

Fig.3.Histograms of distribution densities: pulses - $W(\tau_1)$, pauses - $W(\tau_0)$ and sums - $W(\tau_2)$.

and I_1 have been found to exist indicating the non-Markovian nature of the impact multiplication process responsible for the MP current [4,5].

In the same range of voltages $V \cong V_0$ a maximum frequency of pulse generation for MP was observed $f = f_{max}$. Processes governing p_{01} and p_{10} in the regions $V<V_0$ and $V>V_0$ are different. Neglecting the dependence of I_1 on V and assuming I_1=const(V) one obtains for this voltage range a maximum value of the spectral density of MP noise, assumed to depend only on pulse durations. A corresponding expression for the spectral density [9] and its analysis are given in [4]. In this region, the frequency of pulse generation for MP's investigated throughout the temperature range of interest correspond to $f_{max}=(2-5)\times10^5$ Hz. Further, as seen from a comparison of data on temperature variation of pulsation amplitudes (Fig.1) and of turn-on and turn-off probabilities (Fig.2) at temperatures above $T=T_r \cong 360$ K, the spectral density in this particular range of voltages varies in a manner usual for the MP noise because MP pulses are smoothed out, becoming shorter and less high, as was observed on Si [11].

A mechanism of "formation" of pulses and pauses, that is, $\tau_1=\tau_1(V,t)$, $\tau_0=\tau_0(V,t)$, has a complicated character and, in general, follows neither the Champlin model [9], nor the model of McIntyre-Heitz [2], in which properties of the MP fluctuations are interpreted in different ways. In our case the behavior of p_{01} with V is in fact of intermediate character, with the two models representing limiting cases of small V (McIntyre-Heitz model) and large V (Champlin model). Thus, breakdown mechanisms in different parts of the instability region are essentially different. This is

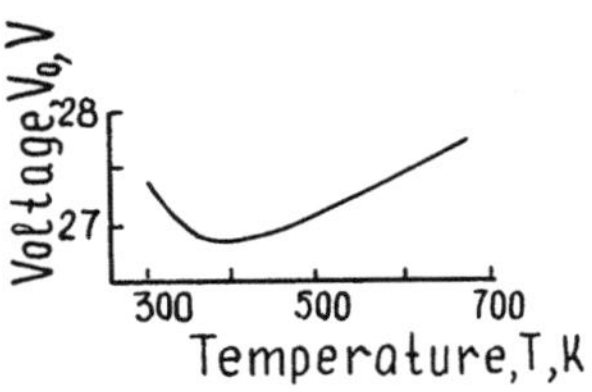

Fig.4. Temperature dependence of the breakdown voltage V_o. The arrow indicates the region of sign reversal of TCBV at $T=T_r$.

substantiated by the complicated behavior of probabilities p_{01} and p_{10} measured in a continuous sampling mode.

From the curves in Fig.2 it is seen that changes in pulse durations and pauses correlate and a relation between the two depends on control parameters (T,V). Initially (low voltages), an MP fluctuation develops in a way determined by pause divergences, pulse durations remaining approximately constant and pulse amplitudes fluctuating noticeably. Above V_o (see Fig.2) pause durations saturate and pulse durations grow and fluctuate.

We think that it is more appropriate to consider V_o as the breakdown voltage instead of V_b, as V_o does not depend on leakage currents and is determined solely by a coincidence of durations of pulses and pauses. In this case all that follows from the models in [2,3,9] (for average values!) is true at $V \geqslant V_o$, in particular, the sign of the temperature increment of V indicates the sign of TCBV for avalanche breakdown. The temperature dependence of V_o is given in Fig.4. It is seen that TCBV changes sign as the temperature is varied: at $T<T_r$ - TCBV is negative, at $T>T_r$ it is positive. Physical processes responsible for sign reversal of TCBV are considered in [10].

References

1. I.V.Grekhov, Yu.N.Serezhkin: "Avalanche breakdown of the p-n junction in semiconductors" (Leningrad, 1980), (in Russian).
2. R.H.Heitz: J.Appl.Phys., 35, (1964), p.1370.
3. V.K.Aladinski: Fiz.Tekhn.Poluprovodn., 6, (1972), p.2034, (in Russian).

4. B.S.Kondratjev et al.: Preprint, Ioffe Physico-Technical Institute, Leningrad, 1988, N 1252, (in Russian).
5. B.S.Kondratjev et al.: Fluctuational phenomena in physical systems - 5th All-Union Conference, Abstracts of papers,Vilnius, 1988,p.75,(in Russian).
6. V.A.Dmitriev et al.: Pisma v ZhTF, 11, 4, (1985),p.238,(in Russian).
7. M.M.Anikin et al.: Fiz.Tekhn.Poluprovodn., 20, 5, (1986), p.133, (in Russian).
8. B.S.Kondratjev, N.A.Sobolev: Fiz.Tekhn.Poluprovodn., 20, 1, (1986), p.132, (in Russian).
9. R.S.Champlin: J.Appl.Phys., 30, (1959), p.1039.
10. B.S.Kondratjev, Popov I.V. and Tutygin V.S.: Fiz.Tekhn. Poluprovodn., 24, 4, (1990), p.647, (in Russian).
11. V.K.Aladinski et al.: Radiotekhnika i elektronika, 2, (1973), p.342, (in Russian).

Positron Diagnostics of Radiation Defects in Silicon Carbide

A.I. Girka[1], *A.D. Mokrushin*[1], *E.N. Mokhov*[2], *S.V. Svirida*[1], *and A.V. Shishkin*[1]

[1]Moscow Engineering Physics Institute, Moscow 115409, USSR
[2]A.F. Ioffe Physical-Technical Institute, Academy of Sciences, Polytechnicheskaya, 26, 194021 St. Petersburg, Russia

Abstract. The defect formations in SiC under different kinds of irradiation and their evolution during subsequent isochronous annealing are studied by Positron Annihilation Technique. The critical influence of nitrogen impurities have been revealed in the appearance of instability in the "matrix-impurity-radiation vacancies" system resulting in formation of vacancy clusters. The size and concentration of these clusters have been estimated.

1. Introduction

The study of radiation defects in Silicon Carbide (SiC) is of fundamental importance. However, the nature of many of these defects and especially the influence of impurities on their evolution mechanics have not yet been understood very well. In order to obtain new information on this matter, the Positron Annihilation Technique (PAT) [1] has been used due to its defect-sensitive and defect-specific properties in atomic scale.

Positrons are particularly sensitive to vacancy-like defects at which positrons get trapped. The lifetime of a positron increases upon getting trapped at such defects (comparing to the bulk lifetime) owing to the lower electron density that is used in positron-lifetime spectroscopy. The PAT sensitivity lies within the concentration range of 0.1÷100 ppm.

The defects have been induced into SiC by electron, ion or neutron irradiation. By specifying the positron-annihilation parameters dependant upon fluences and annealing temperatures, the evolution of primary radiation defects and their interaction with impurities have been studied. For highly nitrogen-doped crystals irradiated to a sufficiently high fluence a stage of "negative" annealing has been found, associated with formation of vacancy clusters.

2. Experimental technique

Single crystals of SiC of the 3C, 4H, 6H and 15R polytypes were grown according to the Lely-method at 2600-2700°C. The nitrogen atom concentration in samples was ranged from $2\cdot10^{17}$ to $1\cdot10^{20}$ cm^{-3}.

Electron irradiation (E=4 MeV) was performed at a linear accelerator at different sample temperatures up to the fluence ranging from $1\cdot10^{15}$ to $3\cdot10^{18}$ cm^{-3}. Irradiation by reactor neutrons (the fraction of neutrons with energies above 10 keV was around 10%) was carried out at temperatures not higher

Springer Proceedings in Physics, Volume 56 **Amorphous and Crystalline Silicon Carbide III**
Editors: G.L. Harris, M.G. Spencer, and C.Y. Yang

than $90^{\circ}C$, up to the fluence ranging from $3\cdot10^{15}$ to $1\cdot10^{21} cm^{-3}$; and irradiation by Xenon ions $^{129}Xe^{+}$ (E=124 MeV) - at temperatures not higher than $50^{\circ}C$ up to fluence ranging from $5\cdot10^{10}$ to $5\cdot10^{14}$ cm^{-2}.

A conventional fast-fast time spectrometer with a time resolution of 380 ps [2] (full width at half maximum (FWHM)) was employed for the positron-lifetime measurements, all of which were performed at room temperature. The sample set was packed in conventional sample-source-sample sandwich geometry. The positron source was 10 µCi ^{22}NaCl salt. The source components were found to have a total intensity of 11%. The isochronal annealing (t=10 min., the step - 50 or $100^{\circ}C$) was made in a pure argon atmosphere in the temperature range of $50-2300^{\circ}C$. The analyses of the positron lifetime measurements were provided by the program POSITRONFIT EXTENDED [3]. The time spectra were analysed by one or by the sum of two decay components when it was possible.

3. Results and discussion

Only one lifetime component has been found in the positron spectra of unirradiated samples of all studied SiC polytypes (3C, 6H, 4H and 15R) independently of nitrogen concentration. It enables us to consider this lifetime to be due to annihilation of quasi-free positrons in the crystal matrix (bulk lifetime) - $\tau_b=(157\pm2)$ps, which coincides with the mean positron lifetime $\bar{\tau}$ in this case. All the subsequent studies were made on Lely single SiC(6H) crystals.

When introducing defects into the system by any kind of irradiation one can observe the growing of $\bar{\tau}$ because of positron trapping. All dose dependencies of $\bar{\tau}$ (for all studied kinds of irradiation and nitrogen concentrations) have regions of saturation, where $\bar{\tau}$ does not change with further fluence increase.

It is well known that the concentration of stable vacancy-type defects is determined by the number of primary displaced atoms (PDA), which is proportional to the fluence of high-energy particle irradiation and can be easily calculated for fast electrons, neutrons and ions [4]. Based on such a calculation in Fig.1 we plotted the dependencies of $\bar{\tau}$ on a number of PDA. For the Xe^{+} ions the values of $\bar{\tau}$ were corrected to the thickness of the damaged layer which is greater than the positron free path [5].

Fig.1 shows that the mechanisms of defect formation are identical for neutrons and ions, but different for fast electrons. In the case of neutrons and ions the mean kinetic energy that heavy particles transfer to the atom (E_a) is much larger than the threshold energy of atom displacement (E_d), which results in formation of displacement cascades - local regions with high PDA concentration. Within such a region the formation of secondary defects containing vacancies is very probable. In [6] we presented arguments that these secondary defects are primarily divacancies which, of course, does not exclude that there cannot exist a smaller concentration of isolated vacancies or even their complexes with impurities or

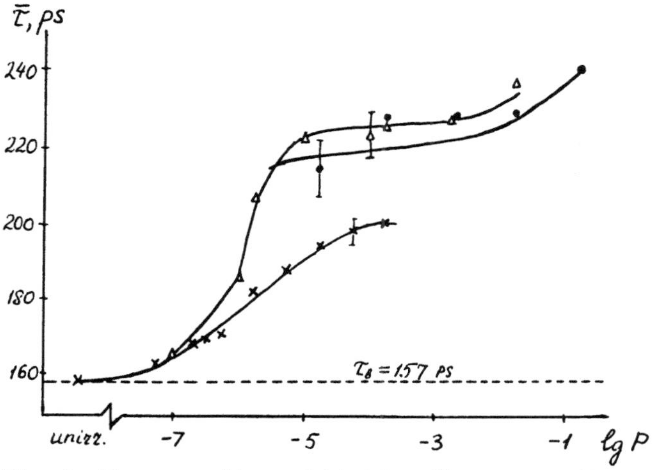

Fig.1. Mean positron lifetime $\bar{\tau}$ versus the concentration of primary displaced atoms P (logarithmic scale) for different kinds of irradiation: (×) - electrons, (▵) - neutrons, (•) - ions.

proper interstitial atoms, as well. Based on the cascade model of defect formation [4] and the diffusion mode of positron capture we were also able [7] to estimate the defect concentration in cascade ($\simeq 3\cdot 10^{20}$ cm^{-3}) and the size of the cascade ($\simeq$4÷6nm), which is in good agreement with the experimental data for other semiconductors.

For electrons with an energy of several MeV, the value of E_a is just a little above E_d, so the Frenkel pairs produced are uniformly distributed throughout the crystal volume. We are bound to believe [8] that there are at least two types of vacancy defects in this case. The defects of the first type, which are stable up to about 150°C, are due to the existence of Frenkel pairs (V-I). The much more stable defects of the second type should be regarded as vacancy-impurity nitrogen atom complexes.

The information about defects interacting with one another and with impurity atoms can be obtained from experiments on isochronous annealing of irradiated samples (Figs. 2,3).

The low temperature stage (ΔT_a=150÷500°C) has been observed in SiC irradiated both by electrons and neutrons. These results are in a good agreement with the optical measurements [9]. According to the data obtained at different temperatures of electron irradiation [8] the low temperature annealing is associated with annealing of genetic Frenkel pairs, which is in agre ement with the ESR data. With the rise of temperature the "close" pairs (with a small spacing between the components) recombine, while the "distant" pairs dissociate into free V and I. This stage does not manifest itself in annealing curves of samples irradiated to high fluences of neutrons ($\Phi_n \geqslant 1\cdot 10^{17}$ cm^{-2}). Apparently, the reason for this is not the absence of Frenkel pairs but the prevailing positron capture into cascades.

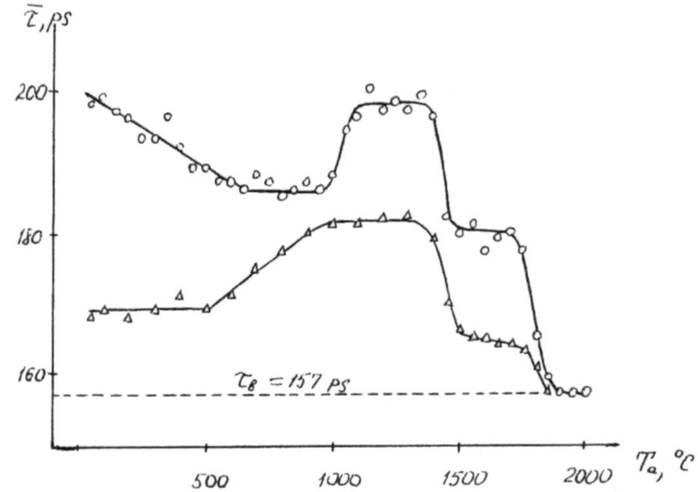

Fig. 2. Mean positron lifetime $\bar{\tau}$ versus the temperature of isochronous annealing of samples irradiated by Xe^+-ions and electrons up to fluence: $\Phi_i = 5\cdot10^{13}$ cm^{-2}; $\Phi_e = 1\cdot10^{18}$ cm^{-2}.

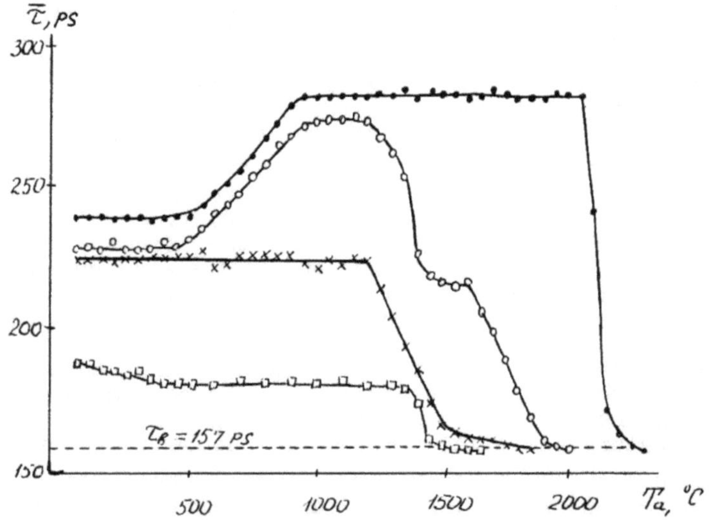

Fig. 3. Mean positron lifetime $\bar{\tau}$ versus the temperature of isochronous annealing of samples irradiated by neutrons up to fluence: (□) – $3\cdot10^{18}$ cm^{-2}; (×) – $1\cdot10^{19}$ cm^{-2}; (○) – $1\cdot10^{20}$ cm^{-2}; (•) – $1\cdot10^{21}$ cm^{-2}.

The stage of "negative" annealing is observed only in samples with the sufficiently high doping level ($N_D-N_A \geqslant 1\cdot10^{18}$ cm^{-3}) and irradiated up to sufficiently high fluences ($\Phi_n \geqslant 1\cdot10^{20}$ cm^{-2}, $\Phi_i \geqslant 5\cdot10^{13}$ cm^{-2}, $\Phi_e \geqslant 1\cdot10^{18}$ cm^{-2}). Undoubtedly, the existence of this stage is due to formation of vacancy clusters in which the lifetime of a localized positron is much greater than in isolated vacancies or divacancies. The component analyses of the spectra enabled us to distinguish a long-lived component at the stage of "negative" annealing of samples irradiated by neutrons, which, in its turn, made it possible to estimate the size of the formed clusters: they consist of 5÷8 and more than 20 vacancies for $\Phi_n=1\cdot10^{20}$ cm^{-2} and $\Phi_n=1\cdot10^{21}$ cm^{-2}, respectively. Concentration of clusters is calculated to be as high as $\simeq10^{18}$ cm^{-3}. Close values of the parameters also characterize the vacancy clusters produced with thermic annealing of SiC samples irradiated by heavy ions. In the case of electrons the formed clusters have smaller size and smaller concentration.

At high temperature stages ($T_a \geqslant 1200°C$) there are either one or two steps in annealing of structural defects depending on the fluence of irradiation. The first step consists in annealing of "loose" clusters, while more compact configurations are annealed at the second stage. For neutron irradiation to the fluence $\simeq1\cdot10^{21}$ cm^{-2} the vacancy clusters formed at the stage of "negative" annealing are much more stable: they are annealed in one step in the temperature range of ΔT_a=2000÷2300°C. This indicates that large clusters consisting of 20 or more vacancies have high thermic stability.

4. Conclusion

The positron Annihilation Technique proved to be very fruitful in analyzing radiation defects in SiC. Under heavy particle irradiation (neutrons and ions) the dose dependencies of the mean positron lifetime are well described by the cascade model of defect formation. The size of displacement cascades, mainly formed by clusters of divacancies with local concentration $\simeq3\cdot10^{19}$ cm^{-3}, is estimated to be ≃4÷6 nm. Under fast electron irradiation Frenkel pairs stable up to about 150°C and, also, the vacancy-impurity complexes are formed.

In the process of annealing of SiC samples doped with nitrogen to high concentration and irradiated to high fluences critical situations take place which lead to instability of the "matrix - impurity - radiation vacancies" system. The evolution of this instability results in phase transformations, i.e. formation of vacancy clusters, the size, concentration and thermic stability of which depend on the type of irradiation and its fluence.

5. References

1. Positron in Solids./Ed. by Hautojarvi P.1979. Springer-Verlag, Heidelberg.
2. Repin I.A., Svirida S.V., Semenihin A.N. // Soviet Pribory i Technika Eksperimenta 6, 28 (1985).
3. Kirkegaard P., Eldrup M. // Comput. Phys. Commun. 4, 401 (1974).
4. Vavilov V.S., Kekelidze N.P., Smirnov L.S. / Deystvie Izlucheniy na Poluprovodniki. 1988 Nauka Moskva (in Russian)
5. Girka A.I., Klopikov E.B., Skuratov V.A., Shishkin A.V. // Soviet Phyzika i Tekhnika Poluprovodnikov 23, 328 (1989).
6. Girka A.I., Mokrushin A.D., Mokhov E.N., Osadchiev V.M., Pogrebnyak A.D., Shishkin A.V. // Proc. 8-th Int. Conf. Positron Annihilation // Ed. by Dorikens-Vanpraet L., Dorikens M. Segers D. Singapore. 1989. P.458.
7. Girka A.I., Mokrushin A.D., Mokhov E.N., Osadchiev V.M., Svirida S.V., Shishkin A.V. // Sov. ZETF 97, 578 (1990).
8. Girka A.I., Kuleshin V.A., Mokrushin A.D., Mokhov E.N., Svirida S.V., Shishkin A.V. // Soviet Physika i Technika Poluprovodnikov 23, 2159 (1989).
9. Yasuda K., Takeda M., Masuda H., Yoshida A. // Phys. Stat. Sol. A71 549 (1982).
10. Loubser J.H.N., de Sousa Balona J.A., van Reneveld W.P. In: Silicon Carbide - 1968. Pergamon Press. N.Y. 1969. P.249.

The Temperature Dependence of Impact Ionization in Silicon Carbide, and Related Effects

A.O. Konstantinov

A.F. Ioffe Physical-Technical Institute, Academy of Sciences, Polytechnicheskaya, 26, 194021 St. Petersburg, Russia

Abstract. The dependences of impact ionization coefficients and of avalanche breakdown voltage on temperature have been studied in α-SiC. The electron ionization coefficient increases with increasing crystal temperature, in contrast with the well known case of a cubic semiconductor. This unusual phenomenon is attributed to the specifics of energy band structure of α-SiC. The thermal instabilities of breakdown current have been studied and simulated. These instabilities are found to be the consequence of the decreasing temperature dependence of avalanche breakdown voltage.

1. Introduction

The high electric strength is one of the most important advantages of semiconductor α-SiC. The avalanche breakdown electric field is anisotropic in α-SiC, and the maximal electric field values are observed in the orientation parallel to the principal axis, $\bar{E}\|\bar{C}$ [1]. The avalanche breakdown in this orientation has a set of anomalies. They are an unusually high breakdown field [1,2], sufficiently exceeding the predictions of SZE [3], unusual temperature dependence [1,4] and avalanche current filamentation observed in abrupt p-n junctions [5,6]. These anomalies are interesting both from the physical point of view and from the point of view of the use of SiC devices.

2. Fabrication of p-n Junctions and Methods of Measurements

The experimental study of impact ionization and avalanche breakdown demands high-quality p-n junctions with uniform breakdown over the area of the diode. These diodes must have no microplasmas or surface breakdown near the periphery. Different methods can be used for making them. We used the modified sublimation "sandwich" method [7] because it allows one to achieve the lowest defect density. Low defect density is obtained by the high temperature of epitaxy and by sublimation etching of substrates directly before growth. The high temperature of epitaxy ($t>2400^{\circ}C$) is required to prevent injection of non-equilibrium point defects while growing highly doped p^+ layer. Crowding of these defects around crystal imperfections causes the appearence of microplasmas. Sublimation etching in an inverted temperature

Springer Proceedings in Physics, Volume 56 Amorphous and Crystalline Silicon Carbide III
Editors: G.L. Harris, M.G. Spencer, and C.Y. Yang

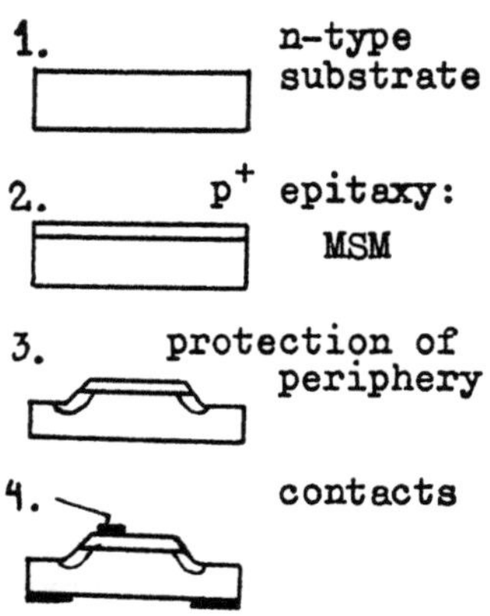

Fig.1 Sequence of diodes fabrication

gradient is used to remove diffused or deposited transition layers. Our substrates were Lely crystals doped with nitrogen to $2\cdot 10^{16}-10^{19}$ cm^{-3}. They provided breakdown voltages in the range 20-1100 V. The profile of p-n junctions was nearly abrupt. The periphery of diodes was protected by proton irradiation or boron diffusion with subsequent oxidation. Ionization coefficients are obtained by measuring electron initiated and hole initiated multiplication. Only the hole initiated multiplication M_p can be derived by photomultiplication measurements on currently available α-SiC p-n junctions [8]. For low voltage samples at moderate temperatures the electron coefficient M_n can be directly measured with the use of multiplication of diode's reverse tunnel current. For higher breakdown voltages (V>100 V) or higher temperatures we used several samples with differing breakdown fields.

3. Experimental Study of Temperature Dependences

The dependence of hole ionization coefficient on electric field $\beta(E)$ at room temperature can be approximated by the exponential function $\beta(E)=\beta_\infty \exp(-E_1/E)$, where $\beta_\infty=2\cdot 10^7 cm^{-1}$, $E_1=1.9\cdot 10^7 V/cm$. The electron ionization coefficient α is sufficiently lower and it has a sharper field dependence. The coefficients ratio β/α changes from 4-6 for $E=(5-7)\cdot 10^6$ V/cm to 30-50 for $E=(3-4)\cdot 10^6 V/cm$. A set of $\beta(E)$ and $\alpha(E)$ dependences derived in 300K-700K temperature band is plotted in fig.2. It should be noted, that β and α have different signs of variation with temperature: the hole ionization decreases with temperature, the electron one increases. It should be noted also that such unusual behavior of impact ionization was never observed in any other semiconductor. Temperature dependences of β and α obtained for fixed values of electric field from fig. 2, as well as the temperature dependence of α obtained for a

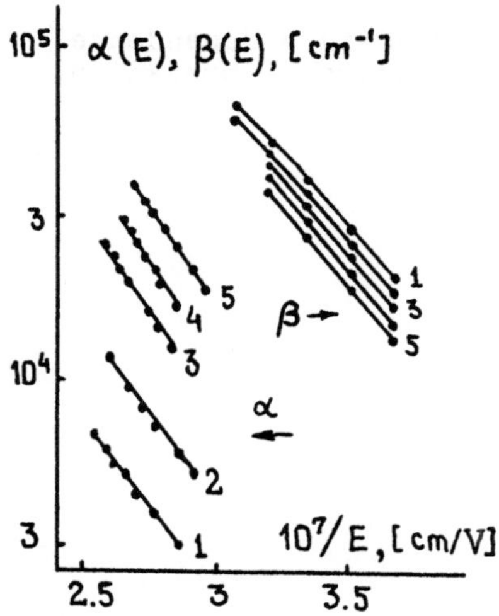

Fig.2 Electron and hole ionization coefficients: 294K#1, 370K#2, 470K#3, 570K#5.

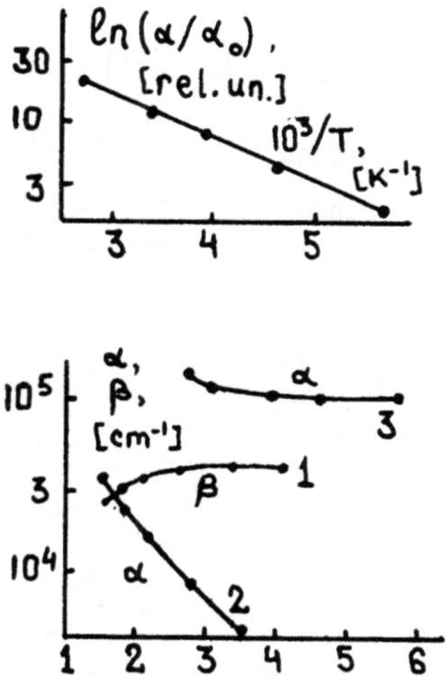

Fig.3 Dependence of impact ionization on temperature for E [10^6 V/cm]: 3.2#1, 3.7#2, 6.5#3. Insert: #3.

strongly doped base sample for a low temperature range are plotted in fig.3. The curve 3 of fig.3 is replotted in the insert in double logarithmic scale. The avalanche breakdown voltage of p-n junctions decreases in the 80K-700K temperature range (fig.4). Strongly nonlinear $V_B(T)$ dependence arises from the competition of hole and electron contributions to avalanche multiplication. Breakdown voltage temperature coefficient of 15 R polytype is lower than of 6H.

The ionization coefficient of a charge carrier is proportional to the probability of reaching the ionization threshold in electric field. In usual cubic semiconductors the ionization coefficient decreases with temperature. The reason for this is simple: as the phonon collision frequency increases, the carrier heating is suppressed. The analysis of impact ionization dependence on temperature has been made by CROWELL and SZE [9]. The high energy asymptotics of hot carrier energy distribution function can be displayed in the form $f\sim\exp(-\hbar\Omega/T_\varepsilon)$, where T_ε is the energetic electron temperature. As the crystal temperature rises, phonon scattering suppresses the electron heating: $T_\varepsilon(T)=T_\varepsilon(0)\mathrm{th}(\hbar\Omega/2kT)$, where $\hbar\Omega$ is the optical phonon energy. At low crystal temperature, $kT\ll\hbar\Omega$, the relative change of ionization coefficient obeys $\ln(\alpha/\alpha_0)\sim T_\varepsilon/T_\varepsilon(0)=-\exp(-\hbar\Omega/kT)$. From the insert in fig.3 we see that the absolute change of electric ionization coefficient in 6H SiC also obeys a similar relation: $\ln(\alpha/\alpha_0)\sim\exp(-\varepsilon_1/kT)$, the characteristic energy ε_1=68 meV is

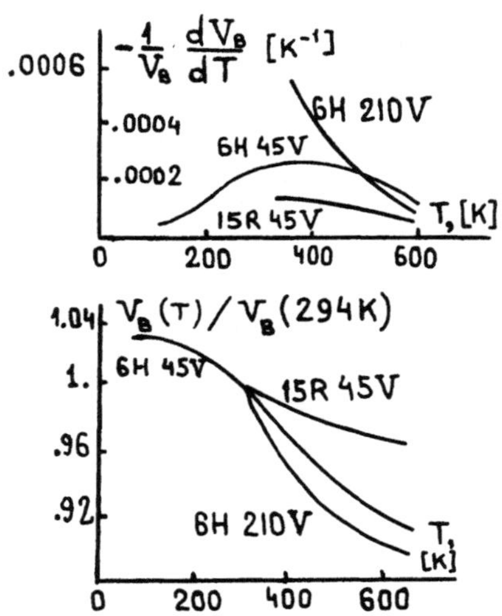

Fig.4 Temperature dependence of avalanche breakdown voltage.

close to the phonon energy. But in our case the ionization coefficient increases with increasing temperature. This result may be considered as the evidence, that phonon scattering enhances charge carrier heating in an electric field.

4. Discussion

We attribute the unusual temperature behavior of electron impact ionization to the energy band structure of silicon carbide. The principal minima of α-SiC conduction band are M-L type valleys. The band structure calculations for wurtzite type SiC (2H polytype) [10] show that the width of the M-L valley for the motion parallel to C_σ axis is about 2 eV, i.e. it is smaller than the band gap energy. Optical absorption studies and energy bands modelling [11] suggest that multilayer poytypes have additional minigap splittings. In a semiconductor with narrow allowed bands the application of strong electric field can result in Wannier-Stark localization of charge carriers. In absence of scattering the electron motion appears to be periodic both in k-space and in real space. Electron drift and heating will be possible only with assistance of phonon scattering and they will be enhanced as the scattering intensity rises. The small width of allowed bands results also in a strong suppression of electron heating and impact ionization.

No direct transition absorption peaks have been reported for the valence band. We suppose that the minigap splittings in valence band are very small or they are located at the energy, exceeding the energy band gap. This is the reason why the hole heating process has no substantial differences from the heating processes in cubic semiconductors. As the crystal temperature rises, the hole heating and impact ionization are suppressed.

The avalanche breakdown anisotropy experiments show that the uniform breakdown voltage of diodes with $\bar{E}\perp\bar{C}$ are lower than for identical diodes with $\bar{E}\parallel\bar{C}$, and the breakdown voltage increases with increasing temperature for $\bar{E}\perp\bar{C}$. At room temperature the anisotropy factor is 1.5-2. Still lower avalanche breakdown field is expected for cubic SiC owing to a smaller energy band gap and simpler energy band structure.

5. Avalanche current filamentation

The decreasing temperature dependence of avalanche breakdown voltage can produce thermal instabilities. In 6H and 4H p-n junctions these instabilities result in high current density filament formation. As the reverse current of p-n junction is increased, the uniformly distributed breakdown luminescence is replaced by a bright spot of current filament. The current density in a filament can exceed 10^4 A/cm^2. We have studied the electric properties of the filament and its transient characteristics. The study of filament relaxation dynamics confirms the thermal model of filamentation process: the relaxation time $\tau=d^2/K$, where K is the thermal diffusivity coefficient. A set of isothermal I-V characteristics of filaments with static currents in the range 20-100 mA was measured with the use of short pulses (fig.5). These I-V characteristics can be used to determine the value of overheating in the filament region, ΔT=200-400K. The value of overheating corresponds to the value calculated for spreading heat from a disk shaped source:

$$T=VI/2d\lambda \tag{1}$$

where λ is the thermal conductivity coefficient. It was found that principal properties of the filament can be explained with the use of a simple analytical model [6]. We

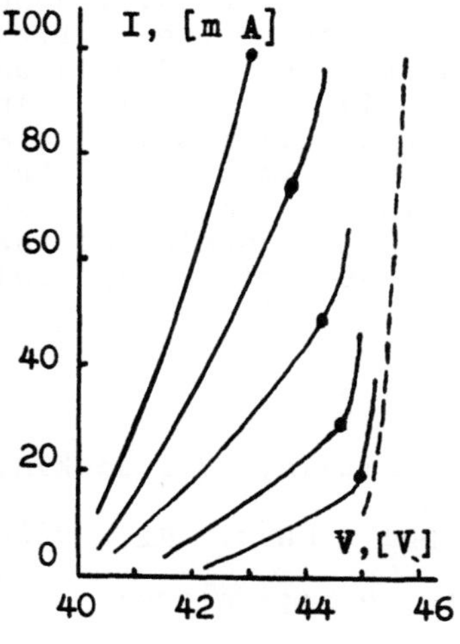

Fig.5 I-V characteristics of uniform breakdown-broken line, filament-solid line, static points are marked.

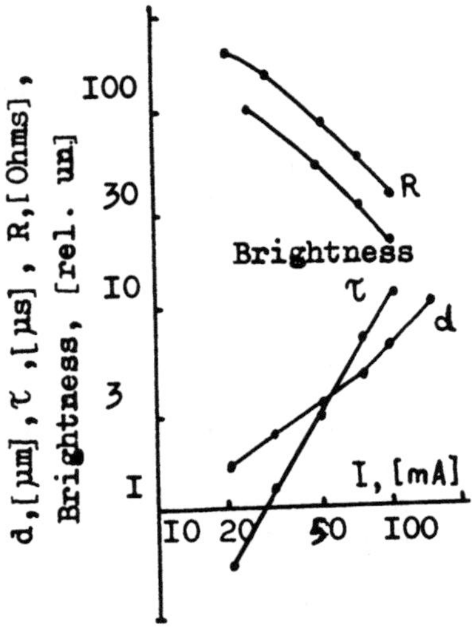

Fig.6 Dependence of filament properties on static current.

suppose that all the current flows through the filament area. As the voltage becomes higher than the breakdown voltage, the current is limited by the spreading resistance in the base of p-n junction: $R=\rho/2d$. The minimal entropy generation principle [12] demands the minimum of the voltage applied to diode. Taking (1) in account, we obtain:

$$V=V_o f(T)+\lambda\rho\Delta T/v_o^2, \qquad (2)$$

where $f(T)=V_B(T)/V_B(294K)$. The minimum of V demands:

$$-f'(T)=\lambda\rho/V_o^2 \; ; \qquad f''(T)>0 \qquad (3)$$

According to (3) the stabilization of filament current density takes place due to the nonlinearity of avalanche breakdown voltage temperature dependence, i.e. f(t) function. The temperature of the filament does not depend on the filament current. According to (1) it results in dependences d~I, j~1/I, R~1/I. The experimental results are in good agreement with these predictions (fig.6). As far as we know, the high current density filament in α-SiC is the only thermal filamentation process in semiconductor devices which can be simulated with the use of analytical modelling.

References

1. A.P.Dmitriev, A.O.Konstantinov, D.P.Litvin, V.I.Sankin. Sov. Phys. Semicond. 17, 686, (1983).
2. C.van Opdorp, J.Vrakking. Journ. Appl. Phys. 40, 2320 (1969)
3. S.M.Sze, J.Gibbons. Appl. Phys. Lett. 8, 111 (1966).

4. A.O.Konstantinov. Sov. Phys. Semicond. 23, 31 (1989).
5. G.B.Lisenkov, V.P.Snegirev, A.S.Tager. Pis′ma Zh. Tech. Fiz. 10, 215 (1984).
6. A.O.Konstantinov. Sov. Phys. Semicond. 23, 616 (1989).
7. A.O.Konstantinov, D.P.Litvin, V.I.Sankin. Sov. Tech. Phys. Lett. 7, 572 (1981).
8. A.O.Konstantinov. Sov. Phys. Semicond. 17, 2124 (1983).
9. C.R.Crowell, S.M.Sze. Appl. Phys. Lett. 9, 242 (1966).
10. F.Herman, J.P.van Dyke, R.L.Kortum. Mat. Res. Bull. 4, 167 (1973).
11. G.B.Dubrovski, A.A.Lepneva, E.I.Radovanova. Phys. Stat. Sol. (b) 57, 423 (1973).
12. B.K.Ridley. Proc. Phys. Soc. 82, 954 (1963).

The Wannier-Stark Effect and N-Shaped Volt-Ampere Characteristics in a Superlattice of 6H-SiC

V.I. Sankin and A.V. Naumov

A.F. Ioffe Physical-Technical Institute, Academy of Sciences, Polytechnicheskaya, 26, 194021 St. Petersburg, Russia

Abstract. A transistor structure N^+-p^--N^+ having the base doped with deep acceptors possesses the properties of an injection transit-time diode at pulse voltages applied to the collector. Carriers leaving the emitter are accelerated in a uniform field F. At $F \simeq 1.5 \cdot 10^5$ V/cm, the current abruptly decreases to a fraction of its initial value. Appearance of the negative differential resistance is explained through the Wannier-Stark effect whose criteria agree properly with the characteristics of 6H-SiC.

1. Introduction

It has been predicted in [1] that the application of the electric field to a solid state having a zone spectrum should lead to interesting quantum mechanical effects.

An active investigation in this field began following the first publications [2,3] on the creation of artificial superlattices (SL) in semiconductors and the possibility of obtaining the negative differential resistance (NDR). This search has been in progress recently [4-9], but NDR has not yet been revealed.

The present study deals with the investigation of volt-current characteristics in the presence of a strong electric field in specially created structures based on 6H-silicon carbide, a semiconductor with a natural superlattice. In the studies dealing with the investigation of impact ionization [10-12] it has been shown that the superlattice determines the properties of transport of electrons (but not holes) in very strong electric fields up to $F=5 \cdot 10^6$ V/cm. According to the estimations made the first mini-band should constitute $E_1=(0.2-0.4)$ eV, while the gap between the first and the second mini-bands should be $E_2=(0.4-0.5)$ eV, which favours the observation of the Wannier-Stark effect in this semiconductor. Special experimental conditions are necessary to avoid the high density of the current J. Besides the Joule heating, high J creates strong electron-electron scattering that induces the gain of the lateral electrons heating and consequently suppresses the Wannier-Stark effect. It is possibly for this reason that the NDR has not yet been revealed.

Editors: G.L. Harris, M.G. Spencer, and C.Y. Yang

2.Results and discussion

An object intended for experimental investigation should possess properties permitting the realization of the controlled electron flow in the presence of a strong electric field. As is known from the analogous study of the Gunn effect, semi-insulating material is required for this purpose. Regrettably, semi-insulating SiC of n-type conductivity is at least a rarity. Therefore an attempt has been made to create an artificial equivalent, and the three-electrode silicon carbide structure, N^+-p^--N^+ has been chosen for the investigation. The base layer of the structure, contrary to known transistors based on silicon carbide [13] is a p-type conductivity layer doped with the deep acceptor Sc, which causes a free-hole concentration as low as $p=10^{10}$ cm^{-3} at the temperature T=300K at the impurity concentration $N_A - N_D = 10^{17}$ cm^{-3}. Besides, the time for recharging Sc is sufficiently long. Due to the circumstances mentioned N^+-p^-(p^--N^+) junctions are turned on only under the action of constant forward bias or long pulse, while the back bias of pulses $\tau_p < 10^{-6}$s is screened at a distance of several thousand μm. Therefore, if the base width in the structure considered is several μm the pulse voltage creates a practically uniform field F=V/W, where V is the voltage and W is the base width.

The distribution of the pulse field in the structure is given in Fig.1. It should be noted that the field direction coincided with the superlattice axis. The same figure also presents a view of the structure and the measurement schematic. In accordance with the transistor terminology, the structure was switched into common-emitter circuit. In compliance with the above, the emitter-base junction was forward biased by a constant voltage U_B, while the back bias of pulses

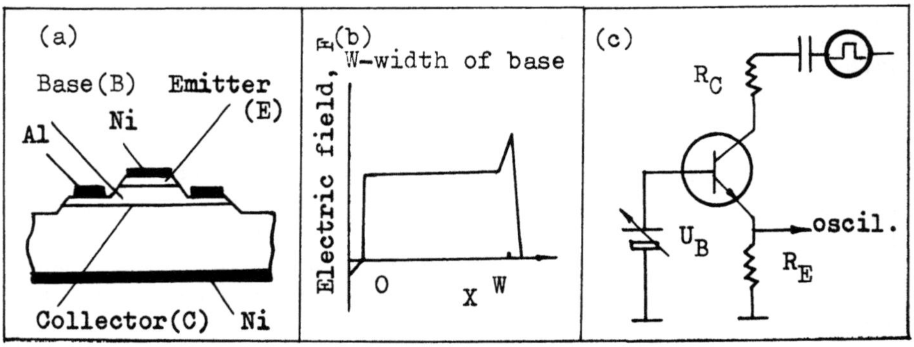

Fig.1 (a) Cross-sectional view of the transistor structure
(b) Distribution of the pulse electric field in the structure base at $U_C<0$ and $U_B>0$
(c) Electric circuit of measurements.

having duration $\tau_p=(2-3)\cdot 10^{-7}$ s was ensured on the collector-base junction.

When the pulse voltage U_C was supplied to the collector, no collector current was measured. This current was detected only when the base received a constant bias of several volts. The base current was $I<10^{-7}$ mA as soon as it was restricted by the large resistance of the base. Abrupt increase of the collector current took place upon increase of the pulse voltage U_C. The change in the collector current under the action of voltage U_B was observed within the range of 2 to 3 volts only. The collector current is determined by the electron injection from the emitter and subsequent drift under the action of the pulse in the base. The action of constant bias U_B leads to straightening of the contact barrier, which is why a voltage of $U_B \simeq 3$V is required, while the injection current caused by this bias is negligibly small and does not have an effect on the collector current. Thus the mechanism of the current transfer in the structure is the injection transit-time mechanism which causes the shape of the volt-ampere characteristic shown in Fig.2. A small effect of U_B takes place within the interval of 2 to 3 volts as the contact barriers straighten. It should be noted that such volt-ampere characteristic corresponds to fields $F<10^5$ V/cm. Measurements covered by this study were made up to the fields as strong as $F=3\cdot 10^5$ V/cm.

Upon increase of the field up to $F=(1.1-1.7)\cdot 10^5$ V/cm the current through the structure decreases jumpwise and then gradually increases (Fig.3). The current is usually reduced upon switching by 3-4 times. With some structures the current was reduced almost to zero and showed no increase later.

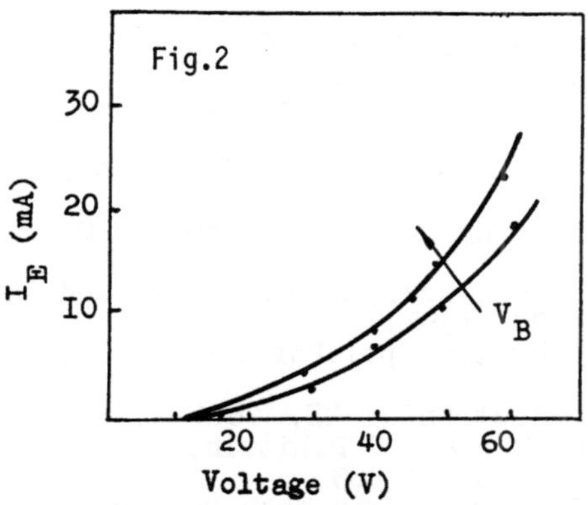

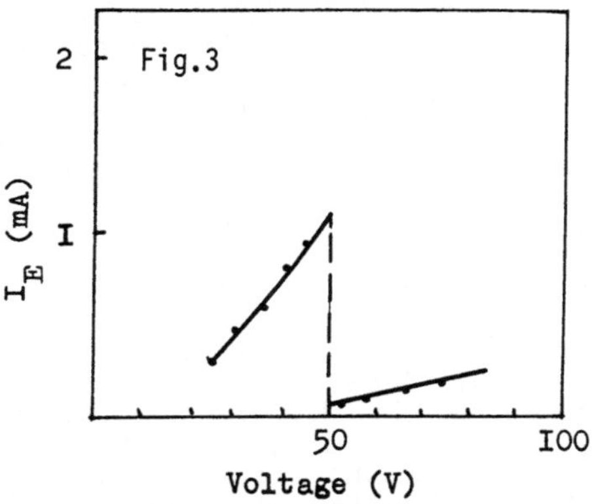

Fig.2 Volt-ampere characteristic of the structure $I_E - V_C$ dependence (without switching).

Fig.3 Volt-ampere characteristic of the structure (with switching).

During the reverse action, i.e. during the decrease of voltage on the structure a current jump takes place at a smaller voltage than during the voltage increase. Thus, some switching hysteresis is observed. A specified electric field spread is determined by the possible inaccuracy in the determination of the base layer width W, which was carried out through the estimation of the size of the luminescence region with the aid of a microscope.

Probably, the N-shape of the presented characteristic may also be explained by some other reason but the quality characteristics of the discussed phenomenon are in good agreement with the criteria of the Wannier-Stark effect [14]. From the relation $\hbar/eFd\tau<1$ at the average field value $F=1.5\cdot 10^5$ V/cm and the superlattice period $d=7.5\cdot 10^{-8}$ cm we obtain $\tau=6\cdot 10^{-14}$ s which is in agreement with the following estimation: $\tau=m^*_{\parallel}\mu_{\parallel}/e$ at $m^*_{\parallel}=m_0$ and $\mu_{\parallel}=100 cm^2/V\cdot s$.

It is known that the recombination instability also contributes to N-shape of the characteristic [15]. However, the time of transit through the active p-layer constitutes less than 10^{-10} s which is considerably less than the recombination time.

3. Conclusion

In our opinion, it can be concluded from the obtained results that this study is the first to establish the quantization of the zone spectrum in 6H-SiC, which is caused by a strong electric field, and to establish N-shaped volt-ampere characteristics in a semiconductor with superlattice at room temperature, which is caused by the Wannier-Stark localization.

References

1. G.H.Wannier: Phys.Rev. **11**, 432 (1960)
2. L.V.Keldysh: Fizika Tverdogo Tela, **4**, 2265 (1962)
3. L.Esaki, R.Tsu: IBM J.Res.Dev., **14**, 61 (1970)
4. K.K.Choi, B.F.Levine, K.I.Malik, J.Walker and C.G.Bethea: Phys.Rev. B, **35**, 4172 (1987)
5. E.E.Mendez, F.Aquillo-Rueda, I.M.Honq: Phys.Rev. **60**, 2426 (1988)
6. J.Bleuse, G.Bastard, P.Voisin: Phys.Rev.Lett. **60**, 220 (1988)
7. P.Voisin, J.Blense, C.Bouche, S.Guillerd, and C.Alibert and A.Regreny: Phys.Rev.Lett. **61**, 1639 (1988)
8. H.T.Grahn, H.Schneider, K.Klitzing: Appl.Phys.Lett. **54**, 1757 (1989)
9. L.Esaki: IEEE J. of Quantum Electronics **22**, 1611 (1986)
10. A.D.Dmitriev, A.O,Konstantinov, D.P.Litvin, V.I.Sankin: Fiz.Tekh.Poluprov. **7**, 1093 (1983)
11. V.I.Sankin, Yu.A.Vodakov, D.P.Litvin: Fiz.Tekh.Poluprov. **18**, 2146 (1984)
12. Yu.A.Vodakov, D.P.Litvin, V.I.Sankin, E.N.Mokhov, A.D.Roenkov: Fiz.Tekh.Polupr. **19**, 814 (1985)
13. W.Munck and P.Hoek: Sol.St. Electronics **21**, 479 (1978)
14. A.Ia.Shik: Fiz.Tekh.Poluprov. **8**, 1841 (1974)
15 A.F.Volkov, Sh.M.Kogan: Uspekhi Fiz.Nauk **96**, 633 (1968)

Physical and Practical Aspects of Electron Heating in Superlattice α-SiC

V.I. Sankin, D.P. Litvin, and Yu.A. Vodakov

A.F. Ioffe Physical-Technical Institute, Academy of Sciences, Polytechnicheskaya, 26, 194021 St. Petersburg, Russia

Abstract. Processes associated with electron heating in α-SiC have been revealed. It's pronounced anisotropy is dependent on the direction of the electric field applied either parallel or perpendicular to the crystal axis C. The purpose of the present study is to show the agreement of the experimental data with the existing concepts on heating of carriers in semiconductors with the superlattice. On the basis of these results we give some recommendations for the development of SiC devices.

1. Introduction

Investigations of the impact ionization in α-SiC [1,2,3] have revealed a number of new properties possessed by this material, and also shown that the impact ionization processes considerably differ depending upon the direction of the electric field (F) in relation to the crystal axis (C). For example, if F is perpendicular to C the electron coefficient of the impact ionization is only 1.2-2.0 times smaller than the hole ionization coefficient while the breakdown voltage increases as the lattice temperature rises. The situation changes abruptly if the electric field direction becomes parallel to the crystal axis. Primarily, the impact ionization is monopolar, which is confirmed by direct measurements of the impact ionization factors. In this work for two directions of the field $F \perp C$ and $F \parallel C$ we give : 1) investigation of the avalanche multiplication noise; 2) temperature dependence of avalanche breakdown; 3) the coefficients of the temperature change in avalanche breakdown for five silicon carbide polytypes; 4) dependence of avalanche breakdown voltage upon the gap width from data for five polytypes only for $F \parallel C$. The observed anomalies we explained by the model of electron heating in semiconductors with a superlattice.

2. Experimental procedure

Unusual peculiarities of impact ionization have been revealed by measuring of avalanche multiplication noise. Results of this investigation are shown in Fig.1 as a dependence $I_s = f(M)$ where I_s is a current of a noise, M is a coefficient of avalanche multiplication. Curve 1 (for $F \perp C$) is described by the dependence $I_s \sim M^{2.9}$, which corresponds to

Editors: G.L. Harris, M.G. Spencer, and C.Y. Yang

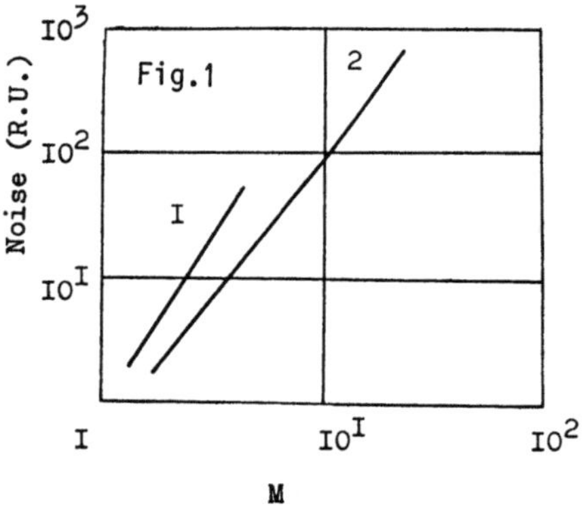

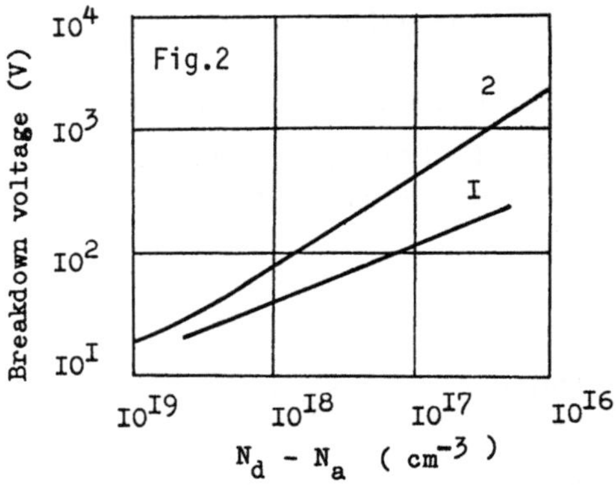

Fig.1 Dependence of spectral density of shot noise upon avalanche multiplication in abrupt p-n junctions in 6H-SiC; (1) for $F \perp C$; (2) for $F \parallel C$. The slopes of the curves are explained in the text.

Fig.2 Dependence of avalanche breakdown (in abrupt p-n junctions in 6H-SiC) upon N_D-N_A concentration in substrate; (1) for $F \perp C$; (2) for $F \parallel C$

a case of the bipolarity of impact ionization. Curve 2 (for $F \parallel C$) is described by the two dependences : $I_s \sim M^{2.2}$ for M <10 and $I_s \sim M^{2.6}$ for M >10. The analysis accounting for the relationship of electron and hole components in the initial current, shows that the relation between the hole ionization coefficient and the electron ionization coefficient is $\beta/\alpha \simeq 1.9$ for $F \perp C$ and $\beta/\alpha \simeq 50$ for $F \parallel C$. The coefficient α is suppressed to the limits only 10-15% smaller than the electric field of the avalanche breakdown. However, this coefficient increases sharply reaching the same order as the hole ionization coefficient at the electric field values equal to 0.98-0.99 of the maximum electric field.

This specific character of heating the electron if $F \parallel C$, leads to anomalous increase of the breakdown voltage. Shown in Fig.2 are the breakdown voltage U_B dependences on the substrate concentrations in sharp asymmetric p-n junctions for two directions of the electric field, i.e. $F \parallel C$ and $F \perp C$ ($U_{\parallel}$ exceeds $U_{\perp}$ by 4-5 times). The temperature coefficient of the breakdown voltage also exhibits an extraordinary anisotropy, unlike the case when $F \perp C$, U_B decreases as the temperature rises if the $F \parallel C$ orientation is used. The $U_B(T)$ dependences for both $F \parallel C$ (1) and $F \perp C$ (2) are given in Fig.3. Temperature coefficients of the breakdown voltage also differ considerably in the polytypes investigated (Fig.4), which appears a non-trivial result since the variation in the energy-band gap (E_g) in polytypes α-SiC does not exceed 10-12%, nor do $E_g(T)$ dependences contain any significant differences. Therefore, the negative temperature

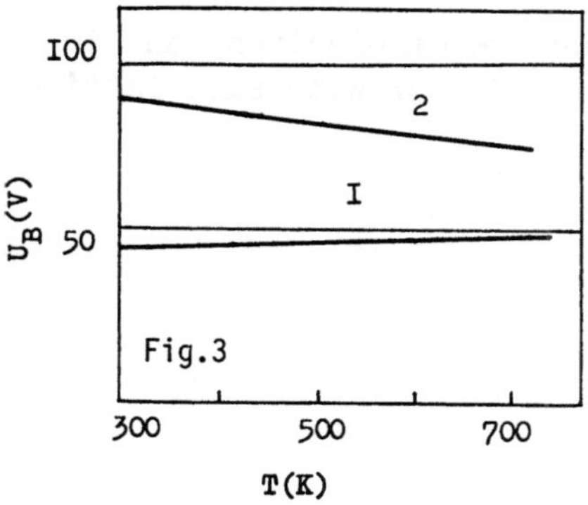

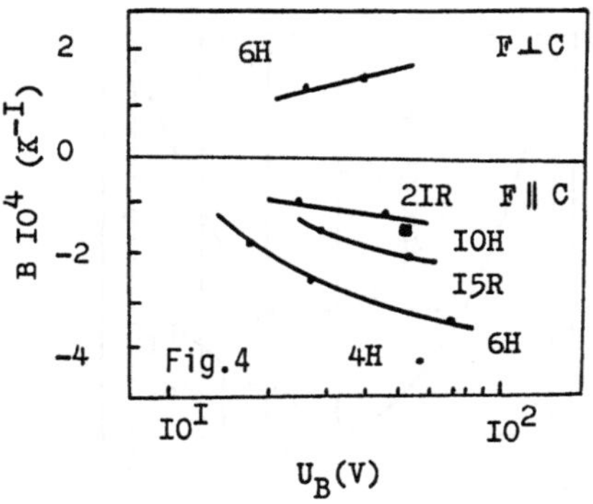

Fig.3 Temperature dependence of avalanche breakdown in 6H-SiC for two abrupt p-n junctions formed on one substrate. $N_D - N_A = 10^{18} cm^{-3}$ (1) for F⊥C; (2) for F∥C

Fig.4 Temperature dependence of change in breakdown voltage (for 5 silicon carbide polytypes)

$$B = -\frac{1}{U_B}\frac{dU_B}{dT}$$

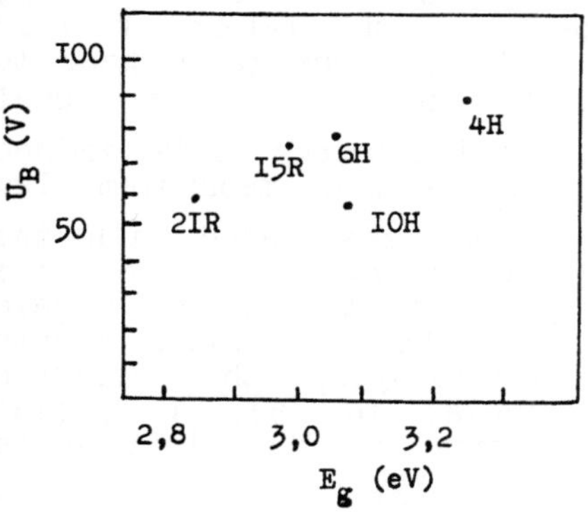

Fig.5 Dependence of avalanche breakdown voltage upon forbidden energy gap for concentration in substrates $N_D - N_A = 1.2\ 10^{18} cm^{-3}$

coefficient of the breakdown voltage cannot be explained through mere band-to-band tunnelling. The $U_B(E_g)$ dependence shown in Fig.5 can also be related to the anomalies observed.

3. Discussion

The features mentioned can be qualitatively explained by the model of heating the electrons within a narrow permitted band (mini-band) by a sufficiently wide forbidden energy band. Such

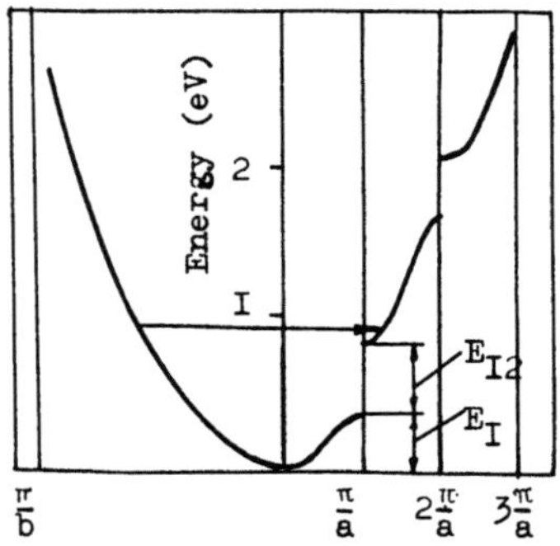

Fig.6 Structure of electron band in semiconductor with superlattice.

a band structure may lie in one crystallographic direction only, and may be caused by the action of the superstructure potential oriented along the crystal axis C. In other words, the observed properties are presumably determined by the minizone structure of the electron spectrum in the direction parallel to the crystal axis C (Fig.6). If the value of the first band is $E_1 \ll Eg$, then at $F \parallel C$ electrons cannot reach the threshold ionization energy in the first band. As the electron moves in the field F parallel to C, the Bragg reflection from the minizone edge takes place. From the calculations made it can be seen [4] that upon dissipation of the electron kinetic energy, e.g. on the phonons, it is pumped into the lateral kinetic energy. Thus the electron gains sufficient energy to pass to the next minizone : $E=E_1+E_{1,2}$. In the case of the usual heating mechanism the threshold ionization energy is determined primarily by the value E_0, while the discussed mechanism also includes some additional pre-conditions associated with the energy required for the interband transition. In this case the energy gain depends to a great degree upon the effective gain of the lateral kinetic energy during the dissipation events taking place on phonons and accompanying the electron transitions between the energy levels. In [5], a special equation for the electronic temperature in the minizone is given:

$$T_e = T \left[1 + \left(\frac{E_1}{\hbar\omega} \right)^2 \right] ,$$

at $eaF \gg \hbar\tau$

where a is the superlattice constant, and τ is the pulse relaxation time.

It can be inferred from this equation that the electron heating does not vary with the electric field, but is determined by the lattice temperature and the relation between the minizone energy and the optical phonon energy. Besides :

$$\frac{dT_e}{dT} \sim \left(\frac{E_1}{\hbar\omega} \right)^2$$

which is why, presumably, the difference is observed between the temperature coefficients of the breakdown voltage in α-SiC polytypes. However independence of T from the field in a wide range of the field values does not explain the anomalous behavior of $U_B(E_g)$. Apparently, it may be explained if we consider values $E_{1,2}$ E_2 etc., in addition to the parameters of the first minizone.

Thus, most of the experimentally observed phenomena associated with the impact ionization in α-SiC agree qualitatively with the concepts of the theory of the energy heating in semiconductors with superlattice.

4. Conclusion

The following information from the specific nature of heating in α-SiC seem to be of practical importance: 1) monopolarity of the temperature avalanche multiplication (α=0) in p-n junctions obtained on the face 0001 direction, permits their use as the avalanche photodiodes in the ultra-violet range. This feature of α-SiC was pointed out by Capasso [6] on the basis of our studies. Preliminary data show that NEP values of such photodiodes are about 10^{-13} W/Hz$^{1/2}$ and can be reduced. 2) anomalously high breakdown fields at F$\parallel$C permit the creation of high-voltage devices (on face 0001) where the avalanche breakdown is not used. 3) negative temperature coefficient for F$\parallel$C which obviates the possibility of preparing p-n junctions on the face 0001 for use in powerful avalanche apparatus. For such apparatus, a face 1210 appears to be preferable since it has a positive value of the temperature coefficient of the breakdown voltage at F$\perp$C.

References

1. A.P.Dmitriev, A.O.Konstantinov, D.P.Litvin, V.I.Sankin: Fiz.Tekh.Polupr.,**17**, 6, 103-1098 (1983)
2. V.I.Sankin, Yu.A.Vodakov, D.P.Litvin, Fiz. Tekh. Polupr., **18**, 12, 2146-2149 (1984)
3. Yu.A.Vodakov, D.P.Litvin, V.I.Sankin, E.N.Mokhov, A.D.Roenkov: Fiz. Tekh. Polupr., **19**, 5, 814-815 (1985)
4. P.Y.Price : IBM J.Res.Dev., v.17, No.1, p.399-446, (1973)
5. Suris R.A., Schamkhalova B.S.: Fiz. Tekh. Polupr., **18**, 7, p.1178-1184 (1984)
6. F.Capasso: Semiconductors and semimetals by R.K.Willardson and Albert Beer (Academic Press, New York, 1985) v.22, p.129

Investigation of Structural Defects in SiC Crystals by Transmission Electron Microscopy

E.N. Mokhov, E.I. Radovanova, and A.A. Sitnikova

A.F. Ioffe Physical-Technical Institute, Academy of Sciences, Polytechnicheskaya, 26, 194021 St. Petersburg, Russia

Abstract. Silicon carbide monocrystals and epitaxial layers (EL) grown by the sublimation method were researched by the transmitting electron microscopy. Second-phase inclusions have been detected. Size and density of the observed structural defects are different in the specimens grown by different methods. In the specimens irradiated with high neutron doses ($F=10^{21}$ cm^{-2}) and annealed at 1900-2600°C, clusters and dislocation loops have been revealed whose size tends to grow to annealing temperatures as high as T_0=2500°C.

1. Introduction

Parameters of SiC-based semiconductor devices are significantly lower than the theoretical estimations [1]. One reason may be concerned with the lack of information on the origin of native defects in grown crystals and their transformation during the subsequent technological operations. Presence of clusters in crystals is likely to be confirmed by the essential deviation of SiC composition from the stoichiometric one. It has been established that SiC crystals contain (3-4)% of excessive silicon [2] which causes a weak effect on the semiconductor properties of the material [3]. On the other hand, certain data prove that the clusterization processes are also active during the heat treatment of crystals irradiated with great doses of neutrons, electrons and other high-energy particles [4-5].

2. Experimental Procedure

The subject of this study is to investigate monocrystal silicon carbide (polytypes 6H and 3C) by the TEM method. Experiments have been run on specimens obtained by the Lely method, and on epitaxial layers grown by the sublimation "sandwich method" [6] in the conditions of relative excess of carbon or silicon. CVD-grown monocrystals 3C-SiC were also used. Specimens for the TEM test were prepared by the chemical etching in the KOH melt.

Editors: G.L. Harris, M.G. Spencer, and C.Y. Yang

3. Experimental Results and Discussion

The study of a large number of monocrystals grown by the Lely method has shown that such crystals contain spherical inclusions 0.1-1 μm in size. The concentration of these defects does not exceed 10^7 cm^{-3}. Fig.1 presents a TEM photo of the spherical inclusion. An origin of such inclusions has not yet been determined, but presumably, this may be silicon.

For the epitaxial specimens grown in conditions of relative excess of silicon, faceted inclusions of unknown composition are more typical. An example of such a defect is given in Fig.2. A specific feature of the crystals grown by the epitaxial method in the conditions of relative excess of carbon are the lined-up deformation centers arranged in chains detected in Fig.3 as light areas due to selective etching. High magnification permits visualization of scaly inclusions of the 2nd phase located in the center of such areas.

A different picture was observed in 3C-SiC specimens. A spatial monocrystal grown by the CVD method did not bear any characteristic defects. Conversely, epitaxial layers grown on the 6H-SiC substrates were less perfect. Individual dislocations with screw-like components of the Burgers vector (Fig.4) were observed on the division boundaries of such crystals.

The subject of the other series of experiments was the transformation of the radiation defects in SiC specimens grown by Lely method. Test specimens were irradiated by a dose $F=10^{21} cm^{-2}$ of reactor neutrons and subsequently annealed at constant temperature T_o in the range 1900-2600°C.

Fig.1

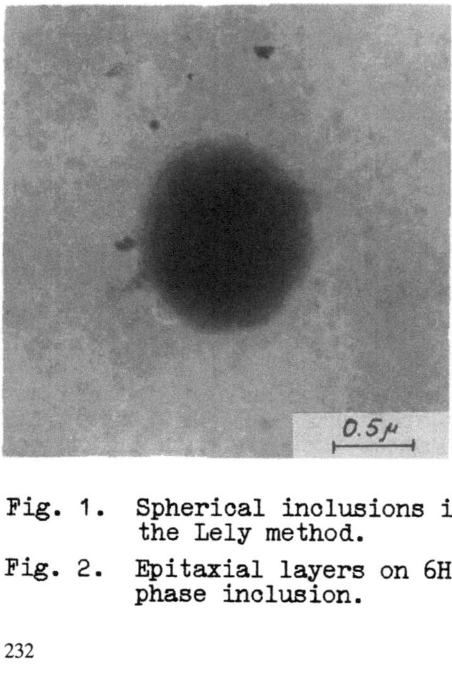

Fig.2

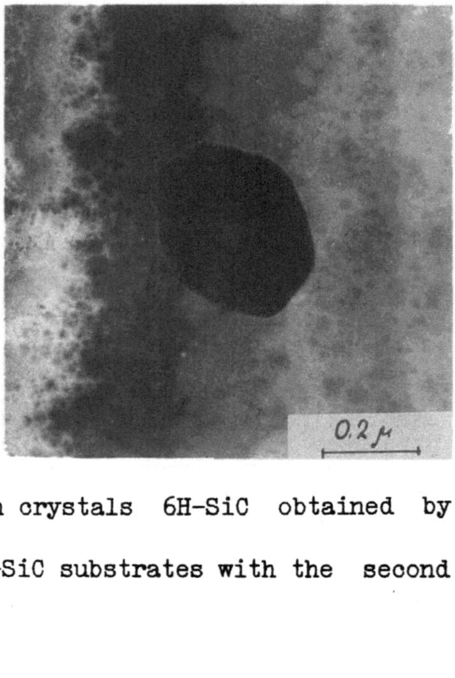

Fig. 1. Spherical inclusions in crystals 6H-SiC obtained by the Lely method.

Fig. 2. Epitaxial layers on 6H-SiC substrates with the second phase inclusion.

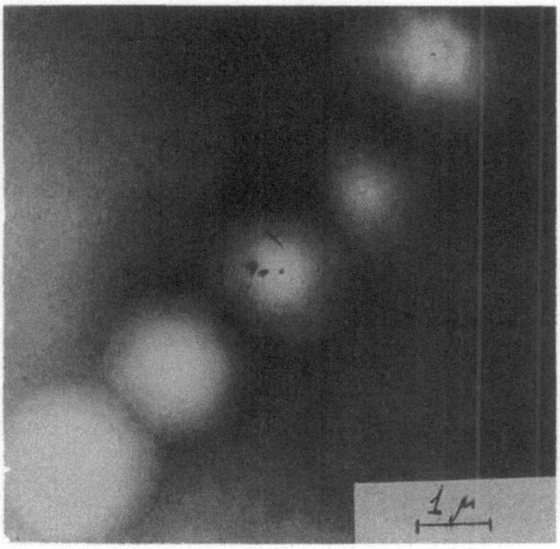

Fig.3

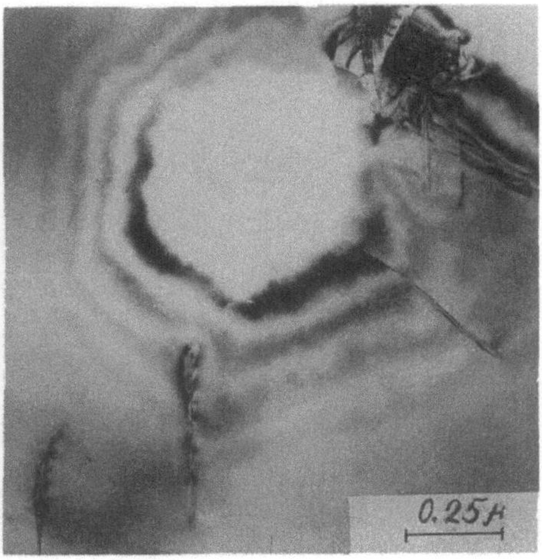

Fig.4

Fig. 3. Chain of deformation centers in EL of 6H-SiC.

Fig. 4. Screw-like dislocations in EL of 3C-SiC near the boundary of 6H-SiC substrate.

The findings of the irradiated crystals study have revealed a relationship between the type of defects and the annealing temperature. For specimens annealed at $T < 2300^{o}C$ we observed the presence of radiational defects which were sufficiently uniform at the microscale. Most of the radiational defects have the form of images with a black-and-white contrast (marked by *A* in Fig.5) which is typical for small clusters of radiational defects in semiconductors. Other defects have the contrast characteristic of dislocation loops located in {0001} planes of crystals (marked by *B* in Fig.5a). The concentration of defects in specimens annealed at $T_o=2300^{o}C$ is as high as $n=(2\text{-}6).10^{12}\ cm^{-3}$. At $T_o>2300^{o}C$ we observed topological changes in the defect structure. Thus, in specimens annealed at $T=2500^{o}C$ we clearly observed localized groups of dislocation loops (Fig.5b). Increasing the annealing temperature up to $T=2600^{o}C$ brings about another change in the defect structure of SiC. Radiational defects were annealed completely (Fig.5) whereas in some crystals isolated loops as large as 0.5-2.5 μm, whose concentration amounted to approximately $n=2.10^{7}\ cm^{-3}$ were observed. The loops lie in planes {0001} and {1101} of SiC crystals. Photograph of these loops is shown in Fig.5d.

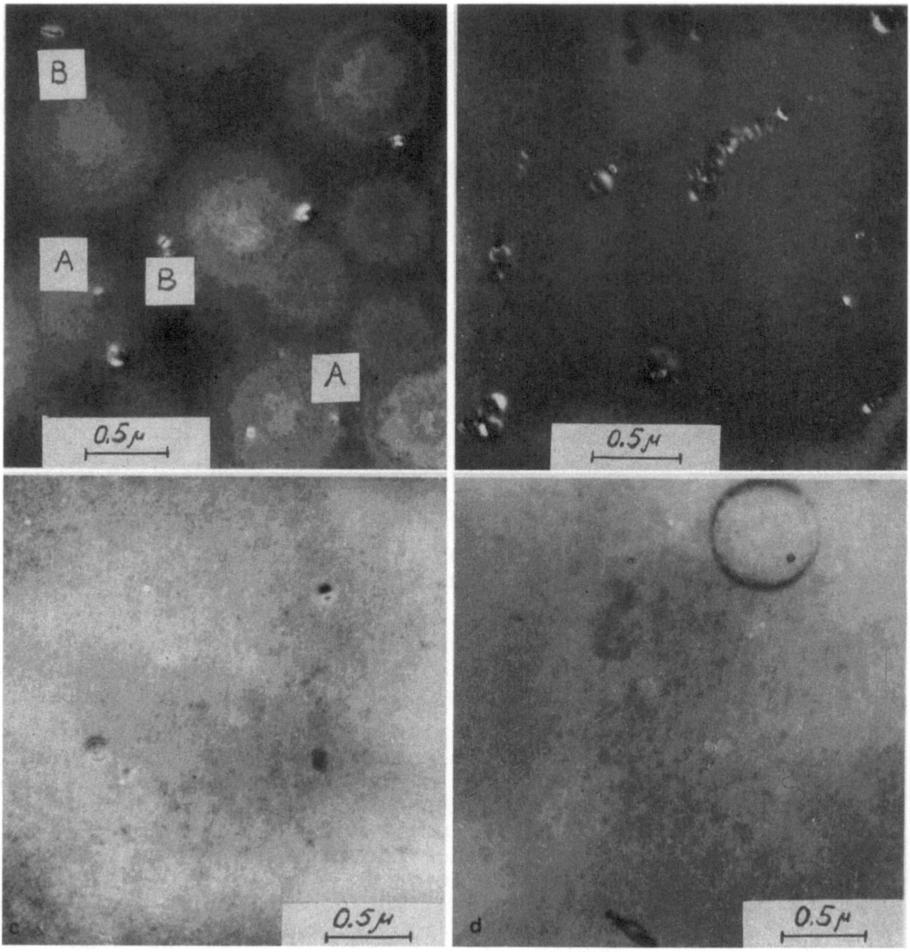

Fig. 5 (a) Radiational defects in 6H-SiC caused by neutron irradiation F= 10^{21} cm^{-2}) and annealed at T_o= 2300°C.

(b) Localized groups of dislocation loops after annealing at T=2500°C.

(c) Residual isolated radiational defects after annealing at T_o=2600°C

(d) Large dislocation loops after annealing at T=2600°C.

4. Conclusion

SiC crystals under study were found to contain structural defects of clusters, dislocation loops and the 2nd phase inclusions whose origin depends upon the conditions of growth, irradiation and heat treatment of SiC specimens. The presence of such defects can explain, in particular, the weak effect of deviation from the stoichiometric composition on the physical properties of the semiconductor.

Acknowledgments

The authors express their gratitude to A.D.Roenkov, A.A.Volfson and A.V.Naumov for their assistance in preparing the specimens and to Yu.A.Vodakov for encouraging our research and for useful advice.

References

1. A.E.Otblesk,V.E.Chelnokov, in "Problems of Physics and Technology of Semiconductors", Leningrad: LIJaF, (1980), p.197.
2. N.D.Sorokin, Yu.M.Tairov, W.F.Cvetkov, M.A.Chernov DAN SSSR, **262**, P.1380,(1980).
3. Yu.A.Vodakov, G.A.Lomakina, E.N.Mokhov Soviet: Fiz.Tverd.Tela, **24**, 1377,(1982).
4. A.I.Girka, A.D.Mokrushin, E.N.Mokhov, W.M.Osadchiev, S.W.Swirda, A.W.Shishkin Soviet: ZETF, **97**, 578, (1990).
5. P.N.Kutt, A.A.Lepneva, G.A.Lomakina, E.N.Mokhov, A.S.Tregubova , M.M.Scheglov, G.F.Yuldashev Soviet: Fiz.Tverd.Tela, **30**, 2606, (1988).
6. Yu.A.Vodakov, E.N.Mokhov, M.A.Ramm, A.D.Roenkov Krist. and Techn. **14**, 729,(1979).

Study of Al/Thermal Oxide/β-SiC MOS Diodes

H. Fuma, M. Kodama, H. Tadano, S. Sugiyama, and M. Takigawa

Toyota Central Res. & Dev. Labs., Inc., Yokomichi 41-1,
Nagakute-cho, Aichi-gun, Aichi, Japan

Abstract. C-V and I-V characteristics of Al/thermal oxide/ß-SiC MOS diodes have been studied using ß-SiC single crystals grown on Si(100) substrates. Oxidation of ß-SiC crystals was performed in a dry or wet O_2 ambient under the following three conditions: at 1050°C for 9h, at 1100°C for 3h and at 1150°C for 1h.

The C-V characteristics of the MOS diodes fabricated by a wet O_2 oxidation were reasonable but those of the MOS diodes fabricated by a dry O_2 oxidation were unreasonable. The equivalent fixed oxide charge density of the MOS diodes fabricated by a wet O_2 oxidation was typically 2×10^{11}-$4\times10^{11}/cm^2$ and it did not depend on the oxidation conditions. Maximum dielectric breakdown field of the oxide layer was about 8MV/cm and a Fowler-Nordheim current was observed prior to breakdown in some MOS diodes. The I-V characteristics did not depend on the oxidation condition and the oxidation ambient (wet or dry).

1. Introduction

Recently, many types of ß-SiC MOS transistors have been reported [1-5]. For MOS transistor performance, electrical properties of a MOS structure are very important. For the electrical properties of thermal oxide/ß-SiC interfaces, Avila et al. reported on the case of using a mercury probe as a metal electrode [6]. Tang et al. also reported about the C-V characteristics of Al/thermal oxide/ß-SiC MOS diodes when using the ß-SiC crystal grown on a just or a 2° off Si(100) substrate [7].

In this paper, we report on the C-V and I-V characteristics of Al/thermal oxide/ß-SiC MOS diodes which were fabricated under several oxidation conditions.

2. Experiment

ß-SiC single crystals were grown on 3-inch Si(100) substrates by chemical vapor deposition (CVD)[8]. CVD was performed using a C_3H_8-SiH_4-H_2 gas system at 1350°C for 3h. The Si substrates were p-type of 100-300 Ω·cm resistivity and about 10 minutes off toward <011>. The grown ß-SiC crystals were 5-8 microns thick without antiphase disorder and showed n-type behavior. From Van der Pauw measurement, the carrier concentration and the electron mobility of the crystals were 3×10^{16}-$1\times10^{17}/cm^3$ and 300-400$cm^2/(V\cdot sec)$, respectively.

Prior to oxidation, ß-SiC crystals were cleaned by a HF solution, rinsed in deionized water and dried. Thermal oxidation of ß-SiC crystals was performed by an electric furnace in a dry O_2(500sccm) or in a wet O_2 ambient (O_2 purging of 150sccm through H_2O in a bubbler). Oxidation conditions were 1050°C for 9h, 1100°C for 3h and 1150°C for 1h. After oxidation, an aluminum film was deposited on the oxide layer by an electron beam evaporation and electrodes were formed by a photolithography and a wet etching of

Editors: G.L. Harris, M.G. Spencer, and C.Y. Yang

aluminum. Finally, the MOS diodes were annealed in a nitrogen ambient at 450°C for 30 min.

The C-V characteristics of the MOS diodes were measured by using a C-V meter (Hewlett Packard 4274A) at a frequency of 100kHz employing 1mm square aluminum electrodes. The I-V characteristics of the MOS diodes were measured by using a picoammeter (Hewlett Packard 4140A) with the 0.1mm square aluminum electrodes serving as the anode. The thickness of the oxide layers were calculated from the capacitances of the MOS diodes at the accumulated condition using 3.82 for the dielectric constant of the oxide.

3. Results and discussions

The thickness of the oxide formed by a wet oxidation in this experiment is nearly the same (see Fig.4.) as those reported by Fung et al.[9]. The thickness of the oxide formed by a dry oxidation is about 100nm for 1050°C 9h, about 80nm for 1100°C 3h and about 50nm for 1150°C 1h. Oxide thickness also depended on the ß-SiC crystals and O_2 gas flow rate in the dry oxidation process. The reproducibility of the oxide thickness was good. A high-resolution-TEM (HRTEM) image of the oxide/ß-SiC interface is shown in Fig.1. The roughness of the interface is about 5Å in Fig.1.

The C-V characteristics of MOS diodes fabricated by a wet O_2 oxidation were reasonable but those of the MOS diodes fabricated by a dry O_2 oxidation were unreasonable, which was probably due to high interface trap density. Typical C-V characteristics of the MOS diodes fabricated by a wet oxidation are shown in Fig.2-a,b. In Fig.2-a, MOS capacitance changes with applied voltages and the shape of the measured curve is the same as the theoretical curve except for the deep inversion region. The theoretical curve was calculated using the values of $6.7cm^{-3}$ for the intrinsic carrier density of ß-SiC at 300K, 3.8 and 10 for the dielectric constant of oxide and ß-SiC and $6\times10^{16}cm^{-3}$ of carrier concentration by Hall measurement. In Fig.2-a, the C-V curve does not show the hysteresis and the formation of an inversion layer. On the other hand, the C-V curve in Fig.2-b shows the hysteresis, where the voltages are applied beyond $\pm10V$.

About this hysteresis, Avila et al. explain that the very low generation and recombination speed of the minority carrier causes the hysteresis. To confirm this explanation, C-V characteristics were measured under illumination as illustrated in Fig.3. In Fig.3, the C-V curve shows that an inversion layer is formed under illumination, while no inversion layer is formed in the dark condition. This change by the illumination supports

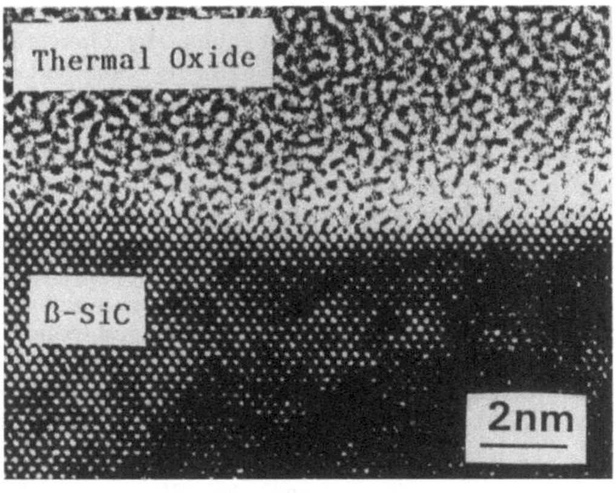

Fig.1. HRTEM image of SiO_2/ß-SiC interface. B=[011]

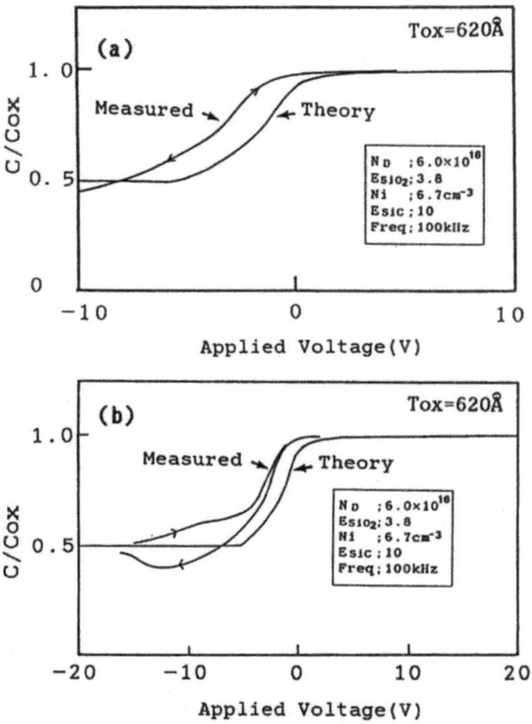

Fig.2. Typical C-V characteristics of MOS diodes fabricated by wet oxidation. (a);appl. volt. ±10V. (b);appl. volt. ±20V.

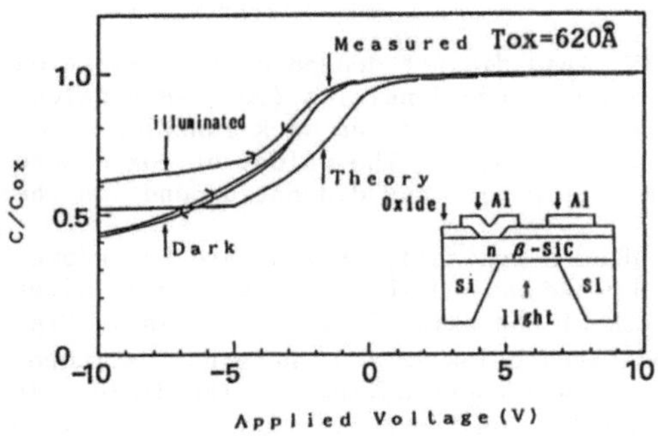

Fig.3. C-V curve under illumination and dark condition.

Avila's explanation. In Fig.3, the C-V curve in the dark condition shows the hysteresis, although the applied voltages are within ±10V. In the samples which did not show the hysteresis at first, hysteresis was observed in the C-V curve even if the applied voltages are within ±10V, after the voltages were applied beyond ±10V which causes the inversion layer. This result shows that the formation of the inversion layer is related to the degradation of MOS diodes.

The flat-band voltages of MOS diodes were obtained from the voltage shift between the measured and the theoretical curves. Flat-band voltages

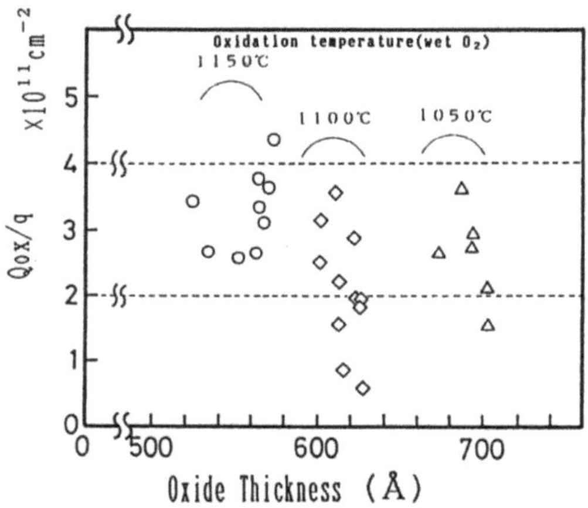

Fig.4. The equivalent fixed oxide charge densities of MOS diodes.

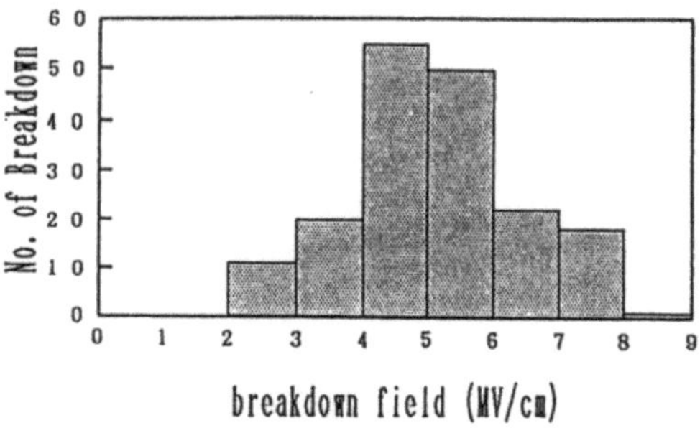

Fig.5. Histogram of breakdown field of oxide.

are in the range from -1.0 to -0.5V, and did not depend on the oxidation temperatures. The equivalent fixed oxide charge densities (Qox) were calculated from the flat-band voltages using 4.2eV for the work function of Al and 4.3eV for the electron affinity of SiC [10]. The values of Qox/q were in the range from 2×10^{11} to $4\times10^{11}cm^{-2}$ (Fig.4), and did not depend on the oxidation temperatures.

The I-V characteristics of MOS diodes typically showed hard breakdown; i.e. no current (below 0.1pA) before breakdown and short-circuit-mode after the breakdown. The dielectric strength of the oxide layers is shown in Fig. 5, where the samples broken at the voltages below 2MV/cm were not included. Maximum dielectric strength of the oxide was about 8MV/cm. The dielectric strength of the MOS diodes mainly depended on the ß-SiC crystal positions in the 3-inch substrate rather than the oxidation conditions. The surface roughness of ß-SiC crystal is considered to decide the breakdown field.

A small current flowed prior to the breakdown in several samples with dielectric strength above 5MV/cm (Fig.6). In the case of silicon MOS diodes, the current prior to the breakdown is well-known as a Fowler-Nordheim current. Therfore, the electric field dependence of this current was analyzed (Fig.7). In Fig.7, $\log(J/E^2)$ vs $(1/E)$ shows the linear region. Supposing that the effective mass of an electron in the oxide is $0.42m_0$(the value in the thermal oxide of silicon), the barrier height is calculated to be 3.1eV from the result shown in Fig.7. The barrier height in Al/thermal oxide/ß-SiC structure is theroretically nearly equal to the energy-level difference between the conduction band of the oxide and the conduction band of ß-SiC;

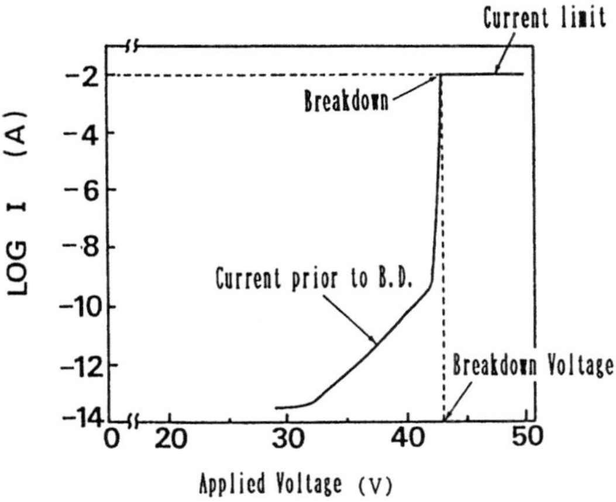

Fig.6. I-V characteristics of MOS diode with high breakdown voltage.

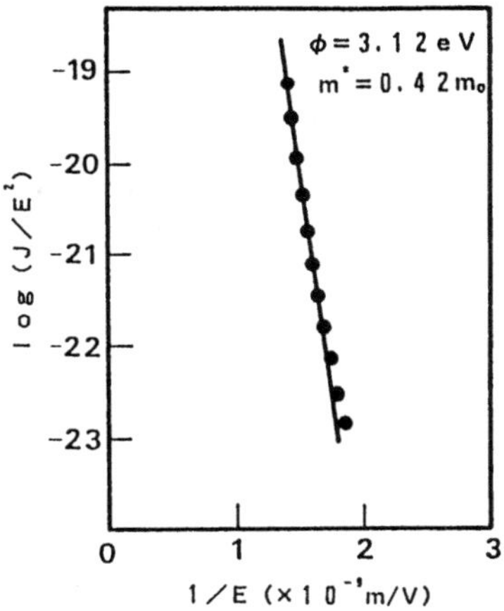

Fig.7. Fowler-Nordheim plot of the current prior to the breakdown.

that is estimated to be 3.4eV. Since the caluculated barrier height of 3.1 eV is close to 3.4eV, the current prior to the breakdown is considered to be a Fowler-Nordheim current. The drop of the barrier height from 3.4eV is considered to be caused by the surface roughness of the ß-SiC crystal, because we estimated that the measured barrier height is lower than the true barrier height when the ß-SiC crystal has surface roughness. However, to decide the barrier height correctly, study of the photoemission-current-dependence on the light wavelength is needed [11].

4. Summary

C-V and I-V characteristics of Al/thermal oxide/ß-SiC MOS diodes have been studied using the ß-SiC crystals grown on Si(100) substrates by CVD. The C-V characteristics of the MOS diodes fabricated by a wet oxidation were reasonable. The equivalent fixed oxide charge density in a wet oxidation was in the range of 2×10^{11}-$4\times10^{11}cm^{-2}$ and it did not depend on the oxidation temperatures. Hysteresis was not observed in the C-V curve when the applied voltages were within ±10V (1.5MV/cm), while hysteresis was observed after an inversion layer was formed by applying high voltages beyond ±10V.

In some samples, dielectric breakdown field of the oxide layer was about 8MV/cm and a small current prior to the breakdown was observed. These currents were considered to be Fowler-Nordheim currents.

References

[1] K.Shibahara,T.Saito,S.Nishino,and H.Matsunami: IEEE Electron Device Lett.EDL-7(1986) 692.
[2] Y.Kondo,T.Takahashi,K.Ishii,Y.Hayashi,E.Sakuma,S.Misawa,H.Daimon, M.Yamanaka,and S.Yoshida: Jpn.J.Appl.Phys.26(1987) 310.
[3] J.W.Palmour,H.S.Kong,and R.F.Davis: Appl.Phys.Lett.51(1987) 2028.
[4] J.W.Palmour,H.S.Kong,and R.F.Davis: J.Appl.Phys.64(4),(1988) 2168.
[5] H.Fuma,A.Miura,H.dadano,S.Sugiyama,and M.Takigawa: proc. of ICACSC'88 [Springer verlag] 178.
[6] R.E.Avila,J.J.Kopanski,and C.D.Fung: Appl.Phys.Lett.49(1986) 334.

[7] S.M.Tang,W.B.Berry,R.Kwor,M.V.Zeller,and L.G.Matus: J.Elec.chem.Soc. Vol 137 No.1 (1990) 221.
[8] S.Nishino,J.A.Powell,and H.A.Will: Appl.Phys.Lett.42(1983) 460.
[9] C.D.Fung,and J.J.Kopanski: Appl.Phys.Lett. 45(7),(1984) 757.
[10] M.A.Kadyrov,N.V.Kunina,A.S.Saidov,Kh.A.Shamuratov,L.V.Sharonova, A.Ya.Shik,and Yu.V.Shamartsev, Sov.Phys.Semicond.14,(1980) 942.
[11] R.Williams, Phys.Rev. Vol.140 No.2A (1965) A569.

Dissipative Structures in Hot Electron-Hole Plasma of Hexagonal Silicon Carbide (α-SiC)

B.S. Kerner, D.P. Litvin, V.I. Sankin, and A.D. Roenkov

A.F. Ioffe Physical-Technical Institute, Academy of Sciences, Polytechnicheskaya, 26, 194021 St. Petersburg, Russia

Abstract. Upon avalanche breakdown in α-SiC the distribution of the electron-hole plasma in the plane of the p-n junction changes from uniform to non-uniform as the current increases and acquires the form of numerous luminous filaments. This stratification of the electron-hole plasma takes place at a certain current density. Increase in duration of a voltage pulse causes the filament diameter to grow, while the number of filaments decreases. In the end the system has one filament only. In this study a model is proposed with the underlying supposition that the rate of the impact ionization is dependent on the concentration of nonequilibium carriers. Such dependence corresponds to the mechanism of heating electrons in a semiconductor with a superlattice. Experimental results have shown good agreement with the conclusions of the linear analysis of diffusion equations describing the discussed model.

1. Introduction

Investigation of the avalanche breakdown and impact ionization in α-SiC shows that this material has a set of anomalies : unusually high breakdown field, negative temperature dependence of breakdown voltage and small value of the electron coefficient of impact ionization. These anomalies are defined by a minizone structure of the electron spectrum in the orientation parallel to the C-axis [1]. In this paper we investigate a new phenomenon of electron-hole plasma stratification upon avalanche breakdown in α-SiC, which is defined by the complicated mechanism of electron heating in the minizone structure of the electron spectrum. In the model proposed, the energy gain of the electron is governed by quasi-elastic scattering of electrons on phonons and electron-electron collisions that result in electron-hole plasma stratification.

2. Experimental procedure

Avalanche breakdown in the p-n junctions of silicon carbide of the hexagonal modification α-SiC, which are formed along the 0001 face, can be characterized by three visually detectable states: 1) uniform electron-hole plasma (EHP) (Fig.1a,b); 2) stratification of EHP (Fig.1c); 3) non-uniform EHP in the form of a single filament (Fig.1d). Breakdowns at voltages lower than avalanche values (U_B), i.e. breakdowns caused by the

Editors: G.L. Harris, M.G. Spencer, and C.Y. Yang

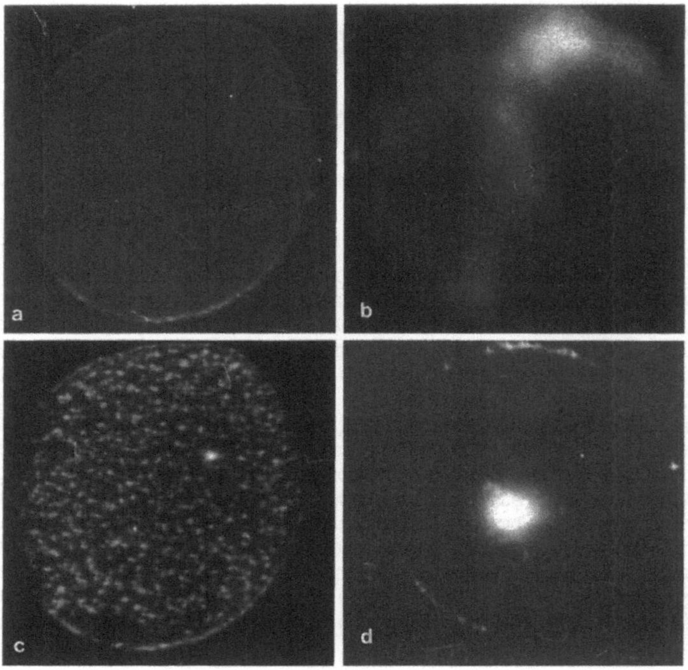

Fig. 1 (a) and (b). Uniform distribution of the electron-hole plasma at avalanche breakdown in an abrupt asymmetric 6H-SiC p-n junction.
$I=5$ mA, $U_B=45$ V, $S=10^{-3}$ cm^{-2}

(c) EHP stratification. $I=10$ mA

(d) EHP distribution in the form of a single filament. $I=25$ mA.

presence of microplasmas are not considered in this study. However, the breakdown picture is often non-uniform with regard to its radiation intensity (see Fig.1b), although no discrete structures can be seen. Such cases will also be regarded as uniform breakdowns. As the current increases the glow picture remains the same until the current exceeds the critical value I_c. As soon as I exceeds I_c, grain structure appears and the night-sky luminance develops (Fig.1c).

To avoid heating of a specimen it is best to observe these pictures using repeated voltage pulses $t_p<10^{-6}$s. The hot electron-hole plasma behaves so extraordinarily only if the electric field is adequately oriented with regard to the axis of the natural superlattice (SL) of α-SiC crystals, i.e. when the directions of the electric field and the axis are close to parallel. Misalignment in excess of 10° results in uniform EHP at any non-destructive currents.

Thus, an assumption was made about the relationship between specific processes caused by the impact ionization in α-SiC [1] and the phenomenon observed. As is known from [1] and [2], heating of the electrons in the field parallel to the superlattice axis is made difficult by the minizone character of the electron spectrum. Energy gain is governed by quasi-elastic scattering of electrons on phonons. As the current density, i.e. the concentration of electrons, increases the electron-electron collisions become more frequent. Such collisions also increase the number of hot electrons. Thus it can be inferred that the ionization rate ν_i becomes higher as the concentration of electrons (n) increases. At the same time the uniform increase in current (at a given voltage on the diode) is suppressed by the increase in voltage drop on the diode base. The situation is different from the avalanche current disturbances which are non-uniform in the plane of the p-n junction. Such shortwave disturbances will build up due to lateral current spreading in the diode base.

3. Theory

The suggested model explaining the stratification effect can be described by a system of two diffusion equations [3], viz the balance equation for the average electron concentration with respect to the thickness of the space-charge region (SCR) :

$$\frac{\partial n}{\partial t} = D \Delta_{\perp} n + n \nu_i (n U_i) - \frac{n}{\tau_n} + G_T \qquad (1)$$

and the continuity equation for the total current in quasi-neutral n or p regions of the structure:

$$C \frac{\partial U_i}{\partial t} = \sigma \tilde{W} \Delta_{\perp} U_i - J + (U - U_i) \rho^{-1}, \qquad (2)$$

where ν_i is the average rate for ionization of carriers (with respect to thickness of the space-charge field) related to the multiplication factor by an obvious relation : $M=(1-\nu_i\tau_n)^{-1}$;

U_i is the voltage drop in SCR of the p-n junction ;

$J=env_n$ is the density of the avalanche current ;

$\tilde{W}$ is the effective thickness of the current spreading in the base of quasineutral p and n regions of the structure;

D and v_n are the diffusion coefficient and the drift velocity of electrons in SCR ;

$\tau_n= w/v_n$ is the time of carrier transit through SCR of the p-n junction ;

w is the SCR thickness;

C is the capacitance of unit area of the p-n structure;

$\Delta_{\perp}= \frac{\partial}{\partial y^2} + \frac{\partial}{\partial x^2}$. The Z axis is chosen square (normal) for the p-n junction plane ;

U is the total voltage drop in the p-n structure;

G_T is the heat generation rate in SCR .

It has been taken into account that in the investigated p-n junctions $\rho_n = W_n/\sigma_n > \rho_p = W_p/\sigma_p$, where σ_n, σ_p and W_n, W_p are the conductivities and thicknesses of quasi-neutral n and p regions, which is why in (2) : $W=W_n$, $\sigma=\sigma_n$, and $\rho=\rho_n$. From (1) and (2) it can be seen that the characteristic length of change in the concentration of carriers in SCR is : $l=\sqrt{D\tau}$, while the characteristic length of change in the voltage drop on SCR (U_i) is $L=\sqrt{\tilde{W}W} \simeq W$.

From the linear analysis (1) and (2) it can be inferred [3] that the uniform distribution of avalanche current at $J > J_c$ loses its stability relative to aperiodic build-up of fluctuations n and U_i with the wave number $k_0 \simeq (lL)^{-1/2}$.

4. Discussion of results

The qualitative picture of build-up of the concentration fluctuations can be presented as follows. Suppose that in the region with $l \ll d \ll L$ an occasional increase in n or J takes place. As long as $d \ll L$, despite a local increase in J, the value of U_i is not reduced due to the large-scale lateral spreading in the diode base. At U=const the value ν_i is dependent only on n, and thus increases with n. At $d \gg l$, lateral diffusion of carriers may be ignored, and the initial local increment of n will grow as ν_i increases. The current distribution changes from uniform to non-uniform, and separate filaments are formed (Fig.2) at a distance of about $t=1/k_0$ from each other. This picture of the current distribution agrees with that obtained experimentally (Fig.1c).

It has been mentioned above that the state marked by the stratified EHP ends in the state of a single aggregate

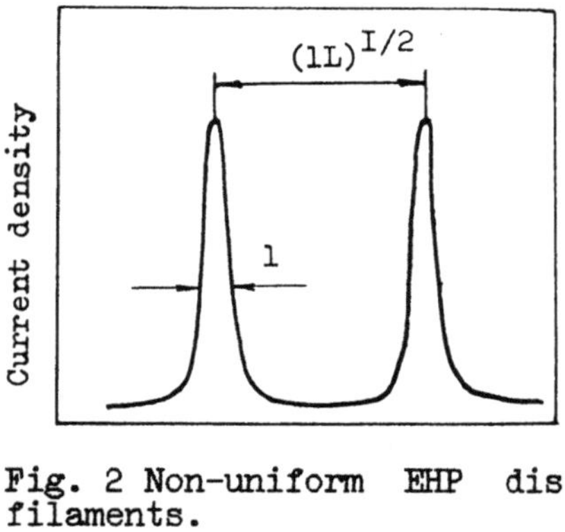

Fig. 2 Non-uniform EHP distribution in the form of filaments.

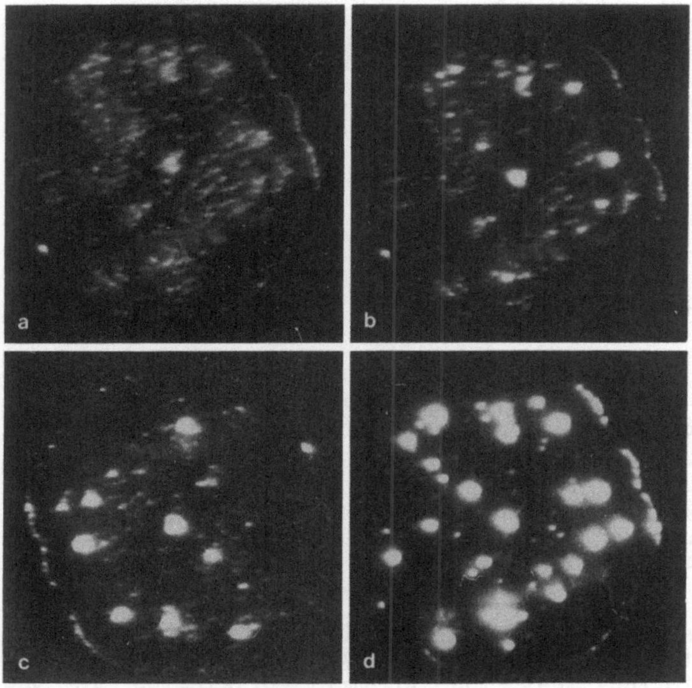

Fig. 3 Evolution of EHP distribution at avalanche breakdown: (a) - t_p=0.2 μs; (b) - t_p=2 μs; (c) - t_p= 4 μs; (d) - t_p=10 μs .

filament. Indeed, as soon the current density reaches I_f the current forms a filament several μm in diameter. It is natural to suppose that the formation of the current filament results from the evolution of stratified EHP. For observation purposes the picture of the EHP evolution was read off at a constant current and at different durations of the feed pulse in the range from 10^{-7}s to 10^{-4}s. The four fragments shown in Fig.3 correspond to pulse durations (t_p) of 0.2 μs, 2 μs, 4 μs and 10μs. It can be easily seen that as the pulse duration grows individual filaments of submicrometer size merge into each other and are converted into larger local regions of several micrometers in size. At these moments the stratification and larger regions coexist. It should be noted that the formation of regions is followed in a certain time t_1 (Fig.4) by the current jump signifying the appearance of negative differential resistance associated, most likely, with thermal fluctuations caused by the lattice self-heating. As is seen from the photograph, within times specified in Fig.3 (b,c and d) new filaments are formed, while the number of the stratification regions decreases. In a certain time (t=50-60 μs for the discussed system) the electron-hole plasma ends in a state

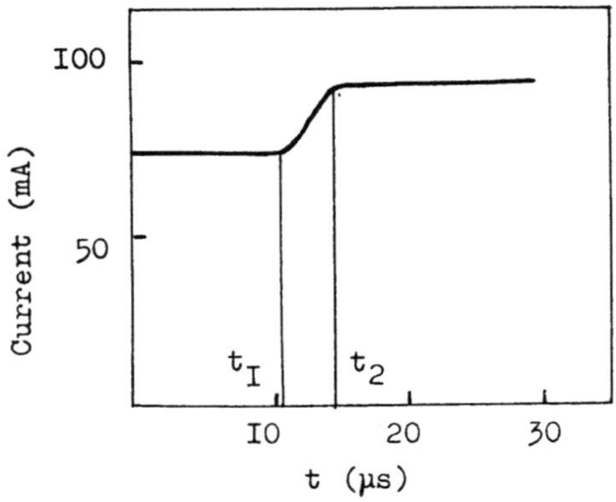

Fig. 4 Transfer characteristic of avalanche breakdown in 6H-SiC p-n junctions.

having one filament only, which is completely different from the situation arising at a thermal instability of the uniform state. The filament is formed within a time period from t_1 to t_2 (Fig.4). We divided this time into smaller periods and observed changes in the filament size in the dynamic range. It was found that the filament diameter increases with time. Thus, it can be inferred that the filament originates in the micro-area where merging of individual stratification features takes place, and then increases its size with the aid of self-heating. The time period t_1-t_2 constitutes 1.5-2.0 μs. It should be noted that the filament formation from the uniform state, as is known from [4], has the reverse dynamics i.e. the size of the column formed decreases with time. It is obvious that the current increase in the filament enables its diameter to grow (which is confirmed experimentally), while the thermal filament originating from the uniform state tends to decrease in diameter as the current increases [5]. These qualitative differences between the ionization filament under consideration and the thermal filament are associated with the fact that the existence of the ionization filament and numerous current filaments is determined by the increasing dependence of the rate of ioniziation of the carriers in SCR upon their concentration. The Joule heating of the lattice plays here a secondary role, causing, as the current increases, a successive transfer of numerous current filaments (Fig.4) into a single filament, i.e. the current column.

Conclusion

Experimental study and theoretical analysis of the electron-hole plasma stratification phenomenon upon avalanche breakdown in the 6H-SiC p-n junction shows that this effect is defined by complicated mechanism of electron heating in the minizone structure of the 6H-SiC. The model proposed is based on the supposition that the rate of the impact ionization is dependent on the concentration of nonequilibrium carriers, which corresponds to the mechanism of electron heating in a semiconductor with a superlattice. Investigations revealed the Joule heating influence on the electron-hole stratification results in transfer of numerous current filaments into a single filament.

References

1. Yu.A.Vodakov, D.P.Litvin, V.I.Sankin, E.N.Mokhov, A.D.Roenkov : Fiz. Tekh. Polupr., **19**, 5, 814-815 (1985)
2. R.A.Suris, B.S.Schamkhalova : Fiz. Tekh. Polupr., **18**, 7, p.1178-1184 (1984)
3. B.S.Kerner, D.P.Litvin, V.I.Sankin : Pisma v JTF, **13**, 13, p.819-823 (1987)
4. B.S.Kerner, A.M.Nechaev, E.A.Rubaha, V.F.Sinkevich : Radiotech. i electr., **25**, 1, 168 (1980)
5. A.I.Baranenkov, V.V.Osipov : Mikroelektronika **1**, 63 (1972)

Characterization of the Electronic and Atomic Structure of Surfaces and Interfaces of SiC/Si Grown at Low Temperature

M. Hirai[1], *M. Iwami*[1], *M. Kusaka*[1], *K. Fukumoto*[1;*], *H. Nakamura*[2], and *H. Watabe*[3]

[1]Research Laboratory for Surface Science, Faculty of Science, Okayama University, Okayama 700, Japan
[2]Osaka Electro-Communication University, Hatsu, Neyagawa, Osaka 572, Japan
[3]Matsushita Electric Industrial Co. Ltd., Twin 21 National Tower, 2-1-61 Shiromi, Chuo-ku, Osaka 540, Japan
*Present address: Sharp Corporation, Abeno-ku, Osaka 545, Japan

We have tried to grow either carbonized layers or 3C-SiC films on Si(100) substrates at lower temperatures. They can successfully be grown at T_s, substrate temperature, $\gtrsim$1230 °C using H_2 gas dilution. A non-destructive analysis of specimens thus obtained is carried out using soft X-ray emission spectroscopy (SXES). The method relies on the Si-$L_{2,3}$ emission band spectra of SiC showing a clear difference from that of Si crystal. Also, we have used the fact that energetic electrons can reach a deeper region in a solid with increasing primary electron energy, E_p. Surface and interface structure of a Si crystal with either a carbonized layer or a SiC on top are clarified by varying E_p up to 4 keV.

1. Introduction

Silicon carbide (SiC) is a candidate semiconductor with applications in several electronic and optical devices, such as the wide-bandgap emitter in Si heterobipolar transistors [1] and the blue light emitting diodes [1,2]. For many devices, the hetero-epitaxial growth of SiC on a silicon substrate has been attracting attention. One of the problems left in this growth process is how much one can lower the growth temperature. Then, it is necessary to have an appropriate tool, hopefully a non-destructive one, which can explore the compositional and/or electronic structure of a specimen with hetero-contacts. For the hetero-epitaxial growth of SiC on a Si substrate, chemical vapor deposition (CVD) may be one of best candidate techniques to lower the growth temperature.

Among methods characterizing surfaces and interfaces, soft X-ray emission spectroscopy (SXES) is a very powerful tool, where photon

Editors: G.L. Harris, M.G. Spencer, and C.Y. Yang

energy ranges from 20 to 1000 eV. SXES has several excellent characteristics, e.g., a clear difference can be observed in spectral shapes of an element in different networks, even when this is not the case for other methods such as Auger electron spectroscopy (AES) and X-ray photoelectron spectroscopy (XPS). In addition, an in-depth structural analysis can be performed non-destructively by SXES [3].

In the present study, we tried to fabricate an epitaxial SiC layer, either a thin carbonized layer or a thick SiC film, on a Si substrate at low temperature. The grown layer is studied using reflection high energy electron diffraction (RHEED), scanning electron microscopy (SEM) and SXES, where the last is carried out under an electron excitation.

2. Experimental

2.1 Sample Preparation

Growth of a carbonized layer and a 3C-SiC film on a Si(100) substrate was carried out by a low pressure CVD method. The preparation chamber was all made of stainless steel, and the base pressure was ~1x10^{-3} Pa. Immediately after the p-Si(100) substrate was etched by hydrofluoric acid, it was inserted in the preparation chamber, which was then evacuated. Then the Si(100) specimen was cleaned by direct Joule heating under a flow of H_2 gas or in vacuum at 800 °C for 10 min; this was followed by cooling for several minutes. Subsequently, the carbonized layer was formed on the Si(100) substrate in a C_2H_2 gas flow with or without H_2 dilution of 1 l/min at T_s, the substrate temperature, =1000~1400 °C for several minutes, where its temperature was measured by

Table I Several preparation conditions for a carbonized layer and 3C-SiC film on Si(100) substrate

Growth Conditions	Carbonized Layer with and without H_2 gas	3C-SiC Film
Substrate Temperature	1000~1400 °C	1100~1300 °C
Total Pressure	5~1000 Pa	1000 Pa
Flow Rate of C_2H_2 Gas	2~100 CCM	2 CCM
Flow Rate of SiH_4 Gas	————	2 CCM
Growth Time	3~10 min	≳3 min

an optical pyrometer. Following this process in some cases, SiH_4 gas was added for 3 min or more to give an epitaxial layer of 3C-SiC. Parameters adopted are shown in Table I.

2.2 Characterization of the Grown Layer

The specimens with the grown layer were characterized by RHEED, SEM and SXES measurements. RHEED studies were carried out using a JEOL JEM-6A electron microscope operated at 100 keV and a SHIMADZU EPMA 8705 was used for the SEM study.

The SXES experiment was carried out in a vacuum chamber with an ambient pressure of $\sim 10^{-6}$ Pa during analysis. Specimens were put in a Henke type X-ray tube, where both exciting electrons and emitted photons are normal to the specimen surface. The electron flux was 1 mA on a region of ~ 0.3 cm^2. Specimens were kept at room temperature by water cooling. Soft X-ray emission spectra were obtained by a grating monochromator with 1200 grooves/mm and a radius of 2 m. Details of the SXES experiments are given elsewhere [4].

3. Results and Discussion

3.1 3C-SiC Film (Buffer Layer and Epitaxial Film) Growth at Lower Temperature

The following results were obtained from RHEED and SEM experiments on specimens with the layer on the Si(100) substrate grown under several growth conditions. The carbonized layer grown without H_2 gas exhibited Debye rings in its RHEED pattern superimposed on some spots at every substrate temperature under all C_2H_2 gas pressures. Particularly, at $T_s \gtrsim 1350$ °C under partial pressure of C_2H_2 gas, P_p, above 200 Pa a structure like a whisker is observed in a partial region of the specimen surface by SEM. Thus, it is clear that the specimens prepared without H_2 gas have poor crystallinity and surface morphology. This fact suggests that the H_2 gas plays an important role in the epitaxial growth of the carbonized layer. On the other hand, the carbonized layer grown with H_2 gas shows the spot pattern for $T_s \gtrsim 1230$ °C under any C_2H_2 gas pressure except for $P_p > 200$ Pa. Further, a 3C-SiC film by CVD growth is obtained at 1300 °C with the conditions as given in Table I.

3.2 Characterization of Carbonized Layer and 3C-SiC(Film) on Si(100)

In Fig. 1 are shown the SXES spectra of the Si-$L_{2,3}$ emission band obtained from (a) 3C-SiC and (b) Si crystal at an incident electron energy, E_p, of 4 keV. From this figure one can easily find clear differences for the structures of the peak around photon energies ($h\nu$) of 90 eV, 86 eV and 97 eV, although such clear differences in Si signal are hard to see in other electron spectroscopies which can give information about the electronic structure of Si, e.g., the conventional first derivative spectra, dN(E)/dE, of Si(LVV) Auger signals as shown in Fig. 2. From these characteristics of the SXES spectra, it can be said that the electronic structure in the valence band of 3C-SiC is different from that of the Si bulk. The Si-$L_{2,3}$ spectrum is considered to reflect the valence band density of states (VB-DOS) with s-like symmetry, because the selection rule works for the electron transition resulting in the photon emission. Furthermore, in combination with the adoption of energetic electrons as a source of excitation of photons, these characteristics of the SXES spectra can be used for a non-destructive analysis of 3C-SiC/Si hetero-structure. The point of this experiment is that an energetic electron can reach deep inside a solid for energies $E_p \gtrsim 100$ eV. Therefore one can gain information from deep inside or from the surface region, using an electron with higher or lower energy, respectively.

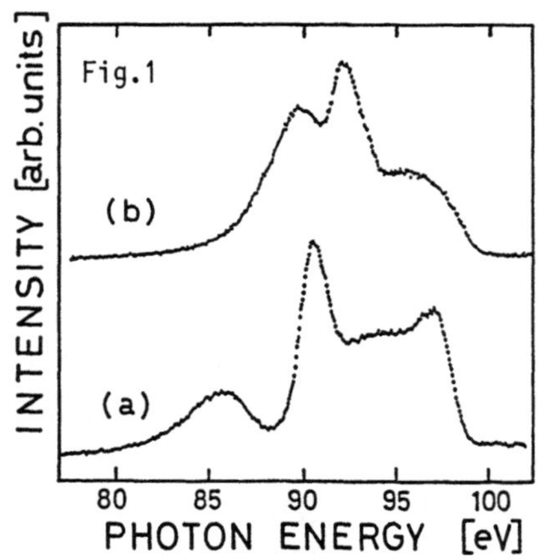

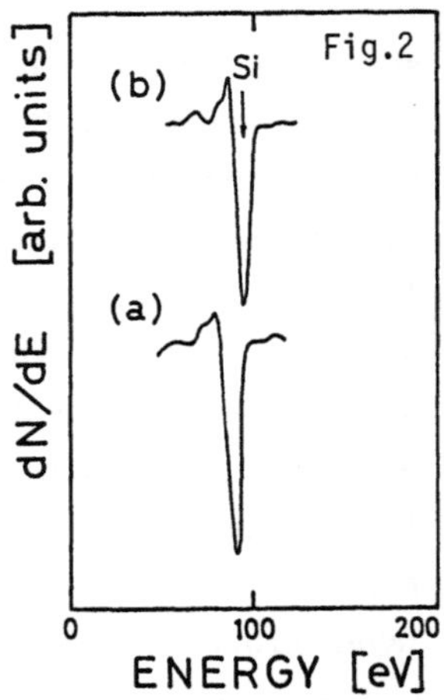

Fig. 1 SXES spectra of Si-$L_{2,3}$ emission band measured at E_p=4 keV: (a) 3C-SiC crystal and (b) Si crystal

Fig. 2 AES spectra of Si-LVV transition: (a) 3C-SiC crystal and (b) Si crystal

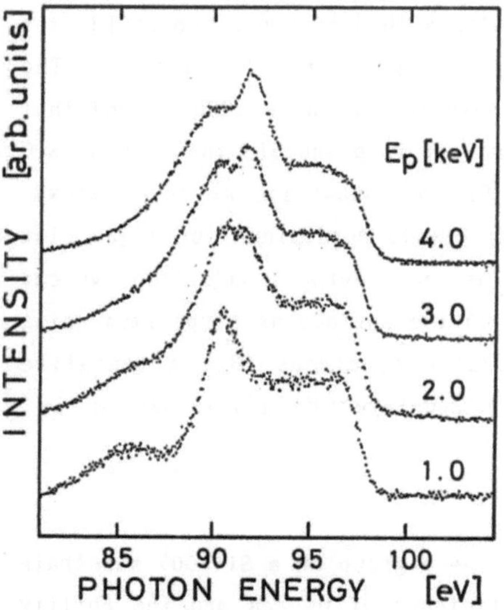

Fig. 3 SXES spectra of Si-$L_{2,3}$ emission band corresponding to several primary electron energies, E_p, obtained from a specimen of a carbonized layer/Si(100)

In Fig. 3 are shown Si-$L_{2,3}$ SXES spectra obtained from a carbonized layer(thin film)/Si(substrate) specimen observed at different incident electron energies (E_p's), where the intensity of the spectra is normalized at the highest peak. In the Si-$L_{2,3}$ emission band spectra measured at different E_p's, one can see clear differences around hν=89-92 eV, ~86 eV and ~97 eV except for details. At E_p=4 keV, the Si-$L_{2,3}$ spectrum has a double-peaked structure around 89-92 eV and varies monotonically with lower photon energy down to ~86 eV; at higher photon energy there is a shoulder at ~97 eV, which appears to correspond to the signal mainly from the substrate Si(100). On the other hand, at E_p=1 keV, the double-peaked structure in the Si-$L_{2,3}$ spectrum changes into the single-peaked one around hν=89-92 eV, a hump appears in the spectrum at hν=~86 eV and the shoulder at ~97 eV changes into a small peak. In comparison with Fig. 1, the Si-$L_{2,3}$ SXES spectrum at E_p=1 keV is considered to be almost the same as the one of the 3C-SiC crystal. With increasing E_p, the Si-$L_{2,3}$ SXES spectrum comes to have an increased contribution of the lower (~89 eV) and the higher (~92 eV) energy components. At intermediate E_p's, the Si-$L_{2,3}$ spectra look like a mixture of the one of 3C-SiC and that of crystalline Si. Therefore, it can be concluded that a 3C-SiC layer is formed on the Si(100) surface during the present carbonization process as before [3]. On the other hand, if we compare these results with a previous report [3] the shape of the Si-$L_{2,3}$ spectra is quite similar except for details. The growth temperature of the specimen as shown in Fig. 3 is 1230 °C and that of the previous one is 1350 °C. Therefore, it is plausible that the 3C-SiC film will be able to be grown on the Si(100) substrate at lower growth temperatures with the selection of appropriate conditions for a gas flow rate, gas composition and pressure during the growth. In addition, we can estimate the thickness of a film formed on the surface of a specimen using the data of penetration depth of energetic electrons [3]. A tentative estimation of the thickness of the carbonized-layer(3C-SiC) is ~20 nm.

4. Summary

The carbonized layer and the 3C-SiC film were grown on a Si(100) substrate by the low pressure CVD method. The efficiency of H_2 gas and the ability of the lower growth temperature are indicated from the results of RHEED and SXES measurements. It should be emphasized that the SXES method is very powerful to characterize the surface and/or interface structure of a thin-film/substrate contact system without adopting any destructive tools. This method can be used even if the same element is included in both film and substrate, because the change in the valence electronic state is reflected in the soft X-ray emission band spectrum, e.g., Si-$L_{2,3}$ for SiC(film)/Si(substrate) specimens. The SXES applied to the carbonized layer/Si contacts clarified the following: (1) A material constructing the surface region of a specimen can be identified. (2) The thickness of the surface and interface layer can be estimated.

References

1. M.M.Rahman, C.Y.-W.Yang, G.L.Harris: In Amorphous and Crystalline Silicon Carbide II (Springer, Berlin, Heidelberg 1989)
2. V.A.Dmitriev, L.M.Kogan, Ya.V.Morozenko, B.V.Tsarenkov, V.E.Chelnokov, A.E.Cherenkov: Sov. Phys. Semicond. 23, 23 (1989)
3. M.Iwami, M.Kusaka, M.Hirai, H.Nakamura, K.Shibahara, H.Matsunami: Surf. Sci. 199, 467 (1988)
4. K.Tsutsumi, O.Aita, K.Ichikawa, H.Nakamura: In Ann. Rept. Inst. Nucl. Study Vol.100 (University of Tokyo, Tokyo 1977)

Two-Dimensional Modeling of New Silicon Carbide Devices

W.R. Curtice[1], *J.R. Szedon*[2], *P.G. McMullin*[2], *F.R. Sutherland*[3], *A.W. Morse*[3], *and J.A. Costello*[3]

[1]W.R. Curtice Consulting, Princeton Junction, NJ, USA
[2]Westinghouse Electric Corporation, Pittsburgh, PA, USA
[3]Westinghouse Electric Corporation, Baltimore, MD, USA

Abstract. Two-dimensional modeling of SiC devices for power microwave applications is presented. Static induction transistor designs were simulated in silicon and silicon carbide using both PISCES and a new two-dimensional analysis that include energy relaxation effects. The simulations show that the silicon carbide structure should be superior in transconductance and frequency response.

1. Introduction

Progress in the realization of silicon carbide devices is being spurred on by the availability of device-quality wafer/epi materials. Effective utilization of powerful device modeling techniques can enhance the selection of potentially best performing device structures for a given application.

For the same donor value, silicon carbide has a higher electron saturation velocity but a lower mobility than silicon. As a result, device structures utilizing low-field mobility, such as the long channel field-effect transistor, (FET), are expected to show poorer frequency response for silicon carbide than that for silicon.

A structure utilizing more the high field property, such as the Static Induction Transistor (SIT) [1] should show improved frequency response for silicon carbide. This occurs because current is controlled by variation of the potential barrier just beyond the source and essentially all current flows at the saturation velocity. The optimum performance occurs for higher resistivity material and at very large voltage biasing. Additional advantages of the SIT are that it is relative simple to construct and a majority carrier device, not suffering from minority carrier storage effects.

The two-dimensional models enable evaluation of both dc and rf characteristics of devices. The dc current-voltage relationships are found from steady-state results and the small-signal rf properties are found from transient simulations using voltage steps on the electrodes. The large-signal properties of rf power capability and power-added efficiency may also be studied, although it is usually more expedient to evaluate the large-signal properties from equivalent circuit models developed from the two-dimensional models [2].

The primary results of this paper contrast the performance of the silicon carbide and silicon SIT structures. It includes the dc and small-signal rf properties of the SIT structure.

2. The Two-Dimensional Simulation Programs

The first program used to study the Static Induction Transistor is called SIT, Version 1.1, or simply SIT/1.1. The second and more detailed program is called SIT, Version 2.3. The first program used a gate electrode and does not include hole currents. The second program includes all details of the implanted gate region. The equations of

Editors: G.L. Harris, M.G. Spencer, and C.Y. Yang

current continuity are solved for both holes and electrons and depletion effects into the p+ gate region are included.

Both SIT programs are based upon the Electron-Temperature Model [4], which includes energy relaxation effects permitting accurate simulation to picosecond time scale. This has been verified by comparison with Yoshii et al. [5] for the case of silicon.

Not all details of the device structure are included in these programs. For example, contact resistance and some parasitics are omitted. For realistic evaluation of performance, these effects would be added as external elements after modeling the internal device carrier effects.

3. The Velocity-Field Relationships

The electron drift velocity relationship to electric field is taken to be the same as used in PISCES with beta = 2.0. For the case studied, the donor value is 6.0E14/cm**3. The parameters used for silicon and silicon carbide are:

Material	Relative Diel. Constant	Low Field Mobility (cm**cm/V-s)	Saturation Velocity (cm/s)
Silicon	11.8	1320.	1.0E7
SiC	9.7	800.	2.0E7

Comparison of the electron velocity-field relationships for these two materials [5] is shown in Figure 1. Notice that for this low doping value, the difference in low-field mobility is not so large as to greatly reduce the velocity (and ultimately the device current) of SiC in the low electric field region. The much larger saturation velocity of SiC will be seen to lead to more current and larger transconductance than for Si.

The drift velocity-electric field relationship for holes in SIT/2.3 is also the same as used in PISCES with beta = 2.0. Because of the large acceptor concentration, a small value of low field mobility is used. The value used is 240 cm**2/V-s for both Silicon and SiC.

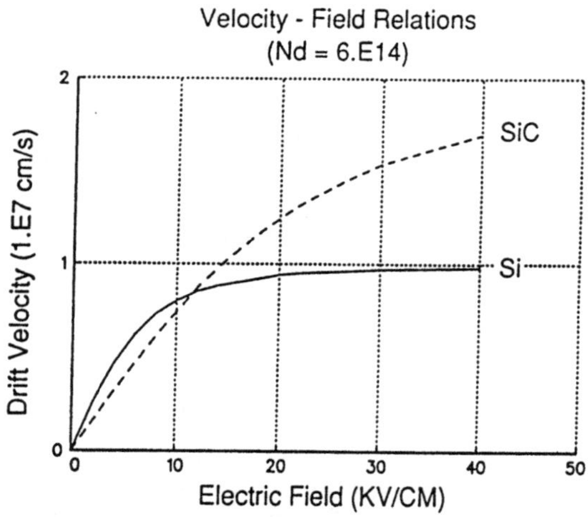

Figure 1. Electron velocity - Electric field relationship for silicon and silicon carbide.

Likewise, the saturation velocity is taken to be 1.0E7 cm/s for both materials. Carrier heating effects are ignored for holes.

The results are presented using the internal gate voltages, not the applied values, which differ by the contact potential. In addition, to avoid confusion, Si and SiC devices are compared at the same current value. This means that quite different external gate-source voltages are applied.

4. Comparison of the 7μm SGSIT for Si and SiC using SIT/1.1

The basic surface gate SIT, or SGSIT, of 7 micrometers gate to gate spacing and described by GTE [6] has been simulated using SIT/1.1. The geometry and other details are in [6] or were obtained directly from GTE. Figures 2 and 3 show the steady-state current-voltage relationships for the structure constructed from Silicon (Figure 2) and Silicon Carbide (Figure 3). Note that for the same internal gate potentials, the SiC device has more current. This is due to the

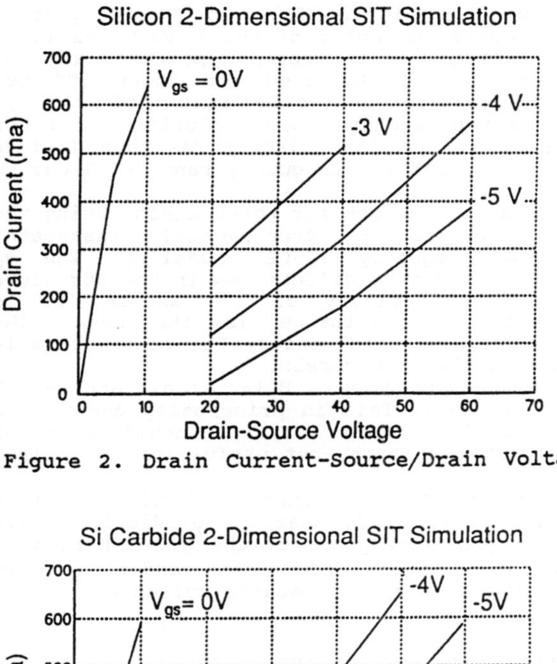

Figure 2. Drain Current-Source/Drain Voltage Curves for the Si SIT.

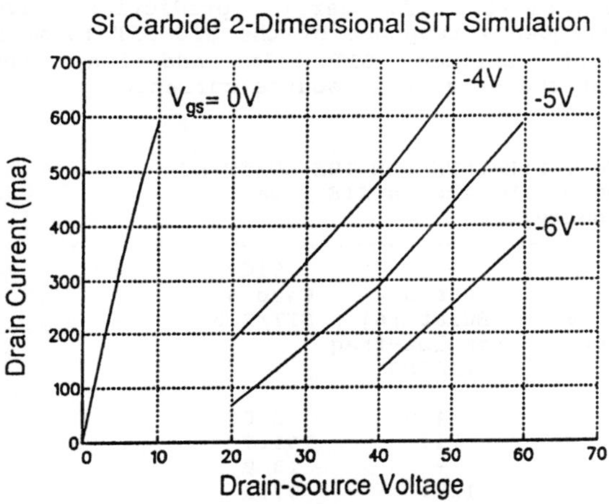

Figure 3. Drain Current-Source/Drain Voltage Curves for the SiC SIT.

higher saturation velocity. Using transient simulations, the transconductance and gate-source capacitance are calculated without difficulty.

A preliminary evaluation of the rf performance was made. When operated at the same current, the SiC structure has similar transconductance but much lower capacitance. This results in much improved F_t, the current cut-off frequency. Thus, one expects better frequency response for the same geometry and material profile. This, obviously, is an interesting result.

However, because holes have not been included in this analysis, one worries that the capacitance may be in error due to the neglect of the p+ depletion region. It is therefore necessary to simulate the structure using SIT/2.3.

5. RF Equivalent Circuits Derived from SIT/2.3

The execution time for SIT/2.3 was longer than for SIT/1.1 due to the inclusion of holes and the greater detail of the implanted gate profile. The results are more realistic than obtained for SIT/1.1.

The procedure used for modeling is as follows. A steady-state condition was found near the operating point of Vds = 60 V and Ids = 300 mA. The gate and drain voltages were stepped separately and the time domain current responses were calculated for about 100 ps. Software for Fourier transformation and calculation of the y-parameters was then used. The y-parameters were evaluated for 1 Ghz to 10 Ghz. A simple small-signal equivalent circuit was then used to extract the average element values for the frequency range of 1 Ghz to 5 Ghz.

Table 1 presents the results for silicon for three cases, using the SIT/2.3 model and reported results [6] from actual s-parameter measurements as well as low frequency PISCES modeling. The g_m difference between case 3 and 1 is due to differences in the structure model, especially detail in the vicinity of the source. It is reassuring that the gate-source capacitance is the same. The capacitance is less affected by these differences. The real device in case 2 has even more capacitance than the models.

A significant difference between results of Bulat et al. and case 1 is that Ft is a factor or two lower. This is principally due to the difference in g_m. The predictions from SIT/2.3 should then be considered optimistic for absolute numbers, but useful for comparison purposes.

Table 1 also shows the results for two-dimensional simulation of the silicon carbide SIT using SIT/2.3. The earlier predictions have been verified, namely, the frequency response of the SiC SIT is much better than the equivalent Silicon SIT. In addition, the transconductance is larger for the same drain source current.

TABLE 1. RF EQUIVALENT CIRCUIT ELEMENTS FOR THE SGSIT AT V_{ds} = 60 V AND I_{ds} = 300 mA (Device is GTE 7-μm SGSIT with 2.0 cm. periphery).

		SILICON			SiC
		Case 1 SIT/2.3	Case 2 Bulat [6] from rf data	Case 3 Bulat [6] from low-freq PISCES	Case 4 SIT/2.3
C_{gs}	(pF)	4.0	6.2	4.0	2.06
g_m	(mS)	155.	92.	75.	201.7
tau	(ps)	45.4	43.	-	13.9
R_{ds}	(ohms)	178.	137.	125.	97.3
F_t	(Ghz)	6.2	2.4	3.0	15.6

6. Conclusion

The two-dimensional simulation of Si and SiC 7-μm SIT devices was successfully accomplished utilizing the Electron-Temperature model. The models show that for the same internal electrode potentials, the SiC SIT has more current. Therefore the SiC device, biased for the same current, has less capacitance and more than twice the current cut-off frequency (F_t). This means that the SiC SIT has potential for higher frequency operation compared with the Si SIT.

References

[1] Nishizawa, J-I, Terasaki, T. and Shibata, J., "Field-Effect Transistor Versus Analog Transistor (Static Induction Transistor)," IEEE Trans. on Elect. Dev., Vol. ED-22, pp. 185, April 1975.

[2] Curtice, W. R., "Intrinsic GaAs MESFET Equivalent Circuit Models Generated from Two-Dimensional Simulations," IEEE Trans. on Computer-Aided Design of Integrated Circuits and Systems, Vol. 8, pp. 395 - 402, April 1989.

[3] Curtice, W. R. and Yun, Y-H, "A Temperature Model for the GaAs MESFET," IEEE Transactions on Electron Devices, Vol. ED-28, pp. 954 - 962, August 1981.

[4] Yoshii, A. et al, "Accurate Modeling for Submicrometer-Gate Si and GaAs MESFET's Using Two-Dimensional Particle Simulation," IEEE Transactions on Electron Devices, Vol. ED-30, pp. 1376 - October 1983.

[5] Sze, S., Physics of Semiconductor Devices, John Wiley and Sons, 1981, D. K. Ferry, Phys. Rev. B, Vol. 12(6), pp. 2361 - 9, June 1975.

[6] Bulat, E. S., Butler, S. J., Regan, R. J., Rock, F. C. and Haugsuaa, P. O., "High Power UHF Transistors," Final Report, AFWAL-TR-86-1140, December 1986.

Optical Properties of 3C-SiC Under High Hydrostatic Pressure

M. Kobayashi[1], *R. Akimoto*[1], *S. Endo*[2], *M. Yamanaka*[3], *M. Shinohara*[3], *and K. Ikoma*[3]

[1]Faculty of Engineering Science, and [2] Research Center for Extreme Materials, Osaka University, Toyonaka, Osaka 560, Japan
[3]Scientific Research Laboratory, Nissan Motor Co., Ltd., 1, Natsusima-cho, Yokosuka 237, Japan

Abstract The pressure coefficient of the energy gap for cubic SiC is determined to be -4.0 meV/GPa at the low pressure limit. The pressure coefficients of the TA(X), LA(X) and LO(X) phonon energies are derived for the first time. In contrast with TA(L) and LA(L) phonons, TA(X) and LA(X) phonons do not exhibit the pressure induced softening. The reported broadening of linewidths of LO(Γ) and TO(Γ) Raman peaks at 10 GPa is found to be ascribable to the non-hydrostatic nature of pressure.

1. Introduction

The pressure coefficient of the optical gap between Γ_V and X_C for cubic SiC was reported to be the smallest in the absolute value among tetrahedrally coordinated semiconductors and insulators ever investigated [1]. In this report, the pressure coefficient of the energy gap is estimated and compared with the theoretical calculation [2]. The pressure dependence of the X-point phonons has not been reported except TO(X) [3] so far. We derive the pressure coefficients of TA(X), LA(X) and LO(X) energies by analyzing the photoluminescence spectra measured at high pressures [1].

3C-SiC is an exceptional substance in a sense that an increase of the LO-TO splitting with pressure was observed [4]. We reexamine the first-order Raman measurement by extending the range of hydrostatic pressure.

2. Experimental

Experimental details for absorption and luminescence measurements under high pressure have been described elsewhere [1]. (111) highly oriented films grown by the chemical vapor deposition (CVD) method were used in Raman measurements in addition to single crystal epilayers with (001) plane. Prior to the measurements, Si substrates were removed. A diamond anvil cell was used to generate hydrostatic pressure. The Raman spectra were excited with the 514.5nm line of an Ar laser. The scattered light in backscattering geometry was focused on a double monochromator (Jobin-Yvon DHR-320S) connected with a spectrograph (Jobin-Yvon HR-640) equipped with a multichannel detector system (Hamamatsu VIM camera and photodiode array PDA).

3. Results and Discussion

In a previous paper [1] we have reported the pressure dependence of absorption spectra. Nonlinear dependence of the optical gap on relative lattice compression or pressure was observed. The mean slope over 15 GPa was -1.9 meV/GPa [1]. In order to compare with the theoretical value, we must determine the initial slope, since the pressure coefficient of the energy gap was calculated for very low pressures [5]. Moreover, one needs the information regarding phonon type involved and its pressure dependence. Figure 1 shows the threshold energy as a function of pressure which was reproduced from Fig. 2 of ref. 1. The best fit curve is drawn in Fig. 1, whose expression is given by

$$E_0 = (2419.6 \pm 0.5) - (3.4 \pm 0.2)P + 0.106P^2, \qquad (1)$$

with E_0 in meV and P in GPa. Strong nonlinearity should be noted. We ascribe the absorption threshold to the indirect transition with TO(X) emission, because a similar threshold feature was observed in diamond and attributed to the transition accompanied with the emission

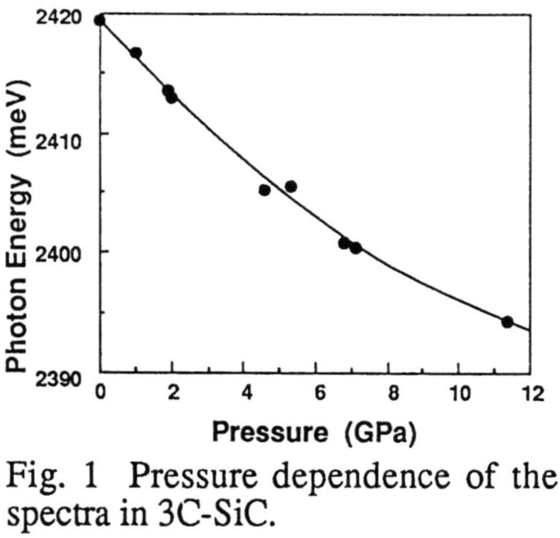

Fig. 1 Pressure dependence of the threshold energy in absorption spectra in 3C-SiC.

of 131 meV phonon [6], which we assign to TO(Δ) phonon according to the data from neutron scattering [7]. Since the mode-Grüneisen parameter for TO(X) in 3C-SiC is reported to be 2.0 [3], we convert it to the pressure derivative of the energy as 0.59 meV/GPa. From the linear coefficient in eq.(1), we derive the pressure coefficient of the indirect energy gap as -(3.4 ± 0.2) - 0.59 = -4.0 ± 0.2 meV/GPa, which is close to the theoretical value of -3.6 meV/GPa [2]. The average of the experimental coefficients for diamond and Si gives -4.2 meV/GPa.

Four luminescence lines observed at low pressures shown in Fig. 3 of ref. 1 were attributed to the recombination radiation of bound excitons trapped by neutral residual donors accompanying emissions of TA, LA, TO and LO phonons at the X point. The energy differences of the TA, LA and LO peaks measured from the TO peak are plotted in Fig. 2 as a function of pressure. LO(X)-TO(X) is almost independent of pressure, while LA(X) - TO(X) and TA(X) - TO(X) show linear dependence as expressed by the solid lines with slopes of -0.36 meV/GPa and -0.43 meV/GPa, respectively. By adding the pressure derivative of TO(X) above described, we derive the pressure coefficients of phonon energies for LO(X), LA(X) and TA(X) as 0.59,

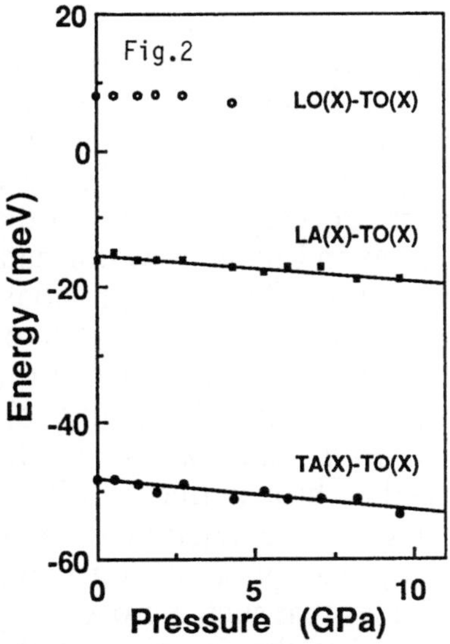

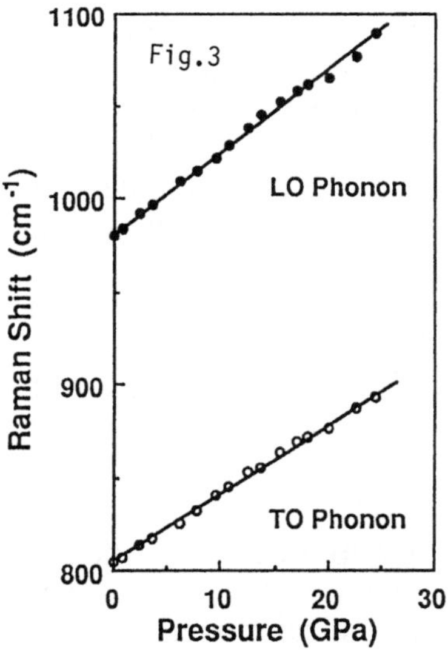

Fig. 2 Energy difference of TA, LA and LO peaks measured from TO peak as a function of pressure taken from Fig. 3 of ref. 1.

Fig. 3 Pressure dependence of LO(Γ) and TO(Γ) phonon energies for (111) highly oriented sample.

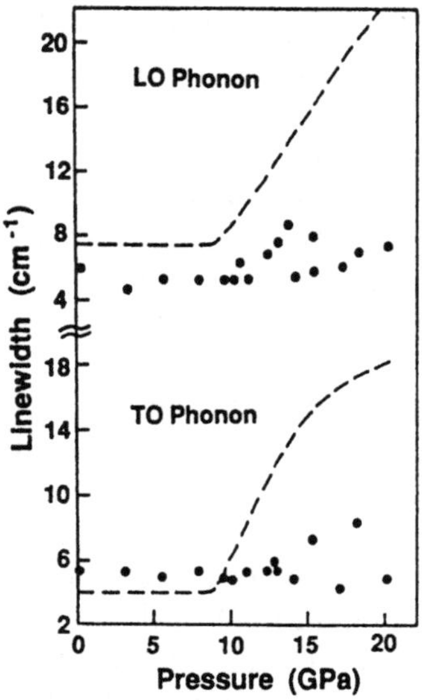

Fig. 4 Pressure dependence of LO(Γ) and TO(Γ) Raman linewidths. Broken lines show data taken from Fig. 9 of ref. 3.

0.23 and 0.16 meV/GPa, respectively. It is to be noted that TA and LA at the X point do not exhibit the pressure induced softening in contrast with the L point [3].

The first-order Raman spectra for (111) highly oriented films have been measured for the first time. Figure 3 shows the pressure dependence of LO(Γ) and TO(Γ) phonon energies, which agrees well with the data obtained for single crystals reported in refs. 4 and 9. We found that the linewidths of Raman spectra are very sensitive to the quality of hydrostatic pressure. Closed circles in Fig. 4 show our data for the pressure dependence of the linewidths for LO and TO phonons. We used a mixture of methanol, ethanol and water in a volume ratio of 16:3:1 as the pressure transmitting medium [8], because it has the wider range of hydrostatic pressure compared with a 4:1 methanol-ethanol mixture which was used by Olego *et al.* [3,4], whose results taken from Fig. 9 of ref. 3 are reproduced by broken lines in Fig. 4. Abrupt increase of the linewidths at 10 GPa was interpreted in terms of an increase in the decay rates of the Γ optical phonons into two acoustical modes [3]. However, Fig. 4 indicates that such a broadening comes not from the intrinsic origins, but from the freezing of the pressure medium.

References

1. M.Kobayashi, M.Yamanaka, and M.Shinohara, J. Phys. Soc. Jpn. **58**, 2673 (1989).
2. K.J.Chang and M.L.Cohen, Phys. Rev. B**35**, 8196 (1987).
3. D.Olego and M.Cardona, Phys. Rev. B**25**, 1151 (1982).
4. D.Olego, M.Cardona and P.Vogl, Phys. Rev. B**25**, 3878 (1982).
5. M.L.Cohen and K.J.Chang, private communication.
6. C.D.Clark, P.J.Dean and P.V.Harris, Proc. Roy. Soc. London A**277**, 312 (1964).
7. J.L.Warren, J.L.Yarnell, G.Dolling and R.A.Cowley, Phys. Rev. **158**, 805 (1967).
8. I.Fujishiro, G.J.Piermarini, S.Block and R.G.Munro, In *High Pressure in Research and Industry*, ed. by C.M.Backman, T.Johannisson and L.Tegner (ISBN, Sweden 1982) Vol.II, p.608.
9. I.V.Aleksandrov, A.F.Goncharov, S.M.Stishov and E.V.Yakovenko, Sov. Phys. JETP Lett. **50**, 127 (1989).

Minority Carrier Diffusion Length in Epitaxially Grown SiC(6H) pn Diodes

M.M. Anikin, A.A. Lebedev, S.N. Pyatko, V.A. Soloviev, and A.M. Strelchuk

A.F. Ioffe Physical Technical Institute, Academy of Sciences, 194021 St. Petersburg, Russia

Only room temperature values for minority carrier (MC) diffusion lengths (DL) have been reported so far for SiC(6H) [1-4]. The basic measurement consisted in recording either the photocurrent value as a function of space charge layer (SCL) width, in a pn junction or a Schottky-barrier structure as it is illuminated with a weakly absorbed intrinsic light (the pC technique) [1-3], or the electron-probe induced current (the EPIC technique) [4]. This paper reports the results of an investigation of MC diffusion length over the 300-800K range in SiC epitaxial layers grown by container-free liquid-phase epitaxy (CFLPE) [5] or open-system sublimation (OSS) [6].

Experimental Structures and Measurement Techniques

Experiments were done on abrupt asymmetric p^+-n-n junctions and Schottky diodes, both grown on n-type substrates. The structures had linear C(V) characteristics in the C^{-2}-V coordinates with the concentrations of uncompensated donors (N_d - N_a) and acceptors (N_a - N_d) of 10^{16} - 3x 10^{17} cm^{-3} and (0.5-1.0) x 10^{19} cm^{-3} respectively. The room temperature values of the cutoff voltage, V_{cutoff}, were 2.5 - 2.6 V for pn and 1.2 - 1.4 V for Schottky diodes. Estimations of the temperature variation of the SCL width were done on the basis of the results obtained in a previous study [7]. Aluminium was employed for the p^- layer ohmic contact. Mesas were etched using the reactive ion-plasma technique as has been described previously [5]. Thermally evaporated Au was utilized for semi-transparent Schotty diodes, which had an area of 3 x 10^{-3} to 2 x 10^{-2} cm^2. The main bulk of the data reported here have been supplied by PC measurements. The EPIC technique was used as a complementary tool for room temperature measurements, following [8]. Both techniques provided excitation of a low level.

In PC technique measurements the light from a mercury or xenon lamp with a prismatic monochromator in front of it could access the respective structure from the top via a window in the aluminium contact or the transparent Au film. The photocurrent was recorded using the simultaneous detection scheme. The measurements were made possible by keeping the load resistance (6 x 10^2 - 10^4 Ω) small compared with the differential resistance of the illuminated structure. Using the simple model with the following limiting conditions which are satisfied in the present case

$$L_\upsilon \gg W; L_n \text{ and } l_p \gg L_n,$$

while ignoring the equilibrium MC concentration in favor of the non-equilibrium

Editors: G.L. Harris, M.G. Spencer, and C.Y. Yang

one, we obtain a linear dependence of the photocurrent, I_p, on the SCL width W, similar to that presented in [9]:

$$I_p = \frac{\beta\phi(1-R)}{L_\upsilon}\left(\exp\frac{-L_p}{L_\upsilon}\right)(L_n + W + L_2)$$

$$L_2 = \frac{L_\upsilon L_p}{L_\upsilon + L_p} \approx L_p \text{ with } L_\upsilon >> L_p$$

where L_υ is the effective length for absorption of light, β is quantum yield, ϕ is the flux of the incident light, R is the light reflection coefficient value at the surface of the sample, l_p is the p-layer depth, L_n and L_p are electron and hole DLs respectively.

Conditions of Uniform Excitation of Minority Carriers

Since L_υ is inversely proportional to temperature, one may expect that the quasi-uniform excitation condition, $L_\upsilon >> W$, L_n, L_p would be violated at elevated temperatures. That this does not happen anywhere throughout the 300-800K temperature range can clearly be seen from the data presented in Figs. 1-3. Fig. la, shows the photocurrent versus SCL width plot normalized to $I_{po} = I_p$ (W_o), where W_o is the SCL width under zero bias, which has been obtained over the temperature range studied for an OSS grown structure in which MC were excited with light having $h\upsilon$ of 3.4 eV. A similar unnormalized, dependence for a CFLPE grown diode is given in Fig. 1b. As can be seen,the linearity of the I_p (W) dependence is maintained up to the highest temperature, thus indicating that the $L_\upsilon >> W$, L_n condition holds over the entire temperature range studied. The reduced angle of inclination seen in the normalized plot of the $I_p(W)/I_{po}$ characteristic is an indication of an increased $(L_p + L_n)$ value. Measurements made at various wavelengths indicate, however, that the I_p value also changes. Fig. 1b, which

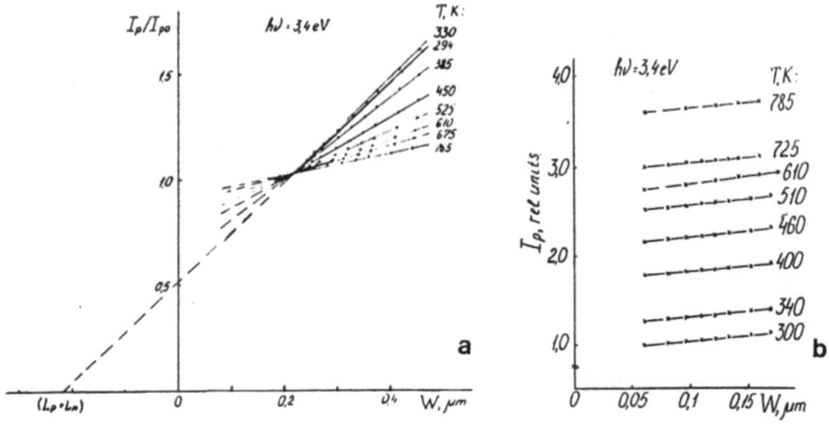

Fig. 1 Photocurrent as a function of SCL width at various temperatures referred to the zero bias photocurrent for (a) OSS pn structure, (b) CFLPE pn structure.

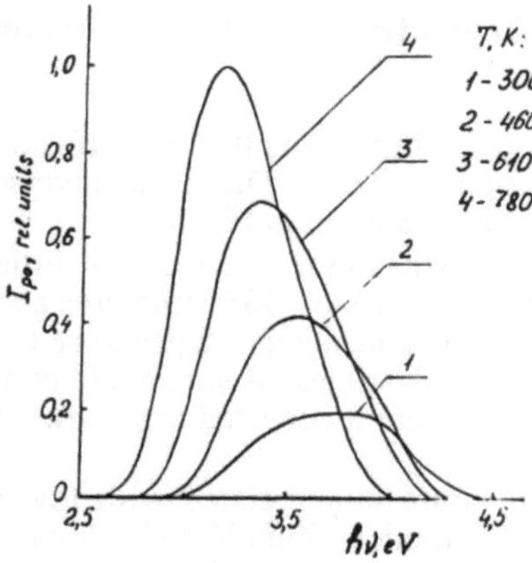

Fig. 2 Spectra of closed-circuit photocurrent for various temperatures.

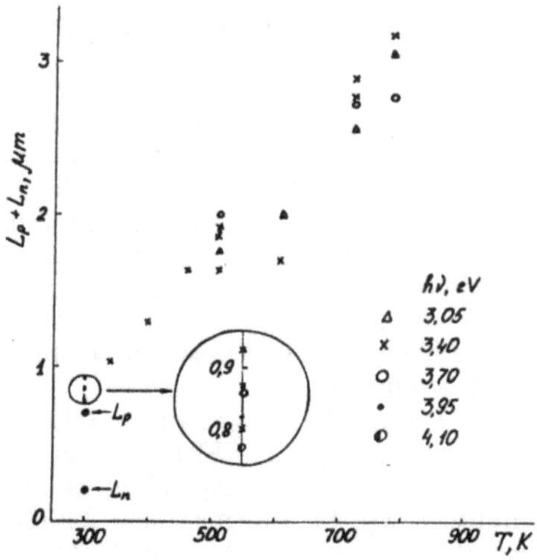

Fig. 3. MC diffusion length ($L_p + L_n$) value obtained for MC excitation light of various wavelengths and at various temperatures.

shows nearly parallel $I_p(W)$ characteristics for different temperatures ($l_p \sim 5$ μm, $h\upsilon = 3.4$ eV), provides clear evidence that the rise in photocurrent is solely due to an increase in the total DL value ($L_p + L_n$).

The temperature variation of the spectral characteristic of photocurrent in pn junctions is shown in Fig. 2. The shape of the spectra is due to the trade-off between two processes that compete with each other as the energy of light is increased, and the absorption coefficient value becomes larger: the rate of generation of pairs in the ($L_n + W + L_p$) region increases while the intensity of the light that reaches the W region is reduced due to the enhanced absorption rate in the p-type layer. The shift of the photocurrent spectra to the long wavelength region with heating is caused by the narrowing of the bandgap, while the integral rise in the photocurrent is due to the increase in the ($L_p + L_n$) value.

Fig. 3 shows that the ($L_p + L_n$) values obtained for MC that have been excited using light of different wavelengths with widely differing absorption coefficients are close to each other at fixed temperatures (within the experimental error), indicating that the excitation at the wavelength of L_p is nearly uniform as well.

Fig. 3 also presents the room temperature values for L_p and L_n measured separately using the EPIC technique. Adding $L_p = 0.70 + 0.05$ μm to $L_n = 0.2$ μm, we obtain the sum which is consistent with the ($L_p + L_n$) value yielded by the PC measurements.

The Temperature and Concentration Dependence of the Diffusion Length

The conclusion that can be drawn from the data of Figs. 1-3 is that the diffusion length in SiC(6H) diodes increases with temperature. The MCDL versus inverse temperature plots for CFLPE and OSS grown diodes are given in Fig. 4.

The plots are seen to exhibit regions where DL displays a decrease or little variation in magnitude: near room temperature in the case of OSS structures and around 600K in the case of CFLPE structures. The same locations happen to be the regions of the largest uncertainty in determination of DL values. A similar effect in $I_p(T)$ dependence was observed in single crystals, where it occurred at somewhat lower temperatures [1].

On the whole, the growth of DL with temperature is associated with lower activation energy in CFLPE diodes than in OSS diodes, 50meV and 120meV respectively. The corresponding values for the activation energy of the respective lifetimes will be 100 meV and 240meV, or slightly larger if corrected for the temperature dependence of the MC mobility.

This is probably correlated with the fact revealed in the analysis of the forward thermo-injection currents in pn structures that recombination can occur via shallow centers [10].

The results of an investigation of DL as a function of N_d - N_a by the C technique in CFLPE and OSS structures at room temperature are presented in Fig. 5. Both Schottky and pn diodes at OSS structures have revealed an approximate correspondence between L_p values, marked in the figure by '0', and the total value of $L_p + L_n$ in a pn junction. For CFPLE structures, where N_d - N_a lies in the range 6 x 10^{16} cm^{-3} to 10^{18} cm^{-3} the value of $L_p + L_n$ has been determined to be 0.4 - 1.5μm. The corresponding value for the OSS structures, which have the N_d - N_a concentration of 6 x 10^{16} to 3 x 10^{18} cm^{-3}, is 0.05 - 0.4 μm. Although it is obvious that both growth technologies are associated with a considerable scatter in

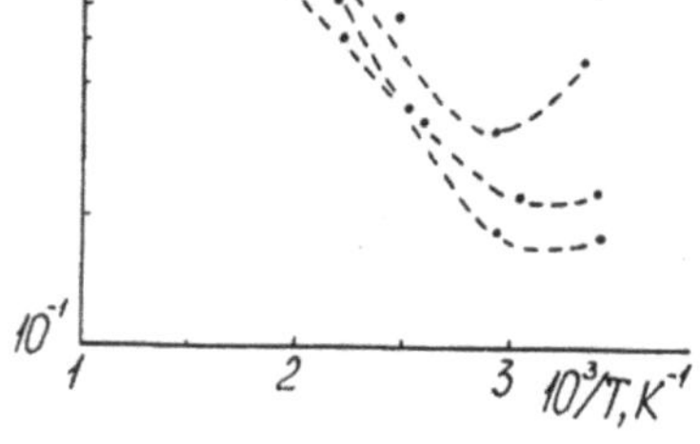

Fig. 4. MC diffusion length in (x) CFLPE and (•) OSS structures ver-versus reverse temperature.

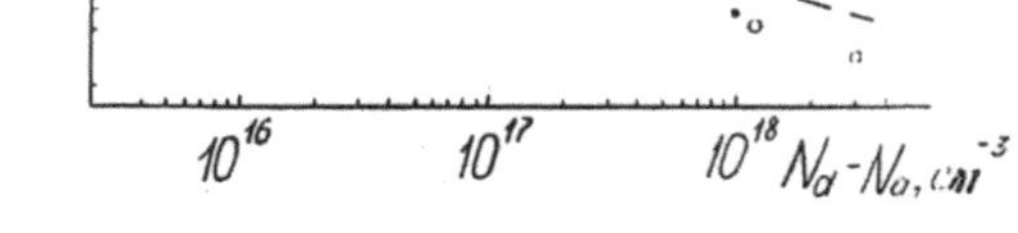

Fig. 5. MC diffusion length ($L_p + L_n$) in (x) CVLPE and (•) OSS structures, and L_p in (o) OSS structures versus (N_d - N_a).

the DL values, it is equally obvious that for the structures it is close to the value reported in [3], where this tendency has been explained by the presence of nitrogen-related centers. Similarily, the larger diffusion length observed at room temperature in structures grown by liquid-phase epitaxy can be accounted for by the ready involvement of centers specific to these structures in the recombination process.

References

1. V.S. Ballandovich, G.N. Violina: In Physics and Technology of Wideband Semiconductors (Leningrad, 1980), p. 293 (in Russian).

2. V.S. Ballandovich, G.N. Violina: J. Semicon. Phys. & Techn. 15, 8(1981), p.1650(in Russian).

3. V.I. Sankin et al: J. Semicon. Phys. & Techn. 16, 7 (1982), p. 1325 (in Russian).

4. M. Ikeda et al.:J. Appl. Phys. 50, 12 (1979), p. 8215.

5. V.A. Dmitirev et al.:J. Theor. Phys. 11, 4 (1985) , p.238 (in Russian).

6. M.M. Anikin et al.;J. Semicon. Phys. & Techn. 22, 2 (1988), p. 298 (in Russian).

7. M.M. Anikin et al.:J. Semicon. Phys. & Techn. 22, 1 (1988), p. 133 (in Russian)

8. K.L. Luke, O van Roos and L. Cheng: J. Appl. Phys. 57, 6 (1985), p. 1978

9. W.W. Gartner: Phys. Rev. 116, 1 (1959), p. 84.

10. M.M. Anikin et al.:J. Semicon. Phys. & Techn. 23, 10 (1989), p. 1813 (in Russian).

Contact Resistance of High-Temperature SiC Metallization

P.G. McMullin[1], *J.A. Spitznagel*[1], *J.R. Szedon*[1], *and J.A. Costello*[2]

[1]Westinghouse Science & Technology Center, 1310 Beulah Road, Pittsburgh, PA 15235, USA

[2]Westinghouse Electronic Systems Group, Baltimore, MD, USA

Abstract: Several candidate metal combinations were examined for suitability as high-temperature contact metallizations for SiC devices of the future. Each metal combination comprised a carbide-forming layer with metals Ta, Ti, and W at the interface with the SiC, and a highly conductive gold layer on the surface. The contact resistance of each metal combination and the sheet resistance of the SiC epilayers on Si substrates were determined utilizing a pattern of rings and dots fabricated in a single photolithography step. The contact resistance varied from about 0.001 to 0.01 ohms-cm^2 depending on the carrier concentration in the SiC layer. The thermal stability of each metal was examined by tracking the measured resistance after repeated thermal stress cycles at 600 to 800°C in a forming gas ambient. A metal system comprising Ti, W, and Au provided the most stable contact system.

1. Introduction

Silicon carbide has long been recognized as a candidate material for semiconductor devices operating at high frequency, high temperature, or high power, due to its greater electron saturation velocity, higher bandgap, higher dielectric strength, and greater thermal conductivity compared to the more familiar semiconductors silicon and gallium arsenide. Realization of these advantages has been made difficult by lack of large single-crystal substrates suitable for conventional device processing. Recently, single polytype boules have been produced at several laboratories worldwide, based on the vapor transport method of Tairov and Tsvetkov [1]. Attention may now turn to the unit processes involved in fabricating high-performance devices. MOSFETs functioning at 650°C have been reported by Palmour et al. [2]. Such devices require formation of gate oxides by thermal oxidation, polysilicon deposition, ion implantation for source and drain formation, and metallization for ohmic contacts. This paper focuses on the formation of low-resistance metal contacts that are stable during high-temperature operation.

Relatively little information is available about the electrical performance of contact metals for SiC. Several combinations of metals and silicides deposited onto 3C SiC were evaluated by Edmond et al. [3]. No information was presented, however, concerning the stability of the contacts during extended exposure to high temperatures. Studies of the interface reactions of SiC with carbide-forming metals were reported by Zeller et al. [4]. Formation of interface carbide compounds was found to promote adhesion, but interdiffusion of Ta through Au

Editors: G.L. Harris, M.G. Spencer, and C.Y. Yang

capping layers caused delamination. Reaction of Ti deposited directly onto 3C SiC and onto surfaces modified to provide excess carbon were reported by Bellina et al. [5]. Enhanced stability under long-term thermal stress was observed when the Ti metal was completely reacted with excess carbon [6]. The structural studies, however, do not contain electrical characterization data.

For high-performance devices, it is essential to achieve low contact resistance and interconnect resistance so that higher frequency of operation and lower power dissipation can be maintained. This may give rise to conflicting requirements. For firm adhesion of the metal, chemical bonding at the surface is helpful. Some diffusion of the deposited metal into the substrate decreases contact resistance if the metal acts as a dopant to increase the carrier concentration and reduce the surface potential barrier. Reaction and diffusion, however, are undesirable for long-term stability at elevated operating temperatures where such processes are thermally accelerated. The lowest interconnect resistance between devices on the same substrate, and the best bond to an external package, may be obtained by a highly conductive metal like gold. Gold is known to be a rapid diffuser and so it must be separated from the interface with the substrate by a diffusion barrier. Therefore, composite metallizations comprising a reactive metal, diffusion barrier, and conductive metal are candidates to combine low contact resistance and thermal stability. In this paper, several such combinations are investigated.

2. Experimental Procedure

In ohmic contacts between semiconductors and metals, the flow of current from the metal conductor to a thin doped layer in the semiconductor is governed by the transfer length, L_t. Beyond about one transfer length from the edge of the contact area, most of the current flows in the metal layer, and a further increase in the dimension of the contact area does not proportionally decrease the contact resistance. The transfer length depends on both the sheet resistance, R_s, of the semiconductor and the contact resistance, R_c, of the metal-semiconductor interface, according to

$$L_t^2 = Rc/Rs.$$

A metal pattern was designed to permit the separate determination of contact resistance and sheet resistance. The pattern is shown in Figure 1 to consist of an arrangement of dots and rings in a metallized field. The resistance measured by probing adjacent electrodes includes a term due to the sheet resistance of the semiconductor in the unmetallized gap, and two contact resistances due to the electrodes. The sheet resistance term depends only on R_s and the geometry, while the contact resistances depend on R_c, R_s, and the geometry. By plotting a normalized resistance against a strictly geometrical term, as shown in Figure 2, the sheet resistance can be found from the slope of the line best fitting the data, and the transfer length can be found

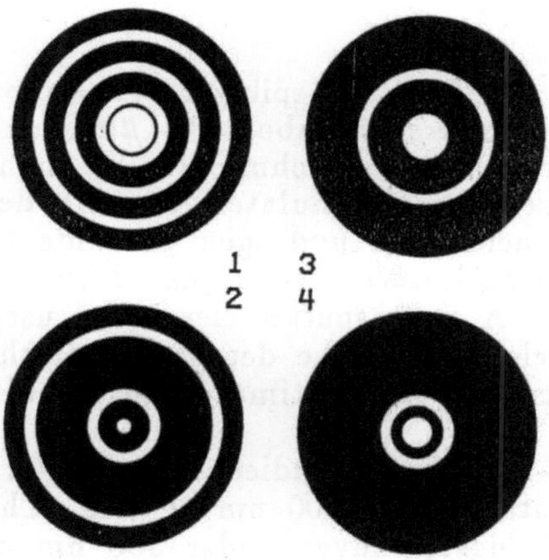

Figure 1. Metal dot and ring pattern for determination of contact resistance and surface layer sheet resistance, requiring only one lithography step.

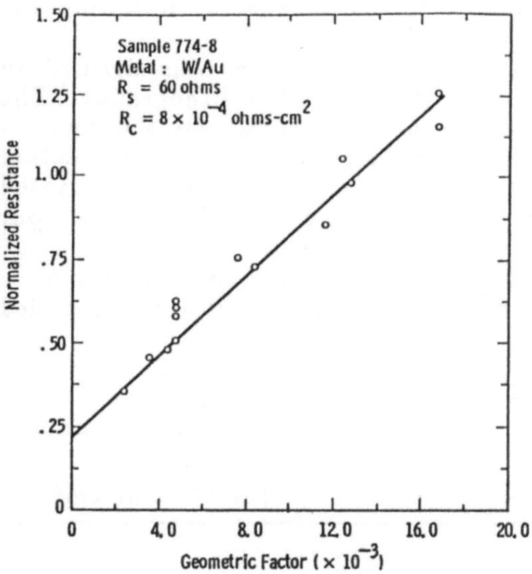

Figure 2. Graphical analysis of measurement data from the ring pattern. The line shows the least-squares best fit to the data.

from the intercept. An analysis of the contact and sheet resistance in circular geometry has been given by Cohen and Gildenblatt [7].

The major advantage of this approach to the measurement of the sheet resistance and the contact resistance is that only a single mask is needed to fabricate the metal pattern. Further, there is no need for a mesa etch step to reduce the edge effects of the rectangular transmission line pattern. This makes the approach suitable for the small and irregular samples encountered in SiC work.

2.1 Sample Preparation

The SiC samples used for this study were 3C SiC epilayers grown on Si substrates at NASA Lewis. The SiC layers are about 10 μm thick and are n-type with resistivities of about 0.1 to 2 ohm-cm. The samples were cleaned with solvents to remove accumulated organic deposits and immersed in hot sulfuric acid with hydrogen peroxide to complete the surface cleaning. The samples were then placed in the vacuum chamber of the sputtering rig. A light sputter clean was used at the beginning of the deposition cycle. During the depositions of the thin metal layers, the sample temperature was estimated to remain below 100°C.

Three metal composite structures have been studied. The Ta/Au metallization comprises 100 nm of tantalum and 300 nm of gold. The W/Au structure contains a 100 nm tungsten layer under 300 nm of gold, and the Ti/W/Au structure has 30 nm of titanium, 100 nm of tungsten, and 300 nm of gold. Sample metallization is identified in Table 1. In addition, one sample was prepared with e-beam deposition of a carbon layer 20 nm thick prior to the Ti/W/Au metallization described above. This sample provides a carbon-rich surface layer to enhance the stability of the interfacial carbide layer, following the results of Bellina [6].

After metal deposition, the dot and ring pattern was defined in negative resist on the surface using standard photolithography. The gold was removed from the surface with aqua regia, and the remaining metals were etched away with hydrofluoric acid. Stripping solution was used to remove the negative resist, and the samples were rinsed in deionized water. All samples were annealed at 800°C for one hour in forming gas to alloy the contacts.

Thermal stress cycles were carried out by placing the samples on a quartz carrier boat and sliding them into a furnace with an extended

Table 1 Sheet and Contact Resistance of Metal Contacts Deposited onto 3C SiC Epilayers on Si Substrates

Sample	Metal	R_s (ohms)	R_c (1×10^{-4} ohm-cm^2)
778-8	Ta/Au	100	20
723-4	W/Au	170	11
774-8	W/Au	60	8
788-8	W/Au	1700	100
720-6	Ti/W/Au	440	10
723-4	Ti/W/Au	200	80
788-2	Ti/W/Au	1900	120
779-9	C/Ti/W/Au	860	90

hot zone at the stress temperature. All thermal stress cycles took place at 600°C in a forming gas ambient atmosphere.

2.2 Measurement Procedure

Newly metallized samples were placed on a probe station and electrically contacted by means of a elastomer ring probe card with 20 contact fingers in a 2x10 pattern with 160 μm spacing between centers. The ring pattern was designed to be compatible with this probe arrangement, and all measurements on a single ring of the patterns can be completed without moving the probes. A curve tracer was used to determine whether the sample characteristics were ohmic at a conduction level of 5 to 10 mA. Resistance data for ohmic samples were taken on the curve tracer. For samples which were not exactly symmetrical about zero, the lower indicated resistance was recorded.

3. Results and Discussion

Each sample was probe tested after metallization to determine if ohmic contact had been obtained. If the sample was ohmic, the sheet resistance and contact resistance were found by graphic analysis as shown in Figure 2. The measurement results are shown in Table 1. If the sample was not ohmic, the resistance measurements were deferred to a later time after thermal stress cycles which tended to form the contacts. The samples with higher SiC layer resistance, 788-8 and 788-2, were non-ohmic as deposited but became nearly symmetric after a short thermal stress cycle at 600°C. The sample that was prepared with the carbon layer at the interface, 779-9, had a rectifying characteristic as deposited and remained non-ohmic through the thermal stress cycles reported here. The resistance values for 779-9 in Table 1 were taken from the lower resistive branch after 13 hours of 600°C thermal stress.

Selected samples from the list in Table 1 were subjected to thermal stress cycles to determine the stability of the contact resistance during high-temperature exposure. The results are shown in Figure 3. In the figure, each data point represents the average of all resistance measurements made on each sample. The resistances are normalized to the initial value, so that the figure shows only changes in the resistance. The initial values are recorded in Table 1. Continuous lines were added to the figure to guide the eye to the associated data.

As shown in Figure 3, the resistance of the contacts on the Ta/Au sample increased steadily with time under the influence of thermal stress at 600°C. The Ti/W/Au samples also show a substantial increase in resistance during the later stage of the thermal stress cycles. In contrast, the data for the W/Au samples remain constant during the thermal stress cycles.

A preliminary interpretation of this data suggests that the presence of the tungsten diffusion barrier metal is crucial for thermal stability.

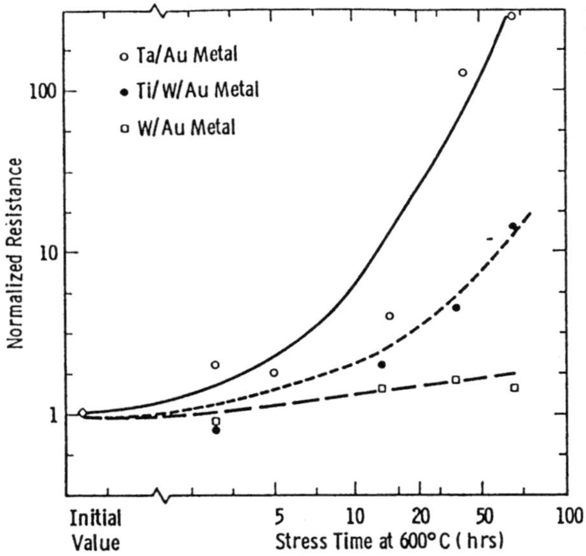

Figure 3. Normalized resistance of contact metal combinations during repeated thermal stress cycles at 600°C.

Both the W/Au and Ti/W/Au composite metal structures exhibited greater stability in thermal stressing than the Ta/Au structure. Since tantalum and gold readily form alloys, while tungsten and gold do not form any alloys, tantalum cannot serve as a barrier to gold diffusion. The lesser stability of the samples with Ti layers at the interface may be due to a continued interface reaction between the deposited Ti and the SiC surface. This would indicate the need for a careful adjustment of the thickness of a deposited carbide-forming material to obtain adequate adhesion without limiting stability by excessive interface reaction.

4. Acknowledgments

The SiC epitaxial samples from NASA Lewis were provided by W. J. Choyke. The expert technical assistance of J. B. McNally for sample processing and measurement and B. A. Blankenship for metal depositions was essential in this work.

5. References

1. Yu. M. Tairov and V. F. Tsvetkov: J. Cryst. Gr. 43, 209 (978).
2. J. W. Palmour, H. S. Kong, and R. F. Davis: J. Appl. Phys. 64, 2168 (1988).
3. J. A. Edmond, J. Ryu, J. T. Glass, and R. F. Davis: J. Elchem. Soc. 135, 359 (1988).
4. M. V. Zeller, J. Bellina, N. Saha, J. Filar, J. Hargreaves, and H. Will: Mat. Res. Soc. Symp. 97, 283 (1987).

5. J. J. Bellina, J. Ferrante, and M. V. Zeller: J. Vac. Sci. Tech. A4, 1692 (1986).
6. J. J. Bellina and M. V. Zeller: Mat. Res. Soc. Symp. 97, 265 (1987).
7. S. S. Cohen and G. S. Gildenblatt: in VLSI Electronics and Microstructure Science, Vol. 13 (Academic Press, New York, 1986), p. 111.

Breakdown in Silicon Carbide pn Junctions

M.M. Anikin, M.E. Levinstein, A.M. Strelchuk, and A.L. Syrkin

A.F. Ioffe Physical Technical Institute, Academy of Sciences,
194021 St. Petersburg, Russia

Silicon carbide (SiC) is a candidate material for avalanche breakdown devices such as IMPATT diodes and avalanche limiters because of its ability to withstand high breakdown fields, its high thermal stability and conductivity.

The tendency for the breakdown voltage to decrease with increasing temperature when the electric field is parallel to the hexagonal C axis was revealed already by the early investigators of SiC(6H) - based pn junctions, who thus established (1) that the temperature coefficient for breakdown voltage (TCBV) given as $\beta = 1U_i :dU_i/dt$ may also assume the negative sign in this material. More recent studies (2-4) further established that the character of breakdown in hexagonal SiC is dependent on the direction of the electric field in the pn structure; with the sign of TCBV being positive just as in other semiconductors at E⊥C and negative at E ∥ C. The authors accounted for the negative TCBV in the terms of the natural superlattice at E C. In practice, this makes the successful operation of avalanche breakdown devices under E ∥ C highly problematic in view of the low thermal resistance of the pn junction to self-generated heat. On the other hand, the choice of the E ∥ C orientation is dictated by the ease with which the devices grown on the Lely substrates can be produced on a large scale.

Consequently, a systematic investigation of the breakdown voltage as a function of temperature was undertaken in SiC(6H/4H) - based pn junctions using p-n-n$^+$ structures that have been fabricated by two essentially different techniques:

A. The n-type layer with the impurity concentration of N_d-N_a = (1-3) x 10^{17}cm^{-3} was grown to a thickness of 3 to 5μm on the n$^+$ substrate having the N_d -N_a concentration of 5 x 10^{18}cm^{-3} by sublimation in an open growth system. The Al-doped p$^+$ layer was fabricated by either ion implantation (Al) or sublimation (A2) and had the respective thickness of 0.1μm and 1.0μm, with the values of N_a -N_d concentration being respectively 2 x 10^{19}cm^{-3} and (5-7) x 10^{18}cm^{-3}.

B. The n-type layer approximately 5μm thick with the N_d -N_a concentration value of 1 x 10^{18}cm^{-3} and the p-type layer of the same thickness having the N_a-N_d concentration in the range 10^{18}cm^{-3} to 5 x 10^{19}cm^{-3} were grown by container-free liquid-phase epitaxy (CFLPE).

In the case of E ∥ C, the substrates were Lely grown (0001)single crystals, while for the E⊥C case we employed (1120) single crystals grown by a method described previously in (5). Mesas were etched using the reactive ion-plasma technique (6).

Investigations on the Al specimens were done in the DC and pulsed modes, as has been described previously in (7). The DC measurements indicated that the breakdown is of the microplasmatic character and that the TCBV can be positive, Fig. 1, as well as negative.

The respective β values were found to be (2-3) x 10^{-4} K^{-1} and 1 x 10^{-1} K^{-1}

Short (70 ns) pulse measurements made on the same samples that displayed negative TCBV for the DC-induced breakdown yielded negative TCBV with β = (2-3) x 10^{-4} K^{-1} for nearly the same values of breakdown voltage, Fig.2.

Editors: G.L. Harris, M.G. Spencer, and C.Y. Yang

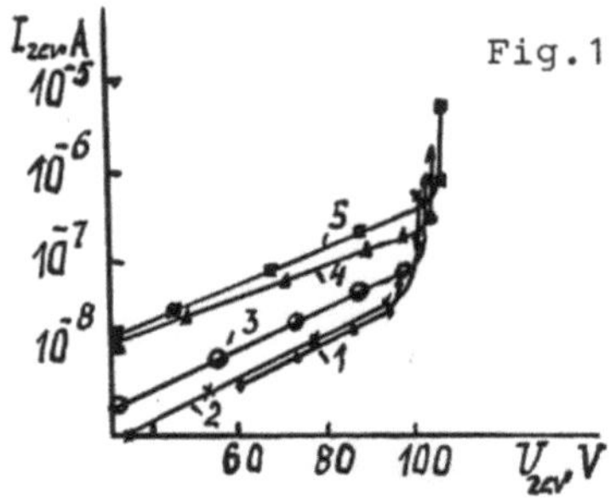

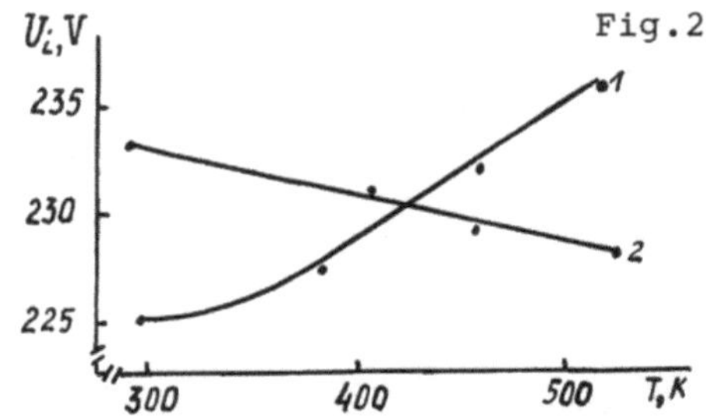

Fig.1 I/V characteristics of Al diodes under reverse bias at 293 (1), 345 (2), 380 (3), 415 (4), and 430 (5) K

Fig.2 Voltage v. temperature plots for pulsed current (1) and DC (2) induced breakdowns

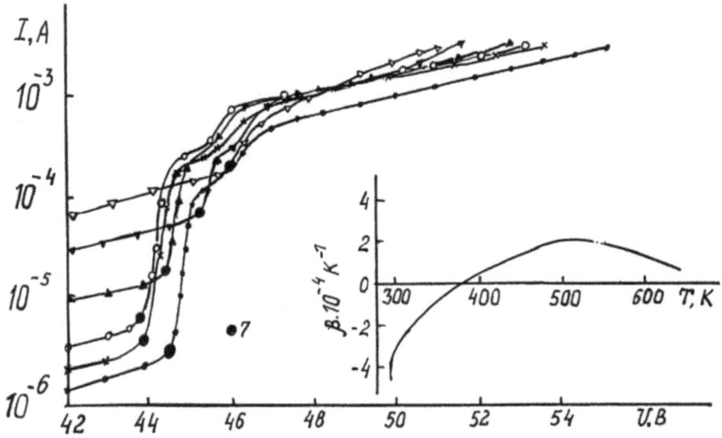

Fig.3. Reverse I/V characteristics of SiC(6H) pn diodes as measured in the breakdown region at E ǁ C for 293 (1), 335 (2), 390 (3), 480 (4), 590 (5) and 670 (6) K, large closed circles mark the onset of microplasma noises. Insert: the corresponding temperature dependence.

The temperature dependences of breakdown and initial microplasma ignition voltage were investigated on the A2 and B specimens. Typical voltage versus current characteristics obtained are shown in Fig.3, where large closed circles mark the earliest occurrence of the microplasma noise signal. The amplitude of microplasma pulses, ΔI, is almost independent of the voltage and has a room temperature value of about 10^{-4} A (curve 1). As the voltage is increased, the rate and duration of microplasma pulses also increase (8), which accounts for the observed rapid increase in the current. Curve 1 shows that there is a steep rise in the differential resistance once the 1 value reaches 10^{-4} A, at which point the microplasma noise ceases, and the first stable microplasma occurs in the sample, clearly seen through the substrate in the microscope.

It is evident that the stability range for the I/V characteristics is about 1 V. As the voltage is further increased, the differential resistance is seen to decline, approaching a point at which the noise signal from the oncoming second microplasma event is recorded. At $I > 5 \times 10^{-4}$A, there are already two stable microplasma strings in the sample. The further rise in termperature, as can be seen in Fig.3, causes the microplasma ignition voltage first to drop

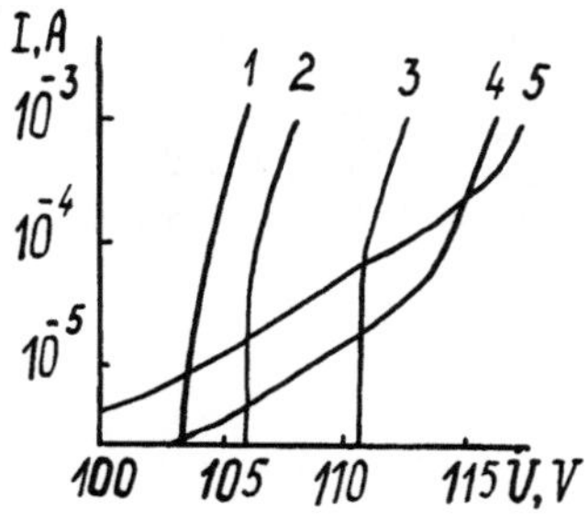

Fig.4 I/V characteristics of SiC pn diodes, taken with E⊥C at 293(1), 405(2), 510(3), 600(4) and 640(5)K

(curves 1-3) and then to rise monotonically (curves 4-6). It, thus, follows that the TCBV of the first microplasma's breakdown and ignition is sign reversible.

The A1 and B specimens were also used for investigation of the breakdown voltage as a function of the microplasma breakdown current. It was found that TCBV is a current dependent characteristic for most of the samples studied (9).

Small currents were found to be associated with the pattern of behavior similar to that presented in Fig.3 in some of the diodes, and with the negative TCBV over the entire temperature range studied in others. For large currents, the TCBV changed sign from the negative to the positive at 450K.

Figure 4 shows the reverse I/V characteristics obtained in the pn junctions for the E⊥C orientation. In complete agreement with the conclusions reached in (2-4), the TCBV is positive, and appears to be independent of the current.

However, the results obtained in this work for the E C case are in disagreement with the hypothesis put forward in (2-4) if, on the strength of the good evidence obtained for the best studied junctions to date- the Si-based pn junctions, it is assumed that the homogeneous breakdown follows closely the microplasma breakdown in the I/V characteristics. A simple possible explanation for our results lies in the influence exerted on the TCBV by the deep centers present in SiC. The change of charge state of the deep centers lying in the space charge region can readily account not only for the different I/V characteristics we obtained for the breakdowns induced by pulsed as opposed to direct currents, but also for such features of the TCBV as its dependence on the current and change of sign with changing voltage for primary microplasma ignition.

The question to which the deep level model does not provide a satisfactory answer is why negative TCBV never occurs with the E⊥C orientation. We believe that the difference in the I/V characteristics between pulsed and direct current induced breakdown, and the fact that TCBV is essentially a sign-reversible parameter are arguments in favor of impurities rather than the superlattice as a basis for the negative sign exhibited by TCBV in SiC at E || C.

References

1. G.Gramber, M.Koniger: J. Sol. St. Electr. 15, 3 (1972), p.285.
2. A.O.Konstantinov, D.P.Litvin, V.I.Sankin: J.Theor. Phys. Lett. 7, 21(1981), p.1335 (in Russian).
3. A.O.Konstantinov: J.J. Semicon. Phys. & Techn. 17, 12 (1983), p.2124 (in Russian).
4. A.I.Dmitriev, A.O.Konstantionov, D.P.Litvin, V.I.Sankin: J.Semicon.Phys. & Techn. 17, 6 (1983), p.1093 (in Russian).
5. Y.M. Tairov, V.F.Tsvetkov: J. Cryst. Growth 52, (1978), p.830.
6. M.M. Anikin, A.A. Lebedev, I.V. Popov, A.M.Strelchuk, A.V.Suvorov, A.L.Syrkin, V.E. Chelnokov: J.Semicon. Phys. & Techn. 20, 5 (1986), p. 844 (in Russian).

7. M.M. Anikin, S.N. Vainshtein, M.E. Levinstein, A.M. Strelchuk, A.L. Syrkin: J. Theor. Phys. Lett. 14, 6 (1988), p. 545 (in Russian).
8. B.S. Kondratiev, I.V. Popov: In Proc. 3rd Int. Conf. (ICACSC'90).
9. M.M. Anikin, M.E. Levinstein, I.V. Popov, V.P. Rastegaev, A.M. Strelchuk, A.L. Syrkin: J.Semicon. Phys. & Techn. 22, 9 (1988), p. 1574 (in Russian).

SiC p–n Junction Photosensitivity in the Long-Wavelength Region

E.Ye. Violin, Yu.M. Tairov, and I.Ye. Jaremenko

V.I. Ulyanov (Lenin) Electrical Engineering Institute,
197022 St. Petersburg, Russia

Abstract. From the analysis of known experimental results, via. the N and B absorption spectra in SiC-6H as well as some properties of p-n junctions produced by boron diffusion, it can be concluded that such p-n junctions should have desirable characteristics for applications in the spectral range $0.5 < E < 2.0$ eV and also, in a number of cases, at $E < 0.5$ eV.

The experimental investigations performed have revealed that SiC < B > diffuses p-n junctions having a p-n-n structure, where n is the high-ohmic compensated region, are capable of operating in the photocurrent injection gain-mode. This mode will drastically increase the photosensitivity of such structures when compared to that of photoresistors. The manner of applying the supply voltage is shown to exert a significant effect on spectral characteristics. The behavior of the pulse-operated SiC < B > p-n-n junctions as well as some technological problems of this photosensitive structure are considered and have been studied in detail.

Part IV

Techniques for Thin Film Growth

Low Pressure Growth of Silicon Carbide on Silicon in a Vertical Reactor

K.G. Irvine[1], *M.G. Spencer*[1], *and M. Aluko*[2]

[1]Materials Science Research Center of Excellence and Department of Electrical Engineering, and [2]Department of Chemical Engineering, School of Engineering, Howard University, Washington, DC 20059, USA

Abstract. We report on the growth of Silicon Carbide on Silicon using the Silane (SiH_4), Propane (C_3H_8), and Hydrogen system in a vertical reactor. Computer simulations predict that for the vertical geometry investigated high flow velocities are necessary to prevent forced convection. Experimental results verify computer prediction and under optimal conditions highly uniform growths are achieved.

Introduction

Silicon Carbide (SiC) is rapidly developing as a viable semiconductor technology. Chemical Vapor Deposition (CVD) is positioned to play a role in this emerging technology. To date there have been limited reports of the low pressure growth of SiC and no reports of growth in the vertical geometry. Low pressure as well as the vertical geometries have been applied with good results in the growth of Silicon (Si) and Gallium Arsenide (GaAs) from the vapor phase. In this paper we report on results obtained with a Low Pressure Chemical Vapor Deposition (LPCVD) Vertical Reactor using Silane (SiH_4) and Propane (C_3H_8) as the reactant gases. The results are reported for SiC on Si system but our ultimate goal is CVD of SiC on SiC substrates.

Experimental

Figure 1 shows the physical geometry of our reactor. This reactor (manufactured by Emcore Corp.) has a stainless steel chamber. The stainless steel chamber allows for the use of standard ultra high vacuum (UHV) components and easy attachment of analytical instrumentation. The substrate is resistively heated by a graphite filament. The filament is located just below a graphite susceptor which holds the substrate. Gases are injected into the chamber at two points. One point is the geometric center of the reactor, where the reactant gases together with hydrogen are injected. The second injection point is called the shroud. At the shroud injection port the hydrogen is uniformly distributed at the top of the chamber before being injected, the gas therefore forms a uniform blanket from a radius of about .625 inches from the center of the the reactor to the chamber wall. This shroud of gas serves to stabilize the gas flow pattern inside the reactor as well as isolate the substrate from the chamber walls. The total inner diameter of the stainless steel chamber is about 8 inches. Studies were undertaken to determine the optimal conditions of flow for this reactor. The growth sequence for deposition of SiC films on Si is similar to those previously reported[1-3]. Optical and electron micrographs were taken to determine surface morphology. Electrical measurements were performed in order to access material quality.

Reactor Simulation

Although this reactor geometry has been used for the growth of GaAs, the conditions for CVD of SiC are quite different. In order to determine the optimal operating conditions for our reactor computer simulations were performed. Our objectives in this study were to find the operating conditions which minimized convention and optimized our temperature profile. The computer program simultaneously solved the continuity equation, momentum balance and energy balance equations. We determined what the optimal operating conditions of the reactor should be by computer simulation and these are shown in table 1. We found that by increasing the neck flow we were able to "compress" the temperature profile and provide a steep gradient

Editors: G.L. Harris, M.G. Spencer, and C.Y. Yang

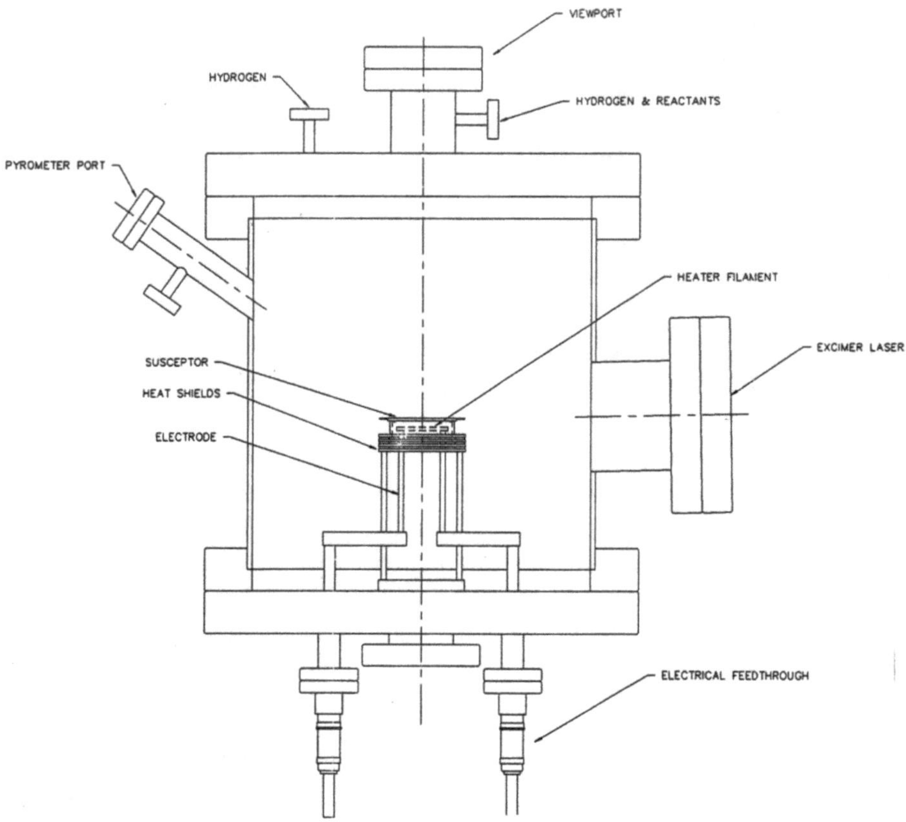

Figure 1. Vertical Reactor Geometry

Table 1. Optimal Operating Conditions for CVD Reactor

Neck Flow (SLPM)	Temperature °K	Pressure (Torr)	Optimal Shroud Flow (SLPM)
5	1623	20	21
5	1623	76	29

at the substrate vapor interface. This is desirable since a steep gradient will act to inhibit pre-nucleation in the gas phase.

Reactor Growths

In order to compare the experimental prediction with actual results we grew several samples of SiC on Si under various flow conditions (figure 2). It was observed that at a growth pressure of one atmosphere it was impossible to obtain smooth surfaces. It is also noted that as

T= 1300 C, P= 20 T, Si/C= .3, Magnification = 413

T=1300 C, P= 76T, Si/C= .3, Magnification =413

T- 1300 C, P- 200 T, Si/C-.3, Magnification -413

T=1300 C, P= 700 T, Si/C= .37, Magnification =413

T= 1350 C, P= 20 T, Si/C= .3, Magnification =413,

T=1350 C, P= 76 T, Si/C= .3, Magnification = 413

T=1350 C, P= 200 T, Si/C= .37, Magnification =413

Figure 2. Growth Morphologies of SiC films grown under different conditions

the pressure is decreased (for a constant growth temperature) the surface morphology improves considerably. However for all growths done without the shroud it was impossible to observe interference fringes on the surface of the wafer. The presence of these fringes requires a surface roughness of less than 250 angstroms[3]. Further on the stainless steel inner wall of the reactor we could observe the presence of deposits in the shape of swirls. These deposits indicated the presence of severe convection. Samples were grown under conditions which were close to optimum as predicted by the computer simulation. One such sample was 23 microns thick and showed the presence of interference fringes indicating a surface roughness of less than 250 angstroms. By counting interference fringes we determined that for this sample our material was uniform to about 7% over a 12 cm^2 area.

Electrical Properties of the Films

Electrical properties of the growths were measured using CV and Hall techniques. Films grown under optimal conditions had carrier concentrations of less than $2x10^{16}cm^{-3}$ at room temperature and hall mobilities of greater than 250 cm^2/v-sec. These values would be considered normal for SiC on Si growths with Silane and Propane at 1350C.

Summary

We have determined the optimal conditions for the growth of SiC on Si in a LPCVD vertical reactor, high flow conditions in the neck and shroud are necessary to produce good quality material. Under optimal conditions films had surface roughness of less than 250 angstroms. Electrical measurements showed carrier concentrations less that $2x10^{16}cm^{-3}$ and mobilities greater that 250 cm^2/v-sec.

Acknowledgements

We gratefully acknowledge the support of the National Science Foundation under MSRCE Grant #NSF-RII8947 537 and NASA Grant #NAG-3887. One of us (M.G. Spencer) acknowledges support under Presidential Young Investigator Program.

References

1. S.Nishino, J.A. Powell and H.A. Will Appl. Phys. Lett. **42** 460 (1983)

2. P. Liaw and R. F. Davis J. Electrochemical Society **132**, 642 (1985)

3. J. A. Powell, L. Matus and M. Kuczmarski J. Electrochemical Society **134** 1559 (1987)

Selective Growth of Cubic SiC on Si by Chemical Vapor Deposition

S. Nishino, K. Takahashi, H. Ishida, and J. Saraie

Department of Electronics and Information Science, Faculty of Engineering and Design, Kyoto Institute of Technology, Matsugasaki, Sakyo-ku, Kyoto 606, Japan

Cubic SiC films were grown on Si substrates by chemical vapor deposition (CVD) using Si_2H_6 and C_2H_2 at normal pressure at a substrate temperature of 900-1150°C. The activation energy for SiC formation was 34.7 kcal/mol. A single crystal of cubic SiC was obtained at a substrate temperature of 1150 °C by adding a small amount of HCl to the reaction gases without employing the buffer layer. Selective growth was achieved in this condition.

1. Introduction

Silicon carbide is a promising material for electronic and optical devices under severe conditions, such as high temperature and high radiation environments, because of its large energy gap, temperature stability, and chemical inertness. Single crystal growth of cubic SiC on Si was difficult for a long time because of large lattice mismatch (20 %) between SiC and Si. Recently, techniques for epitaxial growth of SiC have been established and many device characteristics have been reported [1-7]. The temperature of crystal growth, however, is high, being around 1350 °C. At low temperatures epitaxial growth is necessary to minimize autodoping, the solid-solid diffusion effect and dislocation density [8].

Furumura et al.[9] reported epitaxial growth of cubic SiC on a Si substrate at a substrate temperature of 1000 °C using the $SiHCl_3$ + C_3H_8 + H_2 system in a low pressure CVD. Ohoshita et al.[10] reported selective growth of cubic SiC on Si a substrate at a substrate temperature of 1000 °C using the SiH_2Cl_2 + C_3H_8 + H_2 + HCl system in a low pressure CVD. However, the grown layer is highly oriented polycrystalline SiC. Judging from these two papers, HCl addition seems to be one possibility for growing SiC at a low temperature. We previously reported low temperature growth of cubic SiC using the Si_2H_6 +C_2H_2 + H_2 + HCl system [11]. In this paper, we report successful selective area CVD growth of cubic SiC films on a Si substrate using SiO_2 as the insulating mask. In this CVD growth, single crystalline epitaxial films are formed only on the exposed area of the Si substrate. Areas covered by SiO_2 film result in the growth of polycrystalline films or no deposition.

Editors: G.L. Harris, M.G. Spencer, and C.Y. Yang

2. Experimental

Cubic SiC films were deposited on Si(100) and Si(111) substrates by CVD at atmospheric pressure using Si_2H_6 and C_2H_2 as source materials. The reaction tube was made of clear fused quartz with a water cooled jacket. The susceptor was made of SiC coated graphite and heated by a 15 kW RF generator with a frequency of 300 kHz. The susceptor temperature was measured with an optical pyrometer (CHINO IR-AH) that passed through the chamber end window and into a 15 mm deep hole in the downstream end of the susceptor. The difference between the measured and actual substrate temperature was determined by melting a Si substrate under actual growth conditions. The gases were controlled by mass flow controllers (Nippon Tylan Corp.).

Crystal growth of cubic SiC on Si substrates consisted of two steps: etching and CVD growth. The entire system was evacuated by a rotary pump and backfilled with H_2. After 5 min of H_2 flow, the substrate was heated by RF generator and kept at 1150 ^oC with a HCl flow rate of 20 sccm for 5 min for etching. After the HCl gas was shut off, the temperature was readjusted to 1150 ^oC and left for 2 min. The reaction gases of Si_2H_6 and C_2H_2 were added to hydrogen and crystal growth commenced. Crystal growth was carried out at a substrate temperature of 900-1150 ^oC with the standard gas composition: H_2=3 slm, Si_2H_6=0.6 sccm, C_2H_2=0.1 sccm. The gas ratio of Si/C was kept at 6. The deposition time was usually 10 min. Crystallinity and surface morphology were observed as a function of HCl flow in the CVD growth step.

The thickness of the grown cubic SiC film was estimated by an interference pattern obtained in the reflectance spectra in the visible-light range and the refractive index. The surface morphology of the film was observed by an optical microscope (Nomarski interference contrast). To study the crystal quality of the grown layers, the reflection high energy electron diffraction (RHEED) method was used.

3. Result and Discussion

The layers grown on the Si substrate in the temperature range of 900-1150 ^oC with the standard gas composition showed different colors depending on their thickness. In the above temperature range, the growth rate can be expressed as an exponential function of the reciprocal substrate temperature. The activation energy for the formation of cubic SiC calculated from the slope of the line was about 34.7 kcal/mol.

HCl was added to the reaction gases to control the initial stage of the crystal growth, where the flow rate of Si_2H_6 and C_2H_2 were 0.6 sccm and 0.1 sccm, respectively. By adding HCl to the reaction gases, the growth rate was depressed, as shown in Fig.1. At a HCl flow rate of 30 sccm, no film was grown on the Si substrate. However, by decreasing the HCl flow rate, film formation was observed. The growth rate with a HCl flow rate of 20 sccm was about 20 % of one without HCl addition.

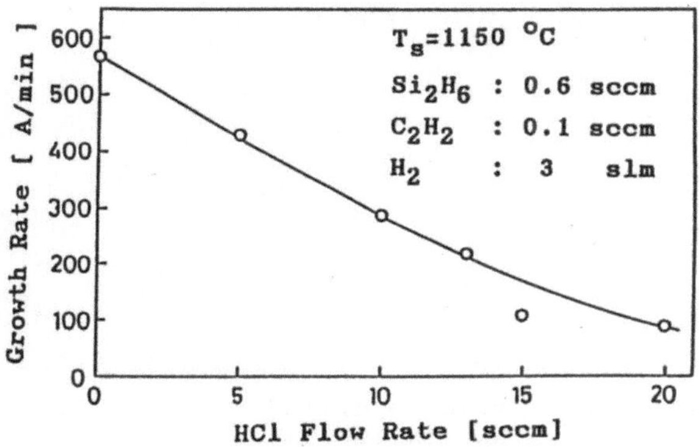

Fig.1. Effect of HCl addition during crystal growth.

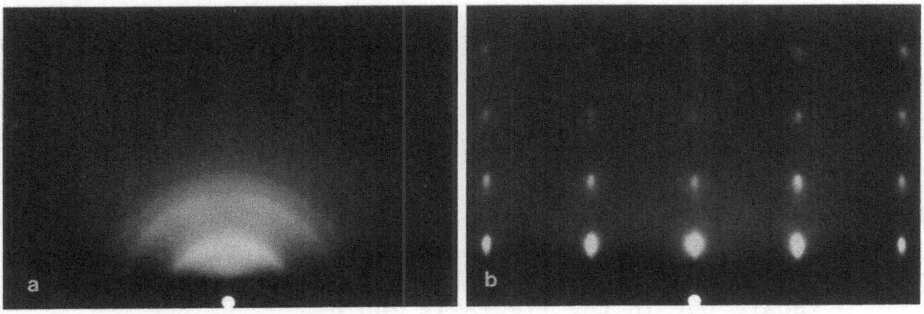

Fig.2. RHEED patterns of the grown layer: (a) without HCl and (b) HCl 20 sccm.

The diffraction patterns of cubic SiC films with and without HCl addition are shown in Fig.2. Debye rings are observed in the layer grown without HCl addition, indicating a highly oriented polycrystalline nature as shown in Fig.2(a). At a HCl flow rate of 20 sccm, the major characteristic pattern consisted of strong spots, which indicates single crystalline growth. From this evidence, appropriate addition of HCl in the reaction gases is necessary to obtain single crystalline cubic SiC on a Si substrate at a low temperature. A Si-rich condition is necessary to grow cubic SiC at a low temperature. Polycrystalline SiC was obtained when the Si/C ratio was 1 even though HCl was added to the reaction gases.

The surface morphology was observed by Nomarski interference microscope. The surface morphology changes with increasing flow rate of HCl as shown in Fig.3. Going from low to high HCl input concentration, the grain size of the crystallites decreased and homogeneous distribution of crystallites occurred. The surface is almost flat and featureless when prepared without addition of HCl, however, when HCl was added, the density of crystallites increased on the surface as shown in Fig.3(b).

Based on these results, we propose two reasons for the low temperature epitaxy of cubic SiC on a Si substrate. One is

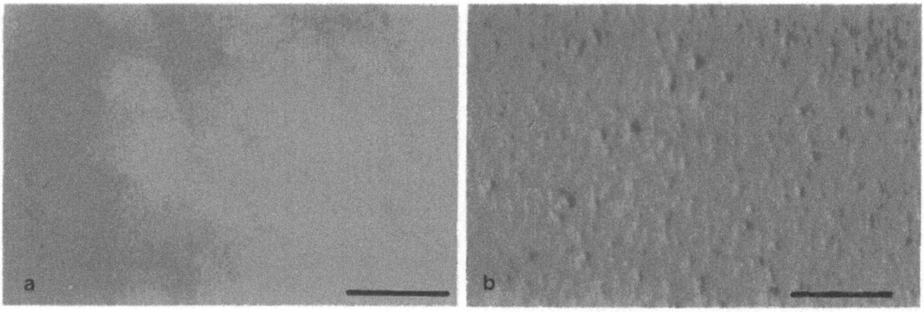

Fig.3 Surface morphology of the grown layer:(a) without HCl and (b) HCl 20 sccm. Scale bar is 100 μm.

Fig.4. Selective growth of 3C-SiC on the patterned substrate: (a) without HCl, (b) HCl 20 sccm and (c) HCl 20 sccm. Scale bar in the figure is 200 μm.

the cleaning effect of the substrate and the other is stabilization of unstable nucleation sites. These two effects are closely related to each other. Since the etching effect of Si is well known in the Si CVD process, SiC growth and etching of Si occurred simultaneously during the CVD process. When HCl is added, etching of the substrate occurred at the initial stage of crystal growth and random nucleation was consequently removed, resulting in uniform distribution of nucleation. When the carbonization process (buffer layer technique) is employed, the Si surface or nucleation site might be stabilized by carbonization at the initial stage of CVD growth. At a higher temperature (1350 ^{o}C), nucleation sites on the Si substrate might be unstable for SiC growth, and the Si surface must be quickly covered with a thin SiC layer which is formed by the carbonization process. In our experiment, the so called carbonization process was omitted, however, single crystalline SiC grew epitaxially on the Si substrate.

Selective growth of cubic SiC on a Si substrate was carried out using SiO_2 as the insulating mask. Field SiO_2 film was thermally grown on Si(100) and Si(111) substrates. The SiO_2 thickness was about 3000 Å. Square openings and a daisy pattern were made by the photolithographic technique. SiC growth on the patterned substrate is shown in Fig.4. Without HCl addition, polycrystalline SiC grew on the SiO_2 mask as

Fig.5. Facet growth of 3C-SiC on the patterned substrate. Scale bar in the figure is 50 μm.

shown in Fig.4(a). Due to the increase of the HCl flow rate, the density of SiC nucleation on SiO_2 was decreased. Perfect selective growth was obtained at 20 sccm of HCl as shown in Fig.4(b). On the daisy patterned substrate, the easy direction of the growth is observed as the dark area, where this direction is <110> and is the fast growth direction. Epitaxial films were grown only on the exposed area of the Si substrate. Areas covered with SiO_2 resulted in the growth of polycrystalline films or no deposition.

As the HCl flow rate was increased, the growth rate decreased and the growth selectivity became better. Similar results were reported for selective deposition of Si in the SiH_2Cl_2-HCl-H_2 system.[12] The mechanism of selective growth of cubic SiC on Si substrates is similar to selective growth of Si. By introduction of HCl in the Si_2H_6-C_2H_2-H_2 system, etching of Si adatoms and SiO_2 mask by HCl occurs and the nucleation density of SiC is different for the SiO_2 mask and Si substrate. Consequently, nucleation on the SiO_2 mask decreases drastically, and selective growth occurs. Facet growth was observed on the small patterned substrate as shown in Fig.5. The faceting plane in this picture is the {111}plane. This is further evidence that the grown layer on the exposed area is a single crystal of cubic SiC. In a selected area, cubic SiC grew laterally on the SiO_2 mask, which suggests a possibility of epitaxial lateral overgrowth on SiO_2; the CVD epitaxial lateral overgrowth technique (ELO) in which an epilayer is seeded through openings in the mask and from these openings grows laterally over the mask. The major problem encountered during the EOL deposition process has been the formation of polycrystalline nuclei over the SiO_2 surface. This problem will be solved by optimizing the SiO_2 thickness and area of uncovered Si.[13]

Al-doped cubic SiC was grown on a patterned n-Si(100) substrate, and heterojunction properties were also studied. Ohmic contacts were made by vacuum evaporation using AuSb for n-Si and Al for p-SiC. The diodes showed good rectification. Current(I)-voltage(V) relations measured at room temperature were expressed by the relation $I \propto \exp(eV/nkT)$, where the value of n (ideal factor) was 2.5, which indicates that the current is predominately generation-recombination. Capacitance(C)-voltage(V) relations were expressed as the relation $V \propto 1/C^2$, which indicates that the junction was abrupt. The diffusion voltage obtained in the C-V curve was about 1.0 volt, and this value was almost the same value as that obtained from the I-V characteristics.

4. Conclusion

Epitaxial growth of cubic SiC on a Si substrate was carried out at the low temperature of 1150 ^{o}C in normal pressure CVD using the Si_2H_6-C_2H_2-HCl-H_2 system. Selective growth of cubic SiC was achieved using patterned SiO_2 masks. HCl addition plays an important role in low temperature epitaxy and selective growth. A possibility of epitaxial lateral overgrowth was observed. A heterojunction of p-SiC/n-Si was made on the patterned substrate and good rectification was oberved.

References

1) S.Nishino,J.A.Powell and H.Will: Appl.Phys.Lett. **42**,460(1983).
2) S.Nishino,H.Suhara,H.Ono and H.Matsunami:J.Appl/Phys.**61**, 4889(1987)
3) K.Shibahara, S.Nishino and H.Matsunami: Appl.Phys.Lett.**50**, 1888(1987).
4) K.Shibahara, T.Saito,S.Nishino and H.Matsunami: IEEE Electron Devices: **EDL-7**,692(1986).
5) P.Liaw and R.F.Davis: J.Elecrochem.Soc.**132**,642(1985).
6) G.L.Harris,K.H.Jackson,G.L.Felton,K.R.Osborne,K.Fekade and M.G.Spencer: Materials Lett.**4**,77(1986).
7) J.A.Powell,L.G.Matus and M.A.Kuczmarski: J.Electrochem.Soc.**134**,1558(1987).
8) H.P.Liaw and R.F.Davis,J.Electrochem.Soc.:**131**,3014(1984).
9) Y.Furumura,M.Doki,F.Mieno,T.Eshita,T.Sugii and M.Maeda: J.Electrochem.Soc.**135**,1255(1988)
10) Y.Ohoshita and A.Ishitani: J.Appl.Phys.**66**,4535(1989).
11) S.Nishino and J.Saraie: Springer Proceedings in Physics, vol **43**,8(1989).
12) K.Tanno, N.Endo, H.Kitajima, Y.Kurogi and H.Tsuya: Jpn.J.Appl.Phys.**21**,L564(1982).
13) L.Jastrzebski, J.F.Corboy, J.T.McGinn and R.Pagliaro: J.Electrochem.Soc.**130**,1571(1983).

Shortwave SiC-LEDs Prepared by the Sublimation "Sandwich Method"

Yu.A. Vodakov, E.N. Mokhov, A.D. Roenkov, V.V. Semenov, R.G. Verenchikova, A.O. Konstantinov, V.I. Sokolov, and V.G. Oding

A.F. Ioffe Physical Technical Institute, Academy of Sciences, 194021 St. Petersburg, Russia

Abstract. The summary of the research of the native defects' influence on the efficiency of the shortwave luminescence and electroluminescence in α-SiC is discussed. Bands of high-temperature boron (BL) and pure defective (DL) luminescence are used as indicators of the state and concentration of the native defects. It is determined that the main part of nonequilibrium native point defects is forming from clusters in SiC. Their size, concentration and temperature stability depend on the specimen obtaining conditions. The specimens grown at the temperatures below 2100-2200°C, especially obtained by sublimation, are enriched with these clusters. During technological procedures at the temperatures 1300-2400°C they generate movable nonequilibrium point native defects. This highly increases the DL bands' intensity in such specimens, thus weakening luminescence bands through the donor-acceptor pairs. The ability to operate the native defects concentration and their state made it possible to obtain the efficient short wave SiC-LED on polytypes 6H and 4H (λ=530;520;485;475;440 nm) by using not traditional liquid epitaxy, but sublimation.

1. Introduction

The present study covers SiC-LEDs with $\lambda_{max} \leq 530$ nm obtained by the sublimation "sandwich method" [1]. The investigation of the behavior of native defects in the SiC specimens obtained by different processes, dependence of intensity of different luminescent bands (L) upon native defects and the doping level of impurities, permitted the manufacture, by sublimation, of blue SiC-LEDs comparable with those obtained by the LPD technique (with respect to efficiency), as well as other types of shortwave SiC-LEDs. Polytype 4H-SiC intended for such LEDs was obtained by specially developed, original and effective process of heteroepitaxial growth of 4H on substrates of other polytypes. A controlled growth of 4H-SiC is based upon the latest studies on the role of native defects in polytypism [2]. In [3-6] it was shown that the following bands may serve to indicate the nature and concentration of native defects in SiC specimens: high temperature boron L (BL), i.e. a known yellow band L ($\lambda_{max} \simeq 585$ nm for 6H) or a green band ($\lambda_{max} \simeq 530$nm for 4H) and defect L via native defects of type (DL) with $\lambda_{max} \simeq 525$ nm in 6H and 475 nm in 4H (type D of the defects is a

Springer Proceedings in Physics, Volume 56 Amorphous and Crystalline Silicon Carbide III
Editors: G.L. Harris, M.G. Spencer, and C.Y. Yang

wider set of native defects giving a green L in 6H-SiC) than that studied in detail in [7,8]. Also of importance is that very promising SiC-LEDs have been developed on the basis of efficient bands BL and DL in 6H and 4H. In [3-6] it was shown, that carbon vacancy V_c included in centers L and DL was presumably a native defect. The generation of movable single V_c takes place intensively in a specific type of specimens at T=1400-2000°C. These temperatures are typical for the formation of EL of p-n structures on SiC.

2. Experimental results and discussion

Fig. 1 presents the dependence of major shortwave bands L (BL, DL and L on the donor-acceptor pairs Ga-N, Al-N) upon the nitrogen doping level as well as upon specific features in preparation of test specimens and subsequent temperature conditions of the specimens treatment. BL also owes to DAP of deep B-N, but the concentration of deep boron is directly concerned with the concentration of carbon vacancies V_c [9,10]. It can be seen that the efficiency of DL and BL bands in low-temperature specimens, i.e. not exposed to high temperatures up to T_g>2100° (let us designate them as LTS) during either growing or further treatment, and high-temperature specimens (HTS) grown at T_g>2300 or subjected to a heat treatment at higher temperatures differ considerably. It can also be inferred from Fig. 2 that blue band L of DAP Al-N (AlL) is dimmed considerably by the defects of D type and it becomes brighter in the specimens having high concentration of native defects only after annealing at a

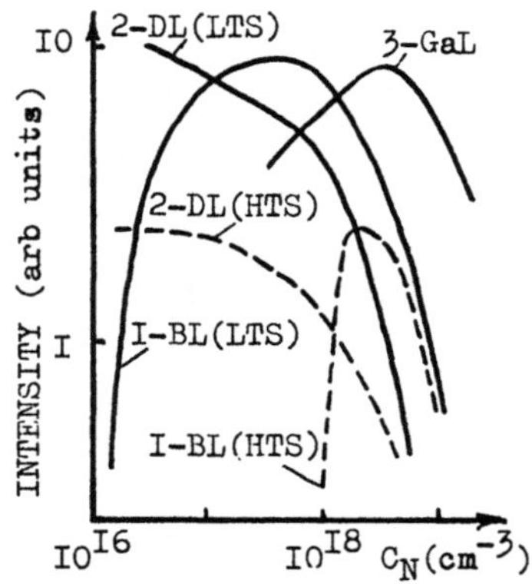

Fig.1

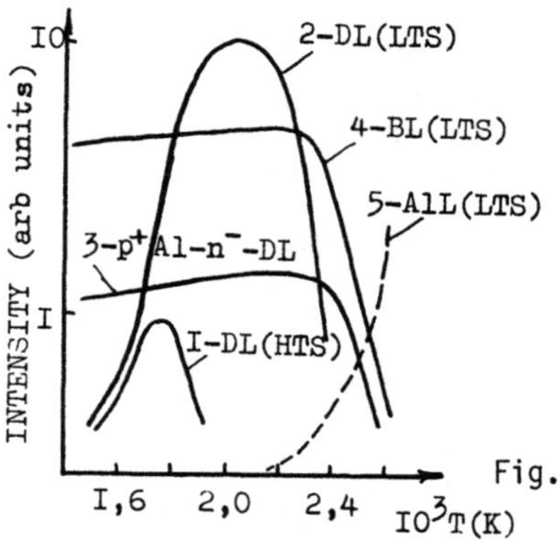

Fig.2

Fig. 1. Dependence of EL band intensity in 6H-SiC upon nitrogen concentration (in different types of specimens). $C_B=2\cdot10^{18}$ cm^{-3}; $C_{Ga}=3\cdot10^{18}$ cm^{-3}.

Fig. 2. Dependence of EL band intensity upon annealing temperature (in different types of specimens).

temperature exceeding 1900°C. Curves 1,2 and 5 in this figure correspond to the specimens with DL after their neutron bombardment followed by annealing (temperatures are given as abscissae); curve 3 corresponds to DL in n-layer adjacent to strongly doped p^+(Al)-layer grown at a rate exceeding 0.2 μm min^{-1}; curve 4 corresponds to BL in LTS, while curve 5 is the blue band AlL. Causes of such peculiarities in behavior of SiC are discussed in more detail in [6]. The explanation is based upon the supposition that native defects are combined in SiC and temperature-resistant microclusters are formed. This formation is accompanied by the creation of regions of local microstresses. Where the stresses approach the plasticity limit, generation of native defects takes place. Defects are accumulated around such microstressed regions and thus screen them. As a result sufficiently strong microlocal formations are created either containing a large amount of excessive native defects or capable of generating them afterwards. In the temperature range 1300-2000°C when a significant restructuring of such clusters takes place [10-12] these regions might be the place of generation of individual nonequilibrium movable point native defects. Their increased concentration aids the existence of DL centers at higher annealing temperatures. Described phenomenon is the most prominent in LTS grown from the vapor phase. The equilibrium vapor over SiC is rich in silicon and contains primarily complex molecules (Si_2C, SiC_2). Nonequilibrium capture of such molecules as well as more complex precipitates formed in the chemisorbate (including excessive silicon) takes place especially intensively at low growth temperatures and may lead to the formation of microclusters in the solid phase. This process is even more effective in the presence of the vapor phase of Al impurity required for the growth of highly doped p^+-layers (with $C_{Al} \simeq 5 \cdot 10^{19}$ cm^{-3}), which results in the formation (near the p-n junction boundary) of a stressed area with higher concentration of native defects giving rise to DL band.

The probability of nonequilibrium capture of complex molecules, clusters, inclusions of the 2nd phase from the vapor phase and chemisorbate increases abruptly in the case of accelerated growth (especially by the sublimation method). This phenomenon can be described through the activity of elementary carbon, which is much less than 1 for the growth from the gaseous phase. On the contrary, in case of growth from solutions or melts, e.g. from liquid silicon, the activity of elementary carbon approaches 1, which facilitates the growth of epitaxial layers (by the LPD methods) having considerably smaller amount of local microstresses. In our opinion, this explains a more easy realization of efficient blue SiC-LEDs based on DAP, as well as purely excitonic (not defective) EL bands if the LPD technology is used.

Emergence of degradation phenomena (more commonly at local spot overheating during passage of working currents) in SiC-LEDs in places of spalling, scratches, microcracks, along steps of growth of epitaxial layer is also explained, in our opinion, by the creation of large local stresses and, consequently, generation of near-by point native defects. Their migration (according to [12] it has a significant rate even at room temperatures) and association causes the appearance of centers L of type DL near the places of local mechanical stresses. If there is a high concentration of boron (primarily non-luminescent) near these places or in the entire specimen, boron interacts with V_c and passes into the luminescent state, and BL emerges.

Appearance and spreading of blue degradation regions with DL near local stresses in p-n structures on 4H-SiC with violet EL indicate convincingly the role of native defects stimulating D-centers, but not the transformation into another polytype as was suggested in [13]. If the B concentration is sufficient ($>10^{17}$ cm^{-3}) in the specimen, the degradation process ends in the transition of stressed regions to DL (yellow for 6H and green for 4H) as the most stable luminescence. Besides the study of effects caused by native defects on the efficiency of L bands, the doping influence was investigated. It is seen from Fig. 1 that DL requires pure n-layers ($C_N<3\cdot10^{17}$ cm^{-3}) but with the higher concentration of native defects in SiC-LTS. On the contrary, nitrogen, native defects and especially local stressed microregions (characteristic for LTS) are undesirable if efficient purely exciton L is required. It could be expected for L on DAP-acceptor that there would be a high intensity in layers strongly doped and compensated with nitrogen and the required acceptor. But the experiments have shown that this supposition is true up to doping levels as high as $(4-7)\cdot10^{18}$ cm^{-3} only. Further increase in concentration and compensation leads to the shift λ_{max} towards the longwave direction and to more time decay L [14], which is concerned with the emergence, upon strong doping, of significant spatial energy fluctuations for the conduction band and the valence band ceiling, and with the growing concentration of the vacancy-impurity complexes. Stretching of the p-n junction enhances the role of the latter mechanisms. It is obvious that more uniform distribution of the doping impurities reduces the adverse action of these mechanisms.

Understanding of above-mentioned processes and their relationship with the specifics of the epitaxial growth of layers of SiC by the sublimation method lets us obtain layers 6H-SiC and 4H-SiC with the efficient luminescence of type BL ($\lambda_{max}\simeq530$ nm for 4H [15], DL (both with help of methods of ion implantation (I.I.) [16] and by introducing the defects after the quality p-n junction has been formed) and aL ($\lambda_{max}\simeq480$ nm for 6H and 440 nm for 4H). For example, the sublimation method has made it possible for the first time to obtain epitaxial layers with a blue high temperature (~300K) L based on DAP Ga-N. The introduction of defects D apart from the creation of the p-n structure permitted to increase linearity and decrease time decay of DL. As a result the SiC-LEDs with different λ_{max} (520-530 nm (6H:D; 4H:B,N); 475-485 nm (4H:D; 6H:Ga, N); ~440nm (4H:Ga, N)) were obtained by sublimation "sandwich method". Complete external quantum efficiency of green LEDs is about 10^{-3}, blue - about 10^{-4} and violet - about $3\cdot10^{-5}$. The EL operation time on DL bands constitutes about 10^{-8} s, on other bands L - about 10^{-7} s. The developed SiC-LEDs stand out for their high stability and

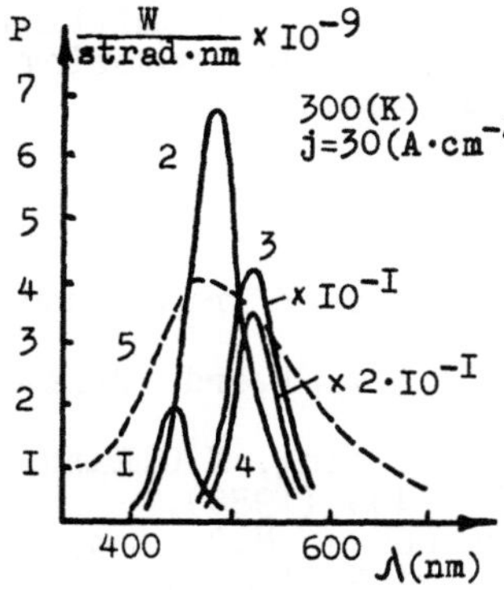

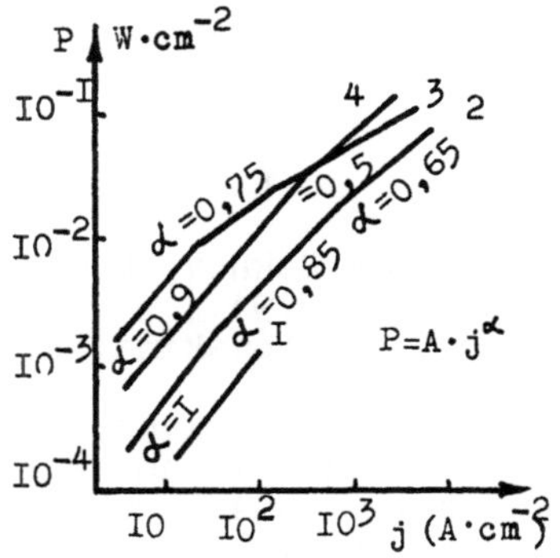

Fig. 3. Unit spectral radiation power of different SiC-LEDs. 1-4 - refer to Fig. 4; 5 - microplasma EL of 6H-SiC.

Fig. 4. Dependence of unit integral power of SiC-LEDs radiation upon current density. 1 - 4H-SiC (Ga, NO; 2 - 6H-SiC (Ga, N); 3 - 6H-SiC (I.I.D.); 4 - 6H-SiC (D).

resistance to external factors and current loads. During 10^5 hours of operation, maximum recorded changes in the diode parameters did not exceed 2%. Spectra of such shortwave SiC-LEDs at current density of 30 A/cm^{-2} are given in Fig. 3 (in absolute chips in the monocrystal matrix, the intensity within 100-120°C is almost constant). Anti-reflection coatings were not used and radiation from end faces was not recorded. Dependence of unit integral power of radiation of such chips within the current density angle of 100°. The dependence has been recorded by means of a calibrated silicon photodetector. In the temperature range of -50°C to +150°C the temperature coefficient of radiation varies from 0.3 to 1% deg.$^{-1}$ for the different types of SiC-LEDs. Understanding of the clusterization processes and their effect or microplasma breakdown in SiC permitted the development of the manufacturing processes for wide-band (curve 5, Fig. 4) SiC-LEDs of special stability with the uniform field of pre-breakdown EL, with the temperature stability better than $2\cdot10^{-4}$ deg.$^{-1}$, time stability of at least 1%, with high linearity of the radiation power-current dependence (all parameters given are for the temperature range T=-50-200°C). Their efficiency is lower than that of other SiC-LEDs but enough to satisfy many light metrological purposes, for use as a reference for stationary and pulse radiation in versatile opto-electronic systems.

3. Conclusion

Thus, it was shown that the sublimation methods can be used to obtain an extensive choice of shortwave SiC-LEDs of various colors. The sublimation "sandwich method" was used to allow a lot of electroluminescence p-n structures having similar parameters, which is typical for the gaseous-phase methods. On the basis of such structures the Designing Bureau of the Ioffe

Physical-Technical Institute manufactures linear light-emitting diode matrices for thousands of elements meant for storing information saved in both analog and digital modes [17].

4. References

1. Yu.A.Vodakov,E.N.Mokhov, M.G.Ramm, A.D.Roenkov //Krist. und Techn. 14, 729 (1979).
2. Yu.A.Vodakov, G.A.Lomakina, E.N.Mokhov // Fiz. Tverd. Tela, 24, 1377, (1982).
3. Yu.A.Vodakov, G.A.Lomakina,E.N.Mokhov, V.G.Oding, M.G.Ramm, V.I.Sokolov // Fiz. Techn. Poluprov. 20, 1433 (1986).
4. Yu.A.Vodakov, G.A.Lomakina, E.N.Mokhov, M.G.Ramm, V.I.Sokolov // Fiz. Techn. Poluprov. 20, 2153 (1986).
5. Yu.A.Vodakov, E.E.Goncharov, G.A.Lomakina, A.A.Maltsev, E.N.Mokhov, V.G.Oding // Fiz. Techn. Poluprov. 21, 207 (1987).
6. Yu.A.Vodakov, G.A.Lomakina, E.N.Mokhov: In Shirokozonnye Poluprovodniki (Mahachkala, 1988) p. 23.
7. V.V.Makarov // Fiz. Tverd. Tela, 13, 2357 (1971).
8. K.L.Patrick, W.J.Choyke // Ohys. Rev. B5, 3257 (1972).
9. S.U.Hagen, A.W.L. van Kamenade // Phys. Stat. Sol. A33, 97 (1976).
10. Yu.A.Vodakov, N.Zumaev, B.P.Zverev, G.A.Lomakina, E.N.Mokhov, V.G.Oding, V.V.Semenov, Yu.F.Simakhin // Fiz. Techn. Poluprov. 11, 373 (1977).
11. V.N.Kutt, A.A.Lepneva, G.A.Lomakina, E.N.Mokhov, A.S.Tregubova, M.M.Sheglov,G.F.Yuldashev //Fiz. Tverd. Tela 39, 2606 (1988).
12. A.I.Girka, A.D.Mokrushin, E.N.Mokhov, V.M.Osadchiev, S.V.Svirida, A.V.Shishkin: ZETF 97, 578 (1990).
13. G.Ziegler, D.Theis: Trans. Electr. Dev. ЕД-28, 425 (1981).
14. Yu.A.Vodakov, E.N.Mokhov, V.I.Sokolov, V.S.Vavilov, A.I.Ivanov, M.V.Chukichev // Pisma v ZTF 15, 60 (1989).
15. A.S.Barash, Yu.A.Vodakov, E.N.Koltsova, A.A.Maltsev, E.N.Mokhov, A.D.Roenkov // Pisma v ZTF 14, 2222 (1988).
16. V.M.Gusev, K.D.Demakov, M.G.Kosaganova, M.B.Refman, V.G.Stolarova // Fiz. Techn. Poluprov. 9, 1238 (1975).
17. A.S.Barash, Yu.A.Vodakov, V.V.Semenov, A.A.Uvarov //Techn. Sredstv Sviazi, OT5, 41 (1982).

Liquid-Phase Epitaxy of Silicon Carbide at Temperatures of 1100°–1200°C

V.A. Dmitriev, L.B. Elfimov, N.D. Il'inskaya, and S.V. Rendakova

A.F. Ioffe Physical Technical Institute, Academy of Sciences, 194021 St. Petersburg, Russia

Abstract. Results of silicon carbide liquid-phase epitaxial growth at 1100-1200°C on SiC wafers are presented. SiC epitaxial layers (EL) of n- and p- conductivity of a thickness 0.1-1.0 μm have been produced. Maximum concentration of aluminium in EL is $3 \cdot 10^{20}$ cm^{-3}. The pn junctions were shown to be abrupt. Electroluminescence is seen in the blue-violet part of the spectrum (300 K). Selective growth of SiC from liquid phase has been realized.

Introduction

Progress has been made recently in fabrication of semiconductor devices based on silicon carbide. Further development of SiC electronics is hindered by the high temperatures involved in the SiC processes.

One way to solve the problem is the development of low-temperature liquid phase epitaxy (LTLPE) of silicon carbide.

The solubility of silicon carbide in Ge, Sn, and Ga melts containing Si additives at temperatures of 1000-1400°C has been studied in [1,2]. Solubility values of 0.02-0.2 mol.% were observed. SiC growth from Sn- and Ga-based melts using LPE has been reported [2].

In our earlier works [3,4] we have reported fabrication of pn junctions in silicon carbide using LPE from Ga and Sn melts containing Al in a range 1200-1350°C. It was found that under forward bias the pn structures emitted in the blue-green part of the spectrum [3]. Under reverse bias structures grown from Ga melt exhibited homogeneous avalanche breakdown [4] which is the first observation of this type of breakdown in SiC structures grown from liquid phase.

The present paper reports data on SiC epitaxial layers (EL) and pn junctions grown at 1100-1200°C as well as on the selective epitaxial growth from the liquid phase.

Experimental

Experiments were performed in an apparatus with a RF heater using a vertical quartz reactor with water-cooled walls in a flow of pure hydrogen. A graphite crucible with compartments for different melts was employed. The "sandwich" consisting of two horizontally positioned plates of silicon carbide held in a graphite clamp was lowered into the melt. Growth proceeded in a temperature dradient ~30°/cm.

Composite melts of Sn-Al-Si, Ga-Al-Si and Ge-Si were used as solvents. As a substrate material we used monocrystalline SiC of 6H polytype. These were commercially available crystals' grown by the Lely method with n- and p-type layers grown onto them beforehand by container-free LPE(CFLPE) [5]. The growth face was (0001)Si.

Results and Discussion

We have produced epitaxial layers (EL) of silicon carbide of n- and p-conductivity types of a thickness 0.1-1 μm. Examination by electron diffraction spectroscopy in reflection revealed the monocrystalline structure of the EL.

In n-type layers grown from Sn- and Ge-based melts the concentration of electrically active uncompensated donor impurities (N_d-N_a) was 10^{17}-$10^{18}cm^{-3}$.

For obtaining SiC of p-type conductivity, we used Sn-Al-Si and Ga-Al-Si melts with components in different ratios. The maximum content of Al in EL's obtained from X-ray spectral microanalysis was $3 \cdot 10^{20} cm^{-3}$, corresponding to a lattice distortion of $\Delta a/a = 5 \cdot 10^{-4}$.

So, in our LTLPE experiments several tyres of epitaxial structures have been grown: n(substrate)-n(LTLPE layer), n(substrate)-p(LTLPE layer), n(substrate)-n(CFLPE layer)--p(LTLPE layer), n(substrate)-p(CFLPE layer)-n(LTLPE layer).

In the pn structures grown mesas 50-300 μm in diameter were etched using reactive-ion etching. Contacts to p- and n-SiC were of alumimium and chromium, respectively.

Capacitance-voltage (CV) characteristics of the structures obtained (Fig.1a) in coordinates $1/C^2$-V were linear, indicating

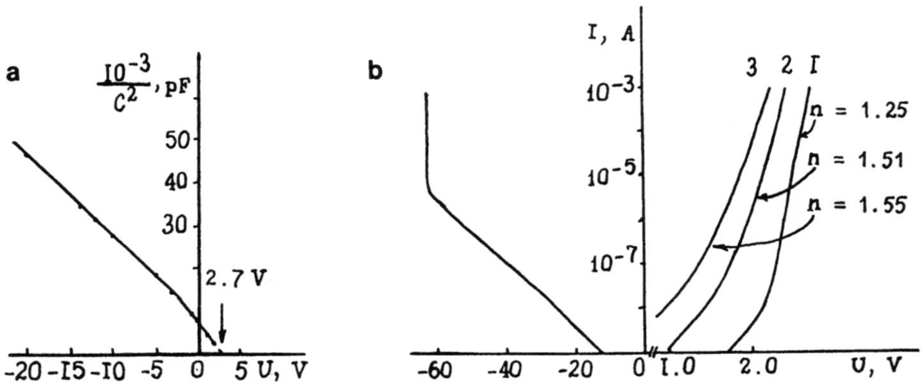

Fig.1 a - Capacitance-voltage characteristic of the SiC pn junction grown from Ga-Al-Si melt (300 K)
b - Current-voltage characteristics of the SiC pn junction grown from Ga-Al-Si melt
(1) - T = 295 K
(2) - T = 430 K
(3) - T = 530 K

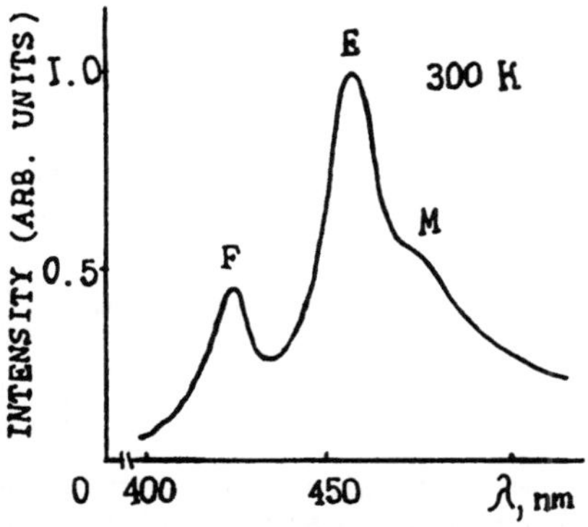

Fig.2 Electroluminescence spectrum of a SiC pn junction (j_{imp}= 100 A/cm^2)

abrupt pn junctions. A kink in the CV characteristic corresponds to a change in the N_d-N_a in the substrate. The zero bias barrier value of 2.7 V is typical of 6H SiC. In the low doped region of the structure, the N_d-N_a deduced from the CV characteristics has a value of $(5\text{-}6)\cdot 10^{17}$cm^{-3}, or about the same as in the substrate. Calculated values of the maximum electric field strength E_m in the junstion and the width of the space charge region W at zero bias are $6.5\cdot 10^5$ V/cm and 0.08 µm.

The best junction electrical properties were found in the n(substrate)-n(CFLPE layer)-p(LTLPE layer), with p-layer grown from the Ga-Al-Si melt. Fig.1b shows the current-voltage characteristic of one of these structures. The breakdown voltage V_{br} was 63 V. The breakdown occurred due to formation of microplasmas. The E_m (at reverse bias $V \approx V_{br}$) calculated from CV data is $3.5\cdot 10^6$ V/cm.

Forward current-voltage characteristics for voltages below 2.5 V measured in a temperature range 295-580 K (Fig.1b) have a portion described with an exponential

$$j = j_o \exp[qV/(nkt)].$$

At room temperature $j_o \approx 3\cdot 10^{-35}$ A/cm^2, n=1.25. To establish a mechanism of carrier transport in this range of bias values further research is necessary.

Under forward bias the structures obtained emit in the short wave length region of the visible spectrum. In Fig.2 the electroluminescence spectrum of the n(substrate)-n(CFLPE layer)-p(LTLPE layer) structure is shown. The spectrum is characteristic of that for pn structures grown from Si melt at temperatures 1600-1700°C.

Thus we have shown that properties of pn junctions grown at temperatures of 1100-1200°C are, on the whole, comparable to those of pn junctions grown by known high temperature techniques.

Lower growth temperatures make possible the use of LTLPE for selective growth of silicon carbide in the openings of carbon masks.

The protective mask for selective epitaxy was deposited on the (0001)Si face of the substrate using the known technique

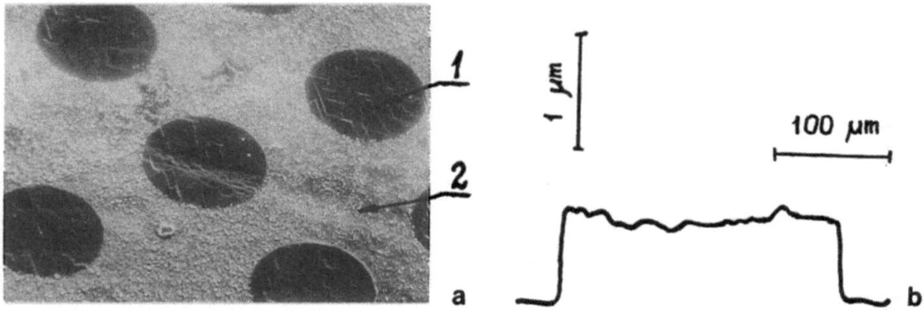

Fig.3 a - SEM of a SiC substrate with selective structure (1) grown from Ga-Al-Si melt inside windows in the carbon mask (2) (x 200)
b - Profilographic picture of the surface of SiC substrate with selectively grown mesa (with mask removed)

for depositing carbon films on other materials [6]. Photolithography was used to make windows of various configurations of linear dimensions 50-200 μm. The thickness of the protective layer was 0.1-1 μm. These substrates were used in epitaxial growth of SiC from the Ga-Al-Si melt.

Auger-spectroscopic investigations prior to the epitaxial process revealed that no other elements besides carbon were present on the mask surface. Investigations carried out on the same samples after contact with the melt revealed compositional changes on the mask surface: silicon was found to be present (up to 30 mol.%) and bonds of Si-C type observed. In unprotected areas both prior to epitaxy and afterwards the ratio of Si to C corresponded to that for silicon carbide. Specrta of Auger-electrons from the surface after removal of the mask coincided with spectra from silicon carbide.

By selective epitaxy silicon carbide layers were grown of a thickness 0.1-2.5 μm. The growth rate was 0.1-1 μm/hour. So, with selective epitaxy the growth rate of SiC is increased. Fig.3a shows a scanning electron micrograph (SEM) of a SiC substrate with selectively grown structures. Light background corresponds to the carbon mask. Fig.3b demonstrates a profile of a selectively grown structure after the mask has been removed. The lateral surface makes an angle of 90° with the substrate plane.

The results presented here demonstrate the feasibility of selective epitaxial growth of SiC from the liquid phase at low temperatures.

Conclusions

SiC layers of n- and p-conductivity types have been grown by the method of liquid phase epitaxy at low temperatures of 1100-1200°C. The epitaxial layer thickness is 0.1-1 μm. The maximum concentration of aluminium in EL is $3\cdot10^{20}cm^{-3}$. The pn junctions were shown to be abrupt. The maximum electric field strength in the pn junction is about $3.5\cdot10^{6}$ V/cm.

Electroluminescence from the pn junctions is mainly in the blue-violet part of the spectrum (300 K). Selective growth of SiC from the liquid phase has been realized.

Acknowledgments

The authors are greatly thankful to A.L.Syrkin for his assistance in preparing mesa structures, to P.A.Ivanov and Ya.V.Morozenko for their help in measuring parameters of pn structures, to V.N.Panteleev for SEM investigations and to M.A.Chernov for X-ray structure investigations.

References

1. Ivanov E.G.: Izvestiya LETI, N281, p.18-22(1981) [in Russian]
2. Tairov Yu.M., Raikhel F.I., Tsvetkov V.F.: Izv. AN SSSR, ser. "Neorg. mater.", v.18, N8, p.1390-1391(1982) [in Russian]
3. Rendakova S.V. In: III All-Union meeting "Physics and technology of wide-gap semiconductors", 10-12 Sept.1986, Makhachkala, Abstracts, p.28-29 [in Russian]
4. Vainshtein S.N., Dmitriev V.A., Levinshtein M.E., Rendakova S.V.: Sov.Tech.Phys.Lett., v.13, N12, p.741-743(1987)
5. Dmitriev V.A., Popov I.V., Chelnokov V.E. In: VII Conference on growth processes and synthesis of semiconductor crystals and films, 9-13 June 1986, Novosibirsk, Abstracts, v.1, p.51-52 [in Russian]
6. Lions A.M., Wilkins C.W., Robbins M.: Thin Solid Films, v.103, p.333-341(1983)

Growth and Characterizaton of 3C-SiC and 6H-SiC Films on 6H-SiC Wafers

J.A. Powell[1], *J.B. Petit*[2], *L.G. Matus*[1], *and S.E. Lempner*[3]

[1]NASA Lewis Research Center, 21000 Brookpark Road, Cleveland, OH 44135, USA
[2]Sverdrup Technology, Inc., Brook Park, OH 44142, USA
[3]Graduate Student, Cleveland State University, Cleveland, OH 44115, USA

Abstract. Single crystal SiC films were grown by chemical vapor deposition (CVD) on vicinal (0001) SiC wafers cut from boules produced by the modified sublimation method. Wafers with tilt angles less than 0.5° yielded 3C-SiC; tilt angles of 3° to 4° resulted in 6H-SiC films. The surface morphology of these films, up to 24 μm thick, were investigated as a function of growth parameters such as the Si/C ratio in the input gases and the presence of dopant materials such as nitrogen and trimethylaluminum.

1. Introduction

Rapid progress is continuing to be made in SiC semiconductor technology [1,2]. A key factor in this progress is the increasing use of the modified sublimation process for growing large boules of 6H-SiC. High-quality 25-mm-diameter wafers are routinely being produced from these boules. The availability of consistently high-quality wafers has benefited the development of device fabrication processes such as chemical vapor deposition (CVD), oxidation, metallization, etc.[3]. This paper will present some recent results of a continuing investigation of CVD for SiC device fabrication. The results were achieved with SiC wafers obtained from sublimation-grown boules.

2. SiC Film Growth

The 6H-SiC wafers used in this work were produced by Cree Research Inc.[4] from sublimation-grown boules [5,6]. The wafers were 25.4 mm in diameter and approximately 200 μm thick with one side polished. With respect to the orientation of the polished face, relative to the (0001) basal plane, there were several types of wafers. For wafers designated as "on-axis" in this paper, the polished face was within 0.5° of the basal plane, and for wafers designated as "off-axis", the tilt angle was in the range 3° to 4°. Also, in this paper, the tilt angle is defined as the angular displacement of the polished surface normal away from the <0001> direction. For the wafers used in this work, no particular tilt direction was selected in sawing wafers from a boule. Precise measurement of tilt angles of the on-axis wafers at Case Western Reserve University showed that the angles were in the range 0.3° to 0.5°. The polished face of all on-axis wafers used in this work was the Si face, and the polished face of most of the off-axis wafers was the C face. So, care had to be taken in comparing results with on-axis and off-axis substrates, because differing results could also be due to the Si-face/C-face difference.

Because of limited availability, each wafer was cut into pieces yielding nine square substrates, 6.4 mm on an edge. No attempt was made to cut the edge of the substrates along any particular crystal direction. Before loading the substrates into the growth chamber, they were scrubbed with liquid detergent, rinsed with 18 MΩ-cm water, and then blown dry with nitrogen.

The SiC films were grown in a horizontal CVD chamber from SiH_4 and C_3H_8 in a H_2 carrier gas at 1 atm. The carrier gas flow was maintained at 3 l/min in the 50 mm I.D. chamber. The CVD system is described in more detail in Ref. 7. The substrates were heated by a rf-heated SiC-coated graphite susceptor. After being loaded into the growth chamber and prior to growth, the SiC substrates were subjected to a 2 min HCl etch at 1200°C. The temperature was then increased to 1450±25°C for the film growth with the SiH_4 and C_3H_8 concentrations at 300 and 120 ppm, respectively. This value of 0.8 for the Si/C ratio in the input gases was about optimum. Typically, a growth rate in the range 3 to 4 μm/hr was achieved, regardless of whether growth was on either the Si face or the C face, or whether a 3C-SiC or 6H-SiC film was being grown.

Both n-type and p-type films were produced by adding nitrogen and trimethylaluminum (TMA), respectively, to the input gases. Since only very limited electrical evaluation has been done on the doped films, this paper will present the effects of these doping gases on the surface morphology. The preliminary electrical evaluation did, however, demonstrate that TMA concentrations of the order of 200 ppm were required to achieve p-type behavior. This TMA concentration level is the same order as the silane and propane! The necessary nitrogen concentration was about an order of magnitude less.

The identificaton of the film polytype was made by transmission electron microscopy (TEM) and low temperature photoluminescence (LTPL) as described previously [3,8]. The surface morphology was investigated by scanning electron microscopy (SEM) and optical microscopy.

3. Results and Discussion

For all cases studied, 3C-SiC films grew on on-axis substrates, and 6H-SiC films grew on off-axis substrates. This includes growth runs with on-axis and off-axis substrates side by side. This result is consistent with previously reported film growth in which 3C films grew on 6H substrates with tilt angles less than 1°, and 6H films grew on 6H subtrates with tilt angles greater than 1.5° [3,8-12]. It is also consistent with a step-controlled growth model proposed by Matsunami et al.[11]. In their model, the density of atomic steps on the vicinal (0001) 6H-SiC substrate determines whether a 3C or 6H film grows. With larger tilt angles (higher step densities), arriving atoms diffuse to steps where growth occurs laterally, i.e., parallel to the basal plane. With small tilt angles (larger terraces between steps), nucleation of 3C takes place. Since two 3C

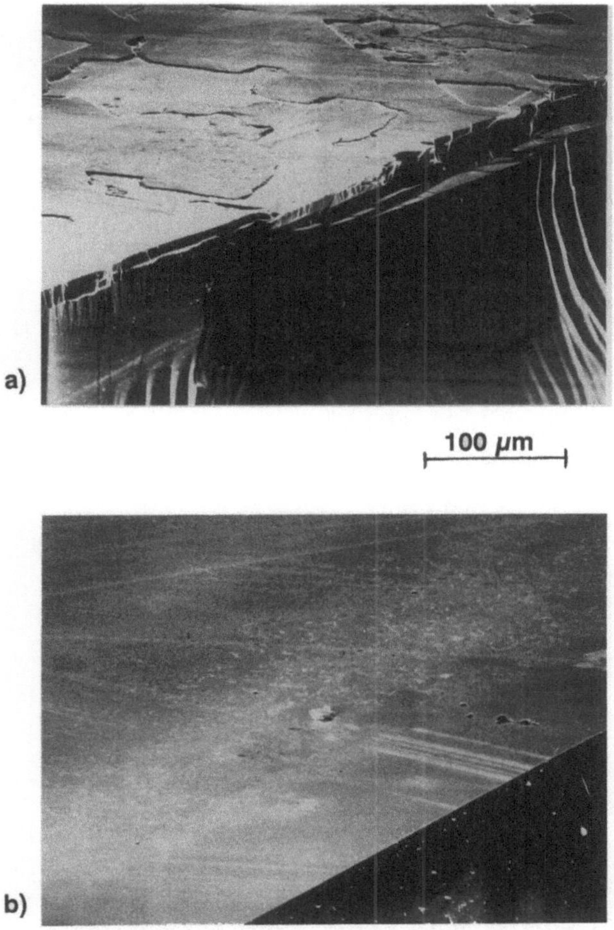

Fig. 1. Scanning electron micrographs of 24-μm-thick SiC films grown on 6H-SiC wafers. (a) 3C-SiC film, (b) 6H-SiC film.

stacking sequences can nucleate, double positioning boundaries (DPBs) can form.

Figure 1 illustrates typical 3C and 6H films. In this figure, the films are about 24 μm thick and were produced in a 6-hour growth run. A significant difference in surface morphology is apparent in these films. The deep grooves in the 3C film are DPBs. Although 3C films with a low density of DPBs have been produced on the as-grown face of Lely crystals [8], all films grown on polished wafers in this work do contain DPBs. The DPB density does vary considerably from sample to sample. The 6H film is nearly featureless; the faint lines seen on its surface are believed to be due to polishing damage in the original substrate.

In Fig.1, both film/substrates were cleaved instead of sawed in order to study the broken edges. The result shown is

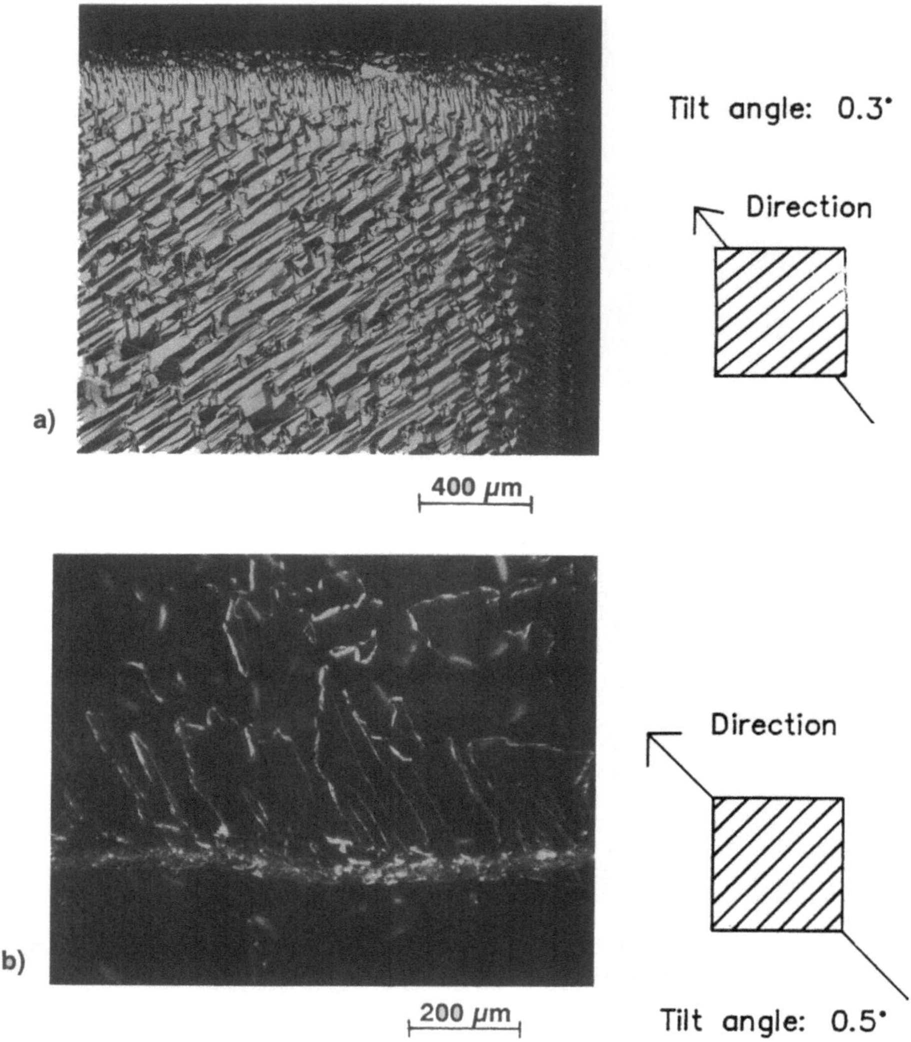

Fig. 2. Correlation of growth features of 3C films with orientation of vicinal (0001) 6H-SiC substrates. (a) Elongated hillocks, (b) Elongated double positioning boundaries (DPBs).

typical. The 6H film/substrates generally broke cleanly and vertically, while the 3C film/substrates tended to break with both vertical and horizontal components near the film/substrate interface. This resulted in the ragged edge seen in Fig. 1(a). We believe that this indicates that the 3C film is under stress near the interface.

The SiC film thickness was determined by viewing thin cross-sectional slices with an optical microscope, equipped with a measuring eyepiece. The 3C film, because of its intrinsic yellow color, was easily distinguishable from the

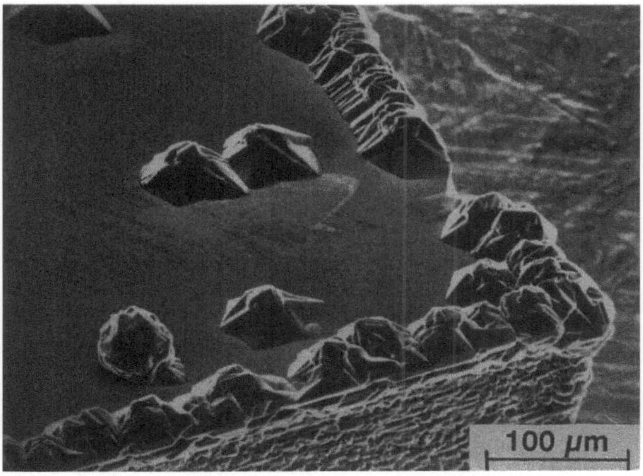

Fig. 3. Scanning electron micrographs of SiC hillocks formed during film growth on 6H-SiC substrates when Si/C = 0.44.

colorless 6H substrate. The 6H film/substrate interface in a cross-sectional slice could not be observed with an optical microscope. Thus, the 6H film thickness was obtained by subtracting the thickness of the original substrate (about 200 μm) from the total film/substrate thickness. This method was applied to films with thicknesses in the 12 to 24 μm range.

Previously, a correlation was reported between growth features in 6H films and the tilt direction of the 6H substrate [3]. In that case, the tilt angles were 3° to 4°. In Fig. 2, there is a representation of this direction for two 6x6 mm^2 samples. Perpendicular to these tilt direction vectors (the arrows), there are parallel lines which represent idealized surface steps. In these drawings, the arrows also point in the direction of the lateral growth one would expect if growth takes place at the steps. For each sample in Fig. 2, the corresponding magnitude and direction of tilt are given at the right. The surface morphology of the sample in Fig. 2(a) consists of elongated hillocks that form parallel lines. The lines are roughly (within the limits of measurement error) parallel to the steps one would expect to be present on the surface. In the sample of Fig. 2(b), the surface was scratched before growth (the broad horizontal line). The irregularly shaped lines are DPBs. As can be seen, there are no DPBs on the lower side of the scratch, but on the upper side, there are DPBs that are elongated roughly in the direction of the tilt. Thus, these two growth features apparently correlate with the tilt direction. In terms of the step-controlled growth model, the presence of the steps, even with small tilt angles, is apparently affecting the net diffusion direction of mobile atoms on the surface during growth.

The Si/C ratio of the input gases was varied over a range of 0.4 to 1.7 in different growth runs. As mentioned in the previous section, the optimum for our system was about 0.8. Generally, the silane concentration was held at 300 ppm and the

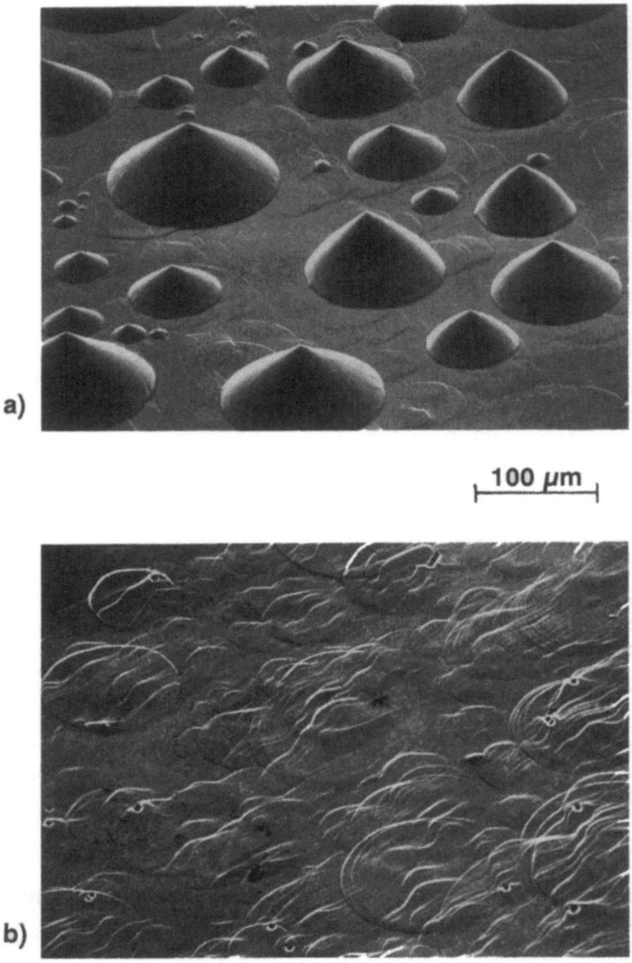

Fig. 4. Scanning electron micrographs of (a) free Si "droplets" formed on SiC substrates during a growth run when Si/C = 1.7, and (b) the surface after removal of the Si with an acid etch.

propane concentration was set to yield the desired Si/C ratio for a given run. For lower values of the Si/C ratio (more propane), hexagonal hillocks formed randomly and along the edge of the substrate as shown in Fig. 3. For higher values (less propane), in the range of 1.0 to 1.7, no growth occurred. Instead, the deposit consisted of droplet-like hillocks of free Si as shown in Fig. 4(a). The Si "droplets" were removed with an acid etch, and patterns remained on the surface, as shown in Fig. 4(b). These patterns indicated that the liquid Si droplets were present on the surface during the growth run and that they migrated during the run. In the case of Fig. 4, the substrate

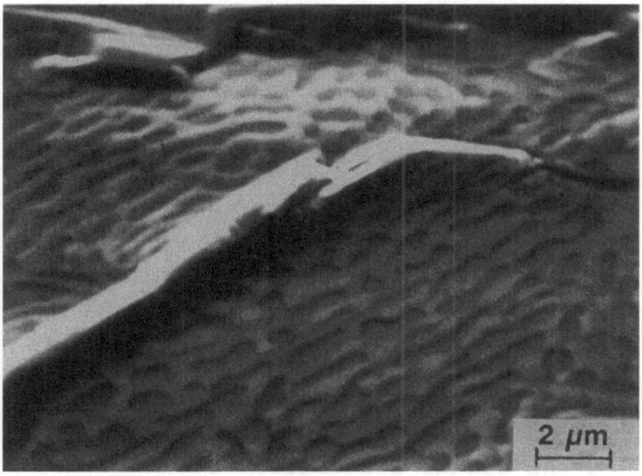

Fig. 5. Scanning electron micrograph of the surface of an Al-doped 3C-SiC film grown on a 6H-SiC substrate.

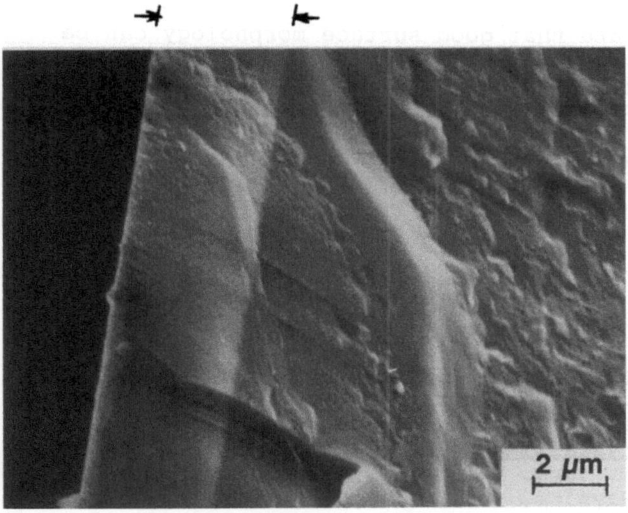

Fig. 6. Scanning electron micrograph of the cross-section of an Al-doped p-type 6H-SiC film (lighter region) on an n-type 6H-SiC substrate.

was an off-axis sample with a C face. For an on-axis sample with a Si face, the result was similar, except there was no evidence of migration of Si droplets. In either case, there was no evidence of any SiC film growth except under some of the Si droplets. This may be explained if we assume that the carbon from the propane is incorporated only in the liquid Si. Then, SiC growth may have taken place under the Si by means of the vapor-liquid-solid (VLS) mechanism [13].

The use of nitrogen and TMA to produce doped samples had the following effect on surface morphology. Nitrogen concentrations up to 40 ppm had no effect on the morphology of either 3C or 6H films. Doping with TMA required concentrations of the order of 200 ppm to achieve p-type behavior. Even at this high concentration, the surface morphology, as seen with an optical microscope, for both 3C and 6H films was not significantly affected. However, when examined with SEM at a higher magnification, a wormy texture, as shown in Fig. 5, was observed. The ridge, seen in this figure, is along a DPB. Normally, the 6H films could not be distinguished from the 6H substrate. However, the Al-doped p-type 6H films could be distinguished from the n-type substrates with SEM as shown in Fig. 6. The contrast is probably caused by the difference in electrical conductivity between film and substrate. This provided an additional method of measuring the thickness of 6H films. Mesa diodes were fabricated from these p-n junction structures of Fig. 6 and the diodes did exhibit rectification and emitted blue light when biased in the forward direction.

4. Conclusions

The results demonstrate that good surface morphology can be obtained by CVD in both undoped and doped 6H-SiC films, up to 24-μm thick, on <u>off-axis</u> 6H-SiC wafers. The morphology of 3C-SiC films, also up to 24-μm thick, grown on <u>on-axis</u> 6H-SiC wafers was not good because of the presence of DPBs. A correlation was shown between growth features on 3C-SiC films and the tilt direction of the 6H-SiC substrate for tilt angles in the range 0.3° and 0.5°.

References

1. <u>Amorphous and Crystalline Silicon Carbide</u>, ed. by G. L. Harris and C. Y.-W. Yang, Springer Proceedings in Physics, Vol. 34 (Springer-Verlag Berlin, Heidelberg, 1989).
2. <u>Amorphous and Crystalline Silicon Carbide II</u>, ed. by M. M. Rahmann, C. Y.-W. Yang, and G. L. Harris, Springer Proceedings in Physics, Vol. 43 (Springer-Verlag Berlin, Heidelberg, 1989).
3. J. A. Powell, D. J. Larkin, L. G. Matus, W. J. Choyke, J. L. Bradshaw, L. Henderson, M. Yoganathan, J. Yang, and P. Pirouz, Appl. Phys. Lett. <u>56</u>, 1442 (1990).
4. Cree Research, Inc., Durham, NC 27713, USA
5. Y. M. Tairov and V. F. Tsvetkov, J. Cryst. Growth <u>43</u>, 209 (1978).
6. G. Ziegler, P. Lanig, D. Theis, and C. Weyrich, I.E.E.E. Trans. Electron Devices <u>ED-30</u>, 277 (1983).
7. J. A. Powell, L. G. Matus, and M. A. Kuczmarski, J. Electrochem. Soc. <u>134</u>, 1558 (1987).
8. J. A. Powell, D. J. Larkin, L. G. Matus, W. J. Choyke, J. L. Bradshaw, L. Henderson, M. Yoganathan, J. Yang, and P. Pirouz, Appl. Phys. Lett. <u>56</u>, 1353 (1990).

9. H. S. Kong, J. T. Glass, and R. F. Davis, Appl. Phys. Lett. 49, 1074 (1986).
10. H. S. Kong, J. T. Glass, and R. F. Davis, J. Mater. Res. 4, 204 (1989).
11. H. Matsunami, K. Shibahara, N. Kuroda, and S. Nishino, in Amorphous and Crystalline Silicon Carbide, ed. by G. L. Harris and C. Y.-W. Yang, Springer Proceedings in Physics, Vol. 34 (Springer-Verlag Berlin, Heidelberg, 1989), pp. 34-39.
12. H. S. Kong, J. T. Glass, and R. F. Davis, J. Appl. Phys. 64, 2672 (1988).
13. R. Wagner and C. J. Doherty, J. Electrochem. Soc. 115, 93 (1968).

The Effects of Impurities on Mass Transport and the Growth Mechanism of SiC Epitaxial Layers Grown by Sublimation

Yu.A. Vodakov, E.N. Mokhov, M.G. Ramm, and A.D. Roenkov

A.F. Ioffe Physical Technical Institute, Academy of Sciences,
194021 St. Petersburg, Russia

Abstract. The growth of high doped SiC epitaxial layers obtained by the sublimation "sandwich-method" was studied. The impurities N, B, Al, Sc were shown to decrease the deposition rate at low supersaturation at high temperatures, which contradicts calculated values. The use of special profiled substrates was proposed to study the effects of the surface orientation and structural imperfections on the growth process. Thus elemental centers of the normal growth may be decorated also. Great anisotropy of the growth rate of the doped SiC layers was observed on the local areas of a singular type of substrate surface. The peculiarities of doped SiC deposition are discussed.

1. Introduction

Device structures based on SiC are primarily formed by the method of epitaxial growth. However, the liquid-phase and CVD techniques used to obtain epitaxial layers (EL) are not flexible enough to study the effect of various factors on the SiC crystallization process [1].

The sublimation "sandwich-method" [2] is, to a considerable degree, free from these drawbacks. In this method, matter is transferred from the SiC source to the substrate by means of sublimation only, i.e. omitting any use of reagents. In such a system the growth of rather perfect SiC layers, both pure and doped with various impurities, is possible within a wide range of temperatures and condensation rates. The subject of the present study is the role of impurities in the kinetics and mechanism of crystallization of epitaxial layers on SiC, polytype 6H, under different external conditions.

2. Experimental procedure

Epitaxial layers were grown at 1700-2600°C in special containers made of air-tight graphite, in the inert gas medium or in vacuum, to which impurities N, Al, B, P, Sc and Ga were added. The impurity source (usually an elementary impurity) was placed inside the container, and the vapor pressure was adjusted by temperature control of the source. Nitrogen was introduced directly into the furnace atmosphere. Monocrystals of SiC grown by the Lely method were used as substrates. The dislocation density was 10^2-10^4 cm^{-2}, while the base surfaces

Springer Proceedings in Physics, Volume 56 Amorphous and Crystalline Silicon Carbide III
Editors: G.L. Harris, M.G. Spencer, and C.Y. Yang

were close to (0001) or disoriented by an angle 0.5-20°. The source-substrate distance (d) was 0.01-5 mm. Supersaturation in the growth region was determined by the temperature drop ΔT between the source and substrate. The temperature drop was varied by changing either the temperature gradient in the working zone, or the source-to-substrate distance, and constituted 0.5-10 K. To ensure adequate interpretation of the test results the growth of non-doped SiC layers was also investigated under the same conditions.

3. Results and discussion

Experimental data show that at low supersaturations ($\Delta T>3$ K) and growth temperatures ($T>2300$°C) impurities decrease the growth rate (Fig. 1). This effect is dependent on the chemical origin of an impurity, and is especially important for the epitaxial growth of SiC in vapors of Sc, B, Al, N and lanthanides. However, the effect is relatively weak in the presence of vapors of Ga, P, In and Sn. The influence of an impurity on the epitaxial growth of SiC increases as the temperature becomes lower, thus the kinetics of the doped layer growth requires apparently higher activation energy than the growth of pure epitaxial layers (Fig. 2). It is possible to compare experimental results with the theoretical values based on the thermodynamic analysis for a number of impurities. Fig. 2 gives information on the SiC-Al system. It

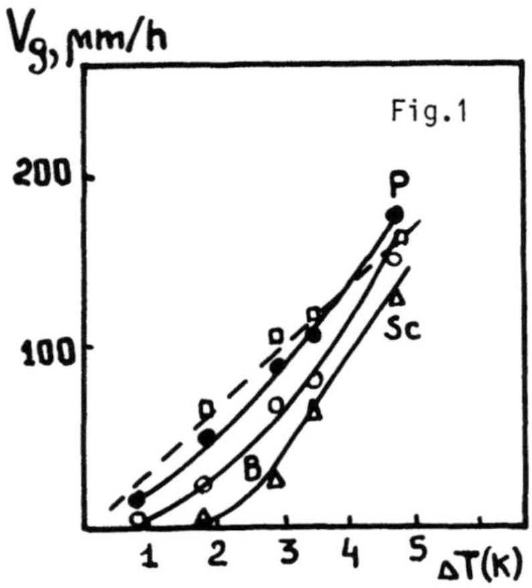

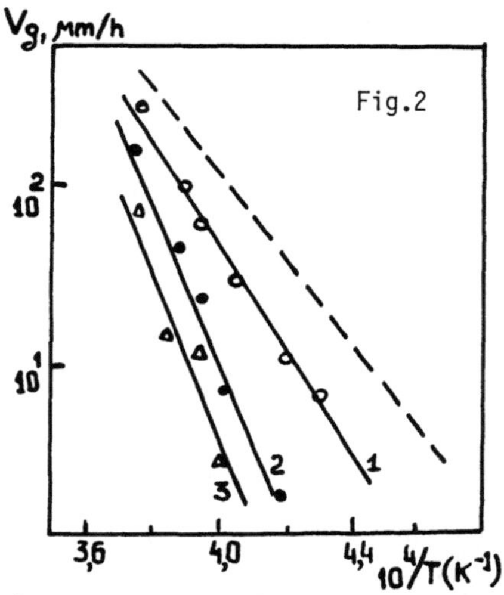

Fig. 1 Dependence of the SiC epitaxial layer growth rate upon source-to-substrate temperature drop in the presence of impurities (solid lines) and without them (dashed line). Growth temperature - 2250°C.

Fig. 2. Dependence of the SiC epitaxial layer growth rate upon temperature at different growth directions. 1 - non-doped EL SiC; 2,3 - Al doped EL SiC. 2 - [000$\bar{1}$]C, 3 - [0001]Si. Dashed line signifies calculated data for SiC-Al system.

can be seen that in accordance with calculation [3], the introduction of Al should accelerate the SiC epitaxial growth, which conflicts with experiment. Analogous results have been obtained in the analysis of the SiC epitaxial growth in the presence of N_2, H_2 and B. Discrepancies between the experiment and calculation grow abruptly as the temperature decreases and supersaturation takes place. At the same time the growth of pure SiC epitaxial layers can be satisfactorily described within the thermodynamical limit [2]. Note that most of the impurities (Al, B, Ga) decrease the growth more effectively in the direction (0001) Si.

Within minor disorientation (1-10°) between the substrate and basic (0001) face, the variation in the growth rate is negligibly small in the presence of such impurities as B and Sc. Conversely, if the growing epitaxial layer of SiC is strongly doped with Al the growth rate on the disoriented face in respect to 0001 is 1.2-1.6 times higher than on the singular surface. In the latter case the thickness of the grown layer often varies considerably over the specimen surface. It should be noted that due to the morphological instability of singular faces these data do not reflect the real orientation anisotropy of growth rate, which may be studied only on small areas of the surface.

The analysis of experimental data shows that the effects associated with the influence of an impurity on the condensation rate at low ΔT are of kinetic origin and concerned with contamination of active areas on the surface of the source or substrate with impurity chemisorbate. This is also confirmed by the fact that the impurities with high bond energy (B, Sc) decrease the growth rate most strongly. The pointed peculiarities of the impurity effect on the epitaxial growth of polar faces (0001) may be explained by different chemisorbtion of impurities on them. Another cause of decrease may be the graphite liberation on the surface of either the substrate or the sublimation source, at the expense of the increased concentration of carbon-containing products in the vapor phase. On the basis of the data obtained a supposition can be made that impurities B, Sc and lanthanides decrease epitaxial growth of SiC primarily due to graphitization of the source surface. Conversely, in the presence of Al vapors (except very low super-saturations) carbon is transported by Al onto the substrate surface, where it may be deposited as graphite. This is clearly seen in the morphological character of the growing surfaces. On the face (0001) Si, co-deposition of graphite may even lead to the growth suppression on a part of the substrate surface or to the formation of numerous pores, which can be excluded by introducing excessive Si vapor into the growth zone. When the growth takes place on the face (0001) C, vapors usually promote considerable modification of the growing surface. At the initial stages of growth, surfaces of singular types often appear and develop into separate areas. As the thickness grows the areas become larger, and thus a relief is formed characteristic for the layerwise growth. Areas of the singular surfaces are separated now by macrosteps from 0.1 to several μm in height. In this case the impurity does not preclude layer wise growth, but makes the transfer of elementary steps difficult, leading to enlargement of the macroscopic size. It is obvious that such a picture

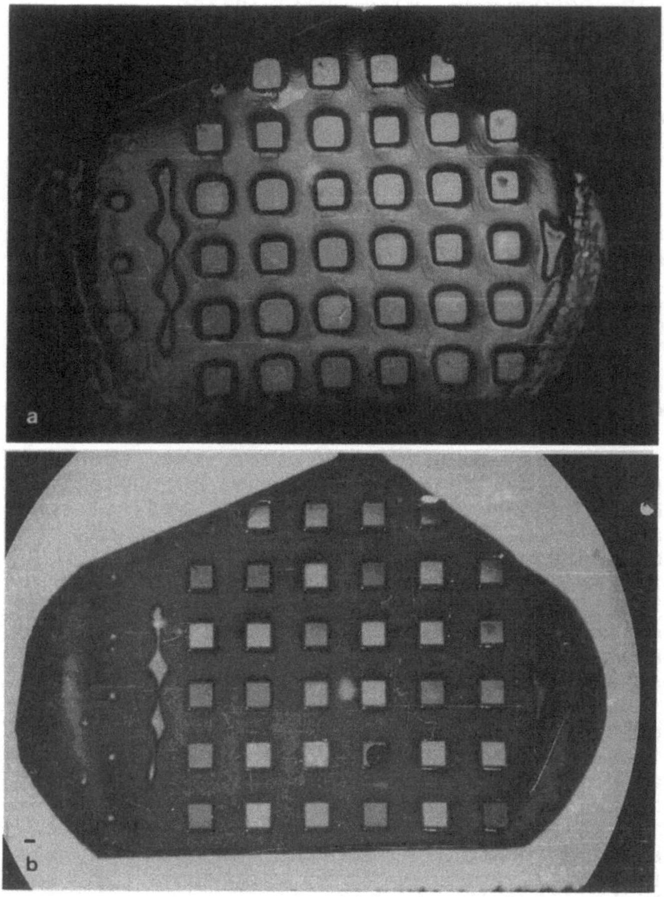

Fig. 3. Photograph of text specimen with Al doped SiC epitaxial layer grown onto the substrate with projected areas shown in deflected (a) and transmitted (b) light. It illustrates normal growth only on a few projections (dark color in transmitted light).

demonstrates a low density of centers of the normal growth, which is confirmed below.

For visualization of these centers the method of selective epitaxy was used, i.e. an epitaxial layer was deposited on the profiled substrate with an individual surface fragment [4]. To do so, projected areas of 0.2-0.3 mm^2 were created on the substrate surface. Then epitaxial layers of SiC were grown on this profile surface. The experiment showed that in the presence of the impurity vapor, if the initial substrate surface is a singular type, normal growth will not take place on most of the projecting fragments and only lateral development of these areas will be detected (Fig. 3).

At the same time the epitaxial growth continues on separate projections and on the remaining part of the substrate surface. Finally, the thickness of the epitaxial layer on different projections varies for times of 10^1-10^2 s or more. This fact can be explained only by the presence of the normal-growth centers on the normally growing fragments of the surface.

To study the origin of these centers, the structural perfection of the crystal was investigated by the X-ray method combined with decorating of the necessary locations by pickling [5]. It was established that such centers could be the screw-like dislocations with the Burgers vector in the [0001] direction. The density of such dislocations does not exceed 10^1 cm^2, which is several orders lower than the density of knot dislocations in the crystal. Other normal-growth centers emerge, apparently, in the initial stage of the SiC epitaxial deposition, and some impurities are found to assist the formation of such centers.

Thus, the impurities considerably influence the kinetics and mechanism of the epitaxial growth of SiC. As a whole they preclude obtaining quality highly doped layers of SiC, and contribute to their polytype non-uniformity. However, by special treatment of the substrate surface and under certain experimental conditions reducing the probability of liberation of incoherent nuclei of other phases, e.g. graphite, on the growing surface, it is possible to obtain SiC layers of both n- and p-type conductivity, with electrically active impurity concentration of about 10^{20} cm^{-3}, and of a rather high structural perfection (comparable with the perfection of the substrate).

On the other hand, the fact that the impurities contribute to the probability of polytype transformation upon growth makes it possible to accomplish controlled preparation of heteroepitaxial layers of rare polytypes of SiC, e.g. 4H, on 6H SiC substrates [6].

4. Conclusion

Impurities Al, B, Sc, N decrease the condensation rate of SiC epitaxial layers especially at low supersaturation. This effect may be associated with the passivation of substrate surface by graphite or impurity chemisorbate. The diminishing density of normal growth centers in the presence of Al, N_2 vapors indicates a macrostep layer condensation mechanism.

5. References

1. E.N.Mokhov. In: Wide-Zone Semiconductors (Machachala, 1988) p.44 (in Russian).
2. Yu.A.Vodakov, E.N.Mokhov, M.G.Ramm, A.D.Roenkov // Krist. und Techn. I4, 729 (1979).
   Yu.A.Vodakov, E.N.Mokhov // Patent Fr. No.7.409.089, prior. 18.3.1973; Patent USSR No.403275, prior. 5.11.1970.
3. Yu.M.Tairov, V.F.Tsvetkov // Neorganich. Mater. I4, 1785 (1978).

4. E.N.Mokhov, Yu.A.Vodakov, A.D.Roenkov, K.L.Matveeva // The Seventh Conference on Procedures of Growth and Synthesis of Semiconductor Crystals and Films. Abstracts. Part 1, p. 277 (Novosibirsk, 1986).
5. Yu.A.Vodakov, E.N.Mokhov, M.G.Ramm, A.D.Roenkov // The Seventh All-Union Conference on Crystal Growth. Abstracts. P. 47 (Moscow,1988).
6. E.N.Mokhov, A.D.Roenkov, A.S.Tregubova: The Seventh Conference on Procedures of Growth and Synthesis of Semiconductor Crystals and Films. Abstracts. Part 1, p. 279 (Novosibirsk, 1986).

Doping Peculiarities of SiC Epitaxial Layers Grown by the Sublimation "Sandwich Method"

Yu.A. Vodakov, E.N. Mokhov, M.G. Ramm, and A.D. Roenkov

A.F. Ioffe Physical Technical Institute, Academy of Sciences, 194021 St. Petersburg, Russia

Abstract. Doping of SiC epitaxial layers (EL) with impurities during the growth process by sublimation in vacuum or inert gas media was studied as a function of the temperature and growth rate, vapor phase composition and the substrate orientation. The dopants were introduced either directly into the SiC vapor source in the sandwich cell or the external vapor phase. Doping level and effective coefficients of the impurity capture were determined. The mechanisms of impurity introduction in SiC-EL are discussed.

1. Introduction

The sublimation "sandwich method" is used to obtain both pure and highly doped monocrystal epitaxial layers (EL) of SiC, suitable for the manufacture of semiconductor devices [1,2]. Advantages of this method are the possibility of growing perfect EL doped with various impurities within a wide temperature range (1600-2600°C) either in vacuum or in an inert gas medium. The subject of this study was to investigate the peculiarities of doping EL of 6H-SiC by the sublimation "sandwich method", with major electrically active impurities (Al, B, Ga, N, P, In), and other impurities used in the SiC technology.

2. Experimental procedure

Heating devices and the design of growth cells were analogous to those described earlier [1]. The doping impurities were placed either directly into the cells (were combined with the source of SiC vapors) or were introduced externally via the vapor phase. In the latter case the required vapor pressure was obtained by proper adjustment of the vapor source temperature. For doping epitaxial layers of SiC with nitrogen a dose of N_2 or its mixture with Ar or NH_3 was introduced into the atmosphere. In a number of experiments the growth was run at an excess pressure of Si vapor. Monocrystals of 6H-SiC oriented with face close to (0001) were used as a substrate. The distance between the sources and substrates varied from 0.5 to 2 mm. The growth rate was 10^1-10^2 μm/h, while the thickness of a grown layer was W-300 μm. The grown EL of SiC had a dislocation density of 10^2 to 10^4 cm^{-3} and did not contain, as a rule, inclusions of other polytypes. The

Editors: G.L. Harris, M.G. Spencer, and C.Y. Yang

determination of an impurity concentration in SiC was made primarily by neutron-activation analytical methods, with the aid of which the total impurity content and character of its spread across the thickness and over the specimen surface were found [3]. The use of an autoradiographic version of the activation analysis permitted the visualization of doping non-uniformity having linear dimensions of up to 1 µm. Concentrations of electrically active impurities (N, B, Al, Ga) were measured by means of different electrophysical methods including the study of the Hall effect, electrical conductivity (in the temperature range 77-1100 K), capacitance of the Schottky barrier, absorbency and EPR.

3. Results and discussion

First, let us consider the analysis of doping L of SiC from the "internal" source. In this case to obtain the correct data, monocrystal sources SiC specially doped with the impurity under test were used. Let us look at the results of theoretical analysis of mass transfer in vacuum, when the free path of a molecule exceeds the distance (d) between the source and substrate. Using the mass transfer model discussed in [4] it is possible to calculate the dependence of an impurity concentration in the layer (C_i) upon the impurity contents in the source and distance (d). Such a dependence is presented for a number of impurities in Fig. 1. As seen in Fig. 1, the experimentally obtained values for the dependence $C_i(d)$ can be satisfactorily described by the calculated curves. From the analysis of the data obtained, an elementary condensation coefficient (K_i) reflecting the probability of capturing the impurity in an elementary collision can also be found. Thus, when EL of SiC are grown in the direction [0001]C at 1950°C, the values obtained for K_i are $5 \cdot 10^{-2}$; $5 \cdot 10^{-3}$; $4 \cdot 10^{-4}$ and $5 \cdot 10^{-5}$ for impurities N, B, Al and Ga, respectively. For acceptor impurities, the coefficient K_i increases as the temperature rises, and the higher values are observed in the growth direction [0001]Si (Fig. 2). These dependences acquire the reverse character for nitrogen.

The value K_i may be regarded as a constant only at a relatively low impurity content in the source ($C_i^s = 1 \cdot 10^{19}$ cm^{-3}). At higher C_i^s K_i decreases monotonically (Fig. 2). In the case of nitrogen this effect can be explained by the fact that the impurity concentration in EL is restricted by its desorption as N_2 [5]. The condensation coefficient for nitrogen depends on the growth rate and Si:C ratio in the vapor phase. At excess pressure of silicon vapor, no orientation doping anisotropy can be observed in EL grown in the polar directions <0001>.

Thus, this method of doping EL of SiC makes it possible to control the impurity concentration both by changing its contents in the source and by variation of distance (d) between the source and substrate. However, it is not possible

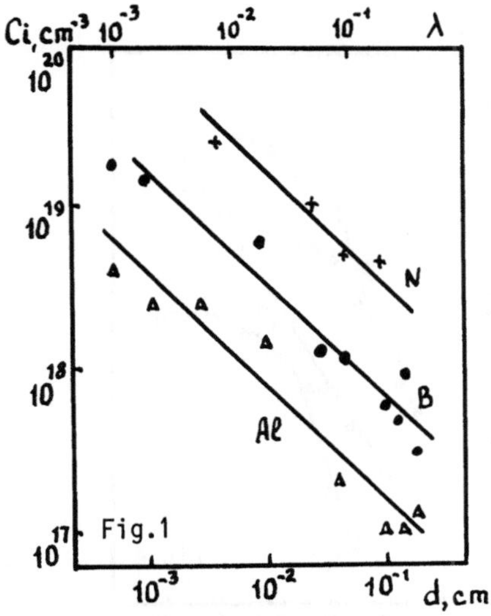

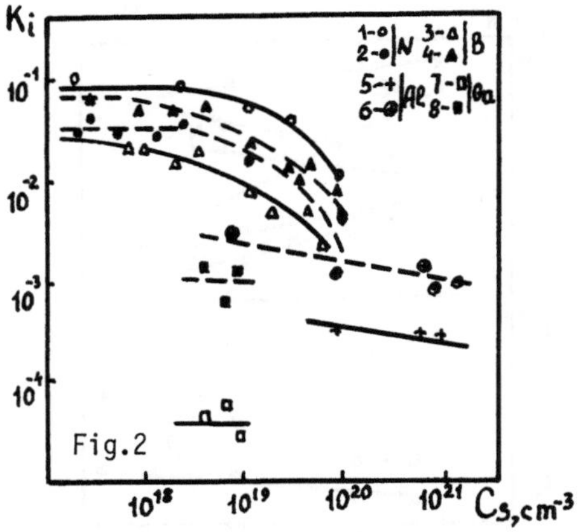

Fig. 1. Dependence of N, Al and B concentrations in EL of SiC upon source-to-substrate distance. Continuous lines are calculated only. Growth direction [000$\bar{1}$]C. T_g=1850°; $C_i^{source}=1\cdot10^{20}$ cm^{-3}.

Fig. 2 Dependence of elementary coefficient of capturing N, B, Al and Ga at SiC resublimation upon impurity concentration in the source. Growth direction: [000$\bar{1}$]C (1,3,5,7); [0001]Si (2,4,6,8). T_g=1850°C.

to reach a high EL doping level because of small values of K_i (10^{18} cm^{-3}). To ensure the growth of strongly doped EL of SiC additional introduction of impurity vapors is required. The dependence of the limit doping level for some major impurities upon the growth temperature is given in Fig. 3. Note the presence of a very strong orientation doping anisotropy of SiC EL grown on polar {0001} faces for the acceptor impurities (Ga, In, Al), which is especially significant at relatively low growth temperatures. In most cases the dependence of the impurity content in EL of SiC may be described by means of the Arrhenius equation. Values of activating energy, effective distribution factors (or conversion factor for Ga, In and Al) are discussed in [3]. Information on nitrogen and phosphorus is given in [5] and [6] respectively. It is worthy of mention that the results given here reflect only the doping level in an almost homogeneous solid solution. Total impurity concentrations are often 1 or 2 orders higher, which is due to the fact that the crystals contain the 2nd phase inclusions located primarily near morphological and structural imperfections [7]. For example, in the case of boron the impurity accumulations may be observed even at low impurity concentrations (10^{16}-10^{17} cm^{-3}). Therefore, an improvement in the structural perfection of grown EL of SiC makes it possible to reduce drastically the density of the 2nd phase inclusions. For Al, B and Ga the results obtained by the

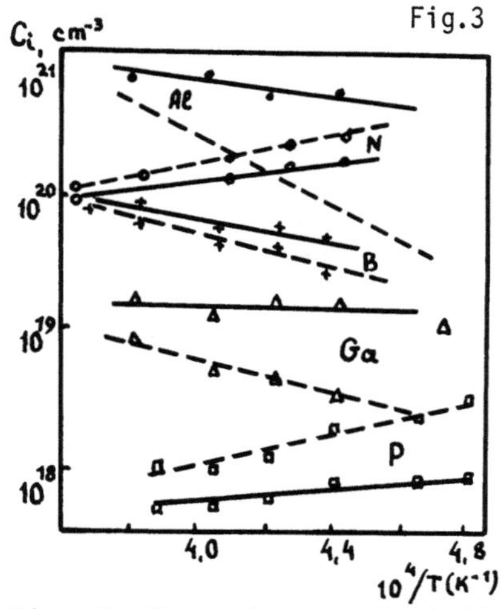

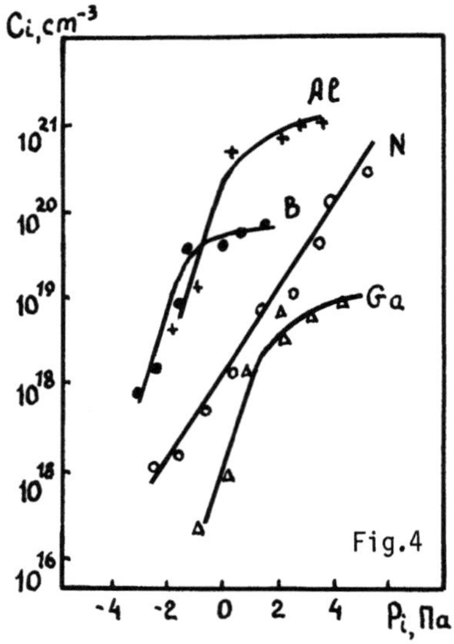

Fig. 3. Dependence of limit level of doping EL of SiC with N, B, Al, Ga and P upon growth temperature. Growth direction: [0001]Si - continuous lines, [000$\bar{1}$]C - dashed lines.

Fig. 4. Dependence of N, B, Ga and Al concentrations in EL SiC upon partial pressure of impurity. T_g=2000°C.

neutron-activation method are in good agreement with the data obtained by the electrophysical measurements of the Hall effect, which confirms that these elements are generally active. Thus, the sublimation "sandwich method" may be employed to obtain EL of SiC, having n- and p-type conductivity, with a concentration of donors and acceptors up to $5 \cdot 10^{20}$ cm^{-3} and $2 \cdot 10^{21}$ cm^{-3}, respectively. Limit concentrations for impurities in EL are given in Table 1.

In some cases the impurity concentration in EL of SiC depends upon the growth rate (V_g).For example,in Ga doping the element concentration decreases at smaller rates of growth. It is of interest that at diffusion saturation (i.e. at V_g=0) the Ga concentration is two orders lower than that obtained in EL of SiC under the same conditions. Independent of the growth rate and method of introduction this impurity is generally in the electrically active state. In the case of nitrogen doping, increase of the growth rate causes a decrease in the nitrogen concentration in EL of SiC. However, this effect can be observed at low concentrations only ($N_D < 10^{18}$ cm^{-3}).

Dependences of impurity concentration in EL of SiC upon partial pressure of impurity vapors (P_i) are given in Fig. 4. The Siverts relationship [10] is true for nitrogen within a very wide range of P_i (10^{-3} - 10^{5} Pa). This impurity does not tend to solubility saturation. Adding acetonitrile to the system of vapors causes only a slight increase of the doping

Table 1. Limit concentrations for impurities in epitaxial layers of SiC, grown by the sublimation "sandwich method"

Element	Concentrat. cm^{-3}	Element	Concentrat. cm^{-3}	Element	Concentrat. cm^{-3}
Li	$1.2 \cdot 10^{18}$	Cr	$3.0 \cdot 10^{17}$	In	$9.5 \cdot 10^{16}$
Be	$8.0 \cdot 10^{20}$	Mn	$3.0 \cdot 10^{17}$	Sn	$1.0 \cdot 10^{16}$
B	$2.5 \cdot 10^{20}$	Cu	$1.2 \cdot 10^{17}$	Sb	$8.0 \cdot 10^{15}$
N	$6.0 \cdot 10^{20}$	Ga	$1.8 \cdot 10^{19}$	Ho	$6.0 \cdot 10^{16}$
Al	$2.0 \cdot 10^{21}$	Ge	$3.0 \cdot 10^{20}$	Ta	$2.0 \cdot 10^{17}$
P	$2.8 \cdot 10^{18}$	As	$5.0 \cdot 10^{16}$	W	$2.5 \cdot 10^{17}$
Sc	$3.2 \cdot 10^{17}$	Y	$9.5 \cdot 10^{15}$	Au	$1.0 \cdot 10^{17}$
Ti	$3.3 \cdot 10^{17}$				

level. From the analysis of the data obtained it may be inferred that the doping process in EL of SiC is primarily kinetic in origin. This is not a surprise if we consider an exclusively low diffusion mobility of impurities in SiC [8], which causes the relaxation processes taking place in the surface layer of the growing crystal to proceed very slowly [9]. It is obvious that the stages restricting the capture of impurities differ depending upon their chemical origin. For nitrogen, doping may be restricted by the desorption of N_2 molecules from the specimen surface; this is why the saturation limit cannot be reached despite the relatively high value of the nitrogen condensation factor K_i. At the same time doping for other impurities, i.e. Ga, Al, B and In, is restricted by the transfer of an impurity from chemisorbate to the subsurface layer. The strong dependence of impurity content in EL of SiC upon many kinetic factors may be the cause of non-uniform doping of crystals and EL up to the appearance of impurity precipitates and the 2nd phase impurities.

4. Conclusion

The impurity doping of SiC EL over the temperature range of 1700-2600°C has a kinetic nature, which is a result of a very slow rate of relaxation processes in SiC crystals. This provides great dependence of the layer doping level on the substrate orientation, vapor phase composition and condensation rate. The sublimation "sandwich method" ensures controlled doping of EL of SiC with different impurities in the range from 10^{16} cm^{-3} up to concentrations close to the solubility limit, and provides sufficiently high structural perfection of the layers grown.

5. References

1. Yu.A.Vodakov, E.N.Mokhov, M.G.Ramm, A.D.Roenkov// Krist und Techn., 14, 729 (1979).
2. E.N.Mokhov, I.L.Shulgina, A.S.Tregubova, Yu.A.Vodakov // Cryst.Res. and Techn., 16, 879 (1981).
3. E.N.Mokhov, R.G.Verenchikova, Yu.A.Vodakov, E.E.Goncharov, G.A.Lomakina, V.G.Oding, M.G.Ramm, A.D.Roenkov,G.G.Ryabova: in Doped Semiconductor Materials, ed. by V.S.Zemslov (Nauka, Moscow, 1985) p. 45.
4. V.F.Dorfman // Crystallography (USSR), 13, 140 (1968).
5. A.G.Zabrodsky, G.A,Lomakina, E.N.Mokhov: in Sixth All-Union Conference on Physical and Chemical Fundamentals of Doping Semiconductor Materials. Abstract of Reports (Nauka, Moscow, 1988) p. 38.
6. E.N.Mokhov, B.S.Makhmudov, M.M.Usmanova, G.F.Uldashev // Inorganic Materials (USSR), 17, 258 (1981).
7. E.N.Mokhov, B.S.Makhmudov, M.M.Usmanova, G.F.Uldashev // Soviet Tech. Phys. Lett. 8, 77, 1982.
8. Yu.A.Vodakov, E.N.Mokhov: in Silicon Carbide - 1973, ed. by R.C.Marshall, J.W.Faust,Jr. and C.E.Ryan (Univ. of South Carolina, Columbia, SC 1974) p. 508.
9. A.A.Chernov: in Modern Crystallography (in Russian) v. 3, ed. by B.K.Vainstein, A.A.Chernov, L.A.Shuvalov (Nauka, Moscow, 1980) p. 7.
10. C.H.P.Lupis: Chemical Thermodynamics of Materials (North-Holland, New York, Amsterdam, Oxford) 1983.

Formation of a Silicon Heterostructure by Germanium Ion Implantation

A. Fukami, K. Shoji, and T. Nagano

Hitachi Research Laboratory, Hitachi Ltd., 4026 Kuji, Hitachi-city, Ibaraki 319-12, Japan

Formation of a silicon heterostructure by germanium ion implantation was investigated. A germanium-implanted layer was grown epitaxially in the solid phase by thermal annealing. Two kinds of crystalline defects were observed. One is a misfit dislocation, and the other is a residual dislocation caused by the ion implantation. The p-n junction fabricated in the SiGe layer had a leakage current which was 3 orders of magnitude larger than that of a pure Si p-n junction. Carbon doping in the SiGe layer improved the crystalline quality and the electrical characteristics.

1. INTRODUCTION

In the 21st century, electronic systems will require higher-speed and higher-density LSIs. To meet these requirements, a very high-speed bipolar transistor operating at lower supply voltage is needed. Therefore, an increase in the cutoff frequency and a reduction in the built-in voltage between the base and emitter junction are essential. These can be accomplished by using a heterojunction bipolar transistor with a narrow-bandgap base, made up of material such as SiGe alloy.

Recently, several studies on the fabrication of a SiGe heterojunction bipolar transistor with various growth techniques have been reported [1-3]. For example, deposition of SiGe with molecular beam epitaxy can optimize control of impurity profiles. However, considering the process compatibility with the present standard Si-LSIs fabrication line,it is difficult and costly to merge any one of these techniques with a standard process. Ion implantation is one of the most conventional techniques, and is expected to be suitable for fabricating SiGe alloy. In this study, SiGe/Si heterostructures are fabricated using germanium ion implantation into silicon and subsequent solid phase epitaxy. The structural properties and the electrical characteristics are also examined. Carbon ions are also introduced to improve the crystalline quality of the SiGe layer, instead of boron doping [4].

2. EXPERIMENTAL PROCEDURES

2.1 Formation of SiGe alloy

The starting material was n-type (100) silicon with a resistivity of 10 Ω·cm. Germanium ions were implanted at room temperature. GeF_4 gas pressure during ion implantation was 4×10^{-5} Torr. Germanium implantation was performed through a thermally grown 10 nm-oxide at an acceleration energy of 50 keV, and a dose of 2.5×10^{16} cm^{-2}. The dose rate was about 1×10^{14} cm^{-2}/min. The peak germanium concentration is about 8×10^{21} cm^{-3}. This value corresponds to a germanium content of 16

%. For some samples, carbon ion implantation was subsequently carried out to compensate the lattice mismatch between the silicon substrate and the implanted layer. Carbon ions were implanted at an energy of 15 keV and a dose of 3.0×10^{15} cm^{-2}. This energy was chosen to yield equal projected ranges for both germanium and carbon ions. The carbon dosage was determined to provide an average lattice constant for the Si, Ge, and C mixture equal to that of Si. These samples were then furnace-annealed in a N_2 ambient at 600°C for 24 h. Rapid thermal annealing (RTA) at 1000°C for 10 s was also performed on some samples.

The impurity concentration profiles were estimated by secondary ion mass spectroscopy (SIMS). The crystalline quality was characterized by cross sectional transmission electron microscopy (XTEM) and also high resolution XTEM lattice imaging.

2.2 Formation of p-n junction diode

In order to evaluate the crystalline quality electrically, a planar p-n junction structure was utilized. Figure 1 shows a schematic of the device structure. The impurity profiles in the n^{++}- and p^{+}-regions are the same as those in the emitter and the base regions of a bipolar transistor. The size of the n^{++}-region is selected to be 1×5 μm^2. Figure 2 shows the designed impurity profiles for Ge, C, B, and As. The n^{++}-p^{+} junction is 30 nm from the surface. The germanium content at this site is around 5 %, whereas the peak germanium content is 16 % at 70 nm from the surface. Therefore, the SiGe bandgap must vary with depth. As the carbon content is about 1/8 of the germanium content, the bandgap of the carbon-added SiGe layer is narrower than that of Si.

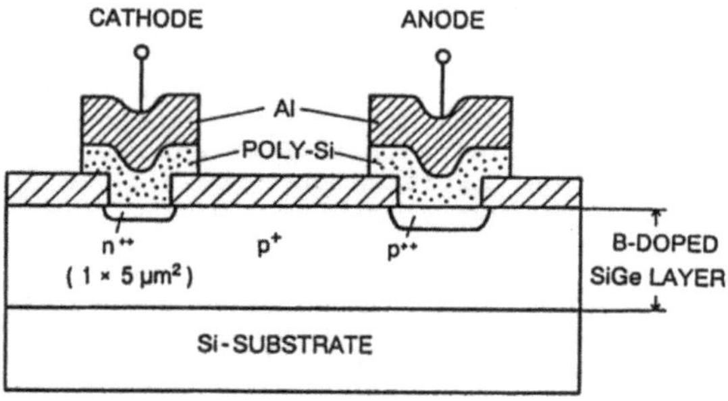

Fig. 1. Schematic cross-section of n^{++}-p^{+} junction diode

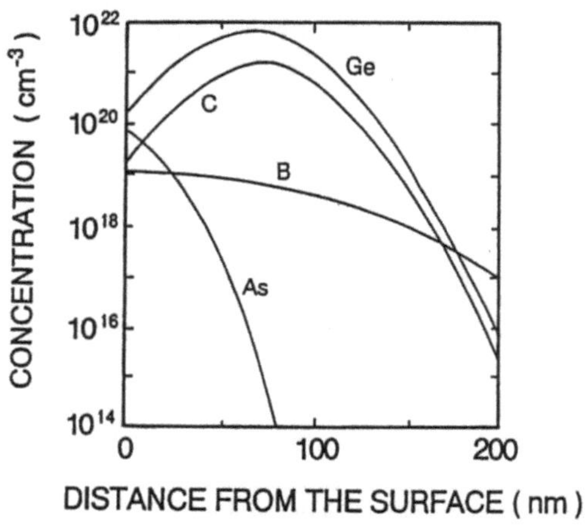

Fig. 2. Designed profiles of Ge, C, B and As in n^{++}-p^{+} junction diode

The starting material was p-type (100) silicon with a resistivity of 10 Ω·cm. Germanium ions were implanted at an acceleration energy of 120 keV, and a dose of 5×10^{16} cm^{-2}. These values are twice that of the previous experiment. The dose rate was about 1×10^{13} cm^{-2}/min. In some samples, carbon ions were also implanted at an energy of 33 keV and the dose range was 3-12×10^{15} cm^{-2}. Boron ions were subsequently implanted to form the p^+-region. Two annealing steps were carried out. First, a low temperature anneal occurred at 600°C for 24 h, followed by a higher one at 950°C for 10 min. Then, a SiO_2 film was deposited, diffusion windows were opened, and a poly-Si film was deposited. Arsenic and boron ions were partially implanted to form the n^{++}- and p^{++}-region respectively. A furnace anneal at 950°C for 10 min was then performed. Finally, aluminum deposition was made to form metal contacts.

3. RESULTS AND DISCUSSION

3.1 Crystalline quality of SiGe layer

Figure 3 shows the germanium concentration profiles for a sample as-implanted and after annealing. As a reference, the dotted line indicates the result of a simulation. The experimental as-implanted result shows a higher concentration at the surface and a deeper region than that of the simulation. This is attributed to a diffusion during the ion implantation step due to beam heating, because of no wafer cooling apparatus in the implanter. Germanium ions diffused about 25 nm due to furnace annealing at 600°C for 24 h.

XTEM micrographs of the as-implanted sample and of the furnace-annealed sample are shown in Figs. 4(a) and 4(b) respectively. Fig. 4(a) shows that an amorphized layer was formed by the germanium ion implantation up to a depth of about 100 nm. The concentration of germanium is more than 3×10^{19} cm^{-3} in this region. As shown in Fig. 4(b), there are some defects under the surface, and at the original amorphous-crystalline interface. The region in between is the recrystallized layer.

To investigate the details of the recrystallized layer, the samples were then examined by high resolution XTEM. Figure 5(a) shows a high resolution micrograph of the interface between the amorphized layer and the silicon substrate in the as-implanted sample, as viewed from the [110] direction. The upper part is totally amorphized

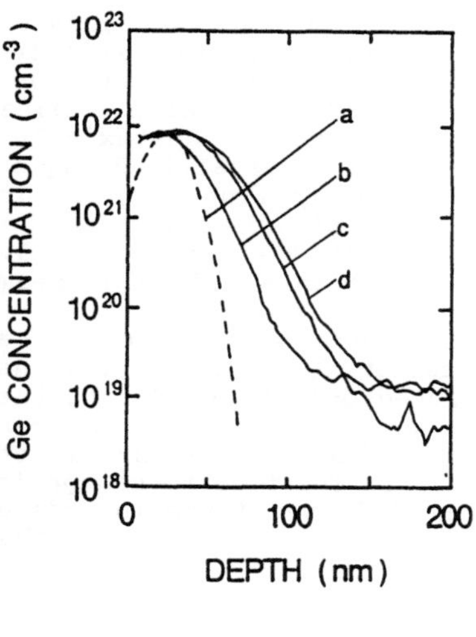

Fig. 3. SIMS profiles of Ge : (a) as-implanted (simulation), (b) as-implanted (experiment), (c) annealed at 600°C for 24h, (d) annealed at 600°C for 24h followed by RTA at 1000°C for 10s

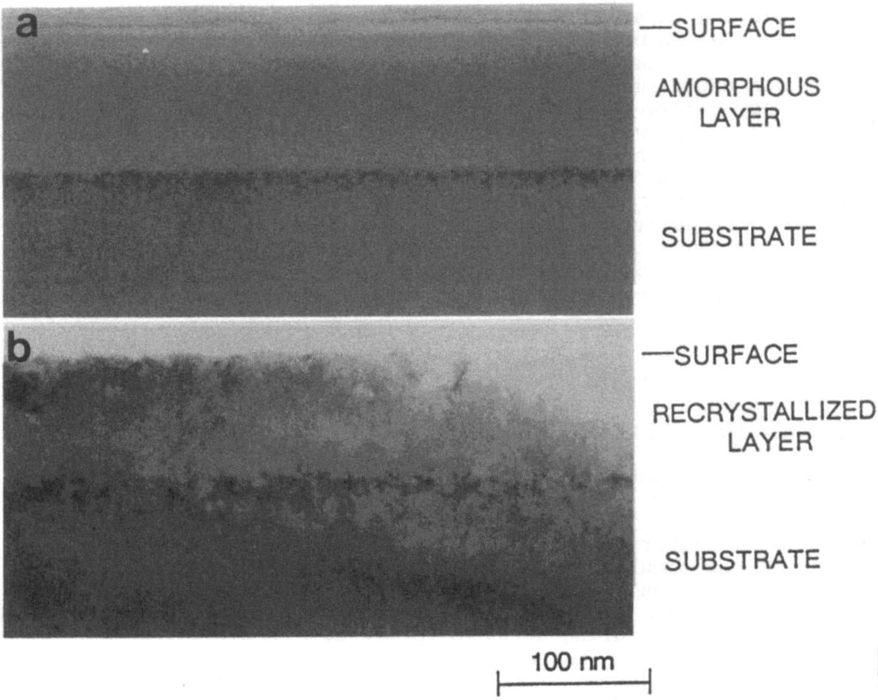

Fig. 4. XTEM micrographs of SiGe/Si structures : (a) as-implanted, (b) annealed at 600°C for 24h

because of no lattice image. In the silicon substrate region there are some strains and defects induced by the germanium ion implantation. Figure 5(b) is a high resolution view of the interface between the recrystallized layer and the substrate in the furnace-annealed sample. There are some defects. These are the dislocations due to the residual germanium ion damage. The lattice image shows that the SiGe layer is epitaxially grown on the silicon substrate. Figure 5(c) is another high resolution micrograph for the furnace-annealed sample, where the surface region is examined. Because there is no ordering of the lattice image under the surface, the top layer is still amorphous even after annealing at 600°C for 24 h. More thermal annealing is needed to grow the entire region epitaxially. Furthermore, this lattice image shows that the recrystallized region is a solid phase epitaxial (SPE) layer and many line and planar defects exist. These defects form an angle of about 60° to the surface. As the concentration is about $8{\times}10^{21}cm^{-3}$ as shown in Fig. 3, these defects resemble the misfit dislocation in a molecular beam epitaxial SiGe/Si heterostructure [5].

The RTA process at 1000°C for 10 s was added to grow the entire region epitaxially and to reduce the dislocation at the interface. As shown in Fig. 3, the germanium atoms diffused only 10 nm during the RTA. Figure 6 shows a XTEM micrograph of the RTA sample. No amorphized layer exists under the surface. That is, the SPE layer reached the top surface by the additional RTA. There is no change in the defects near the surface. However, the defects of the interface changed their shape. Some of the dislocations were removed by the RTA.

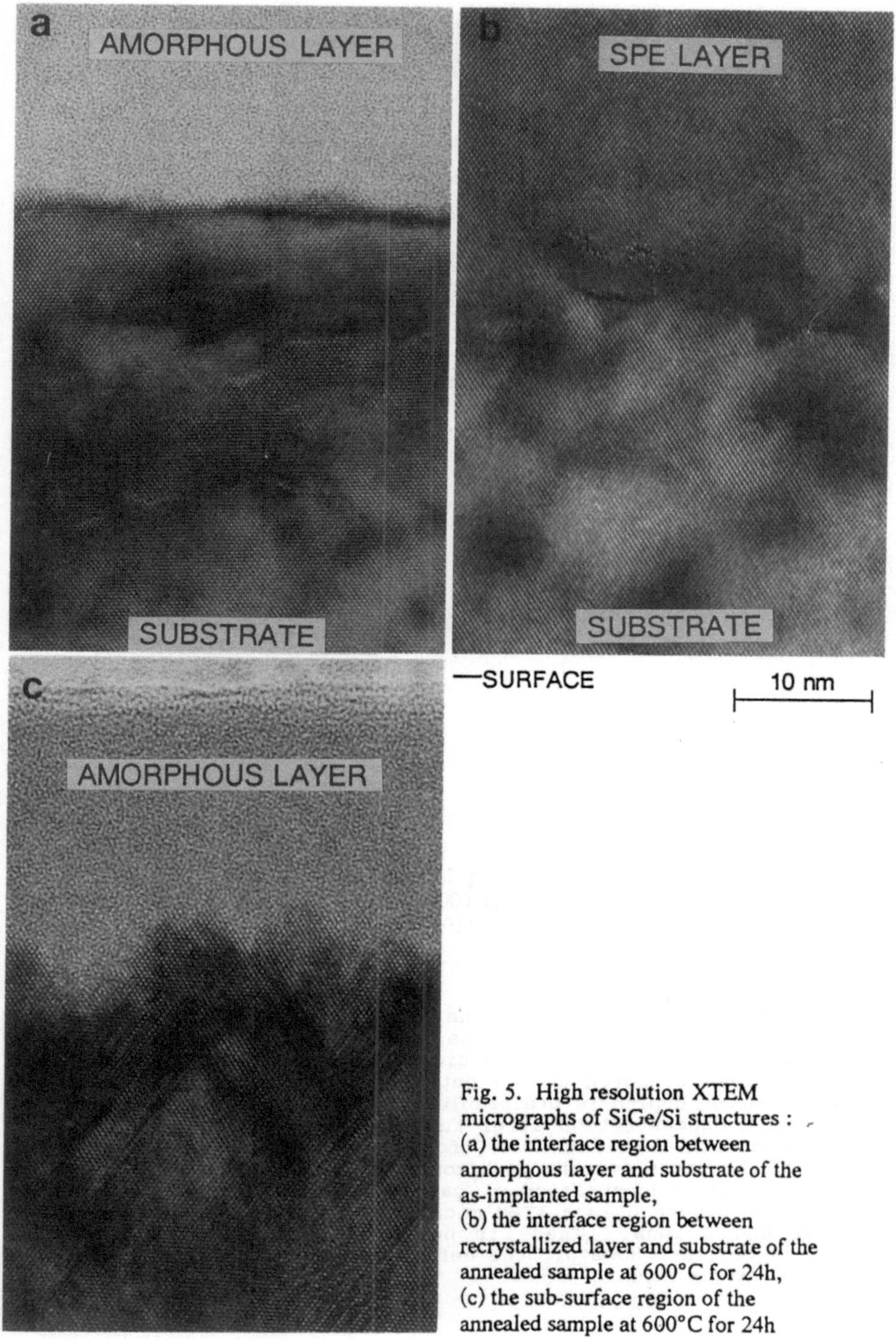

Fig. 5. High resolution XTEM micrographs of SiGe/Si structures : (a) the interface region between amorphous layer and substrate of the as-implanted sample, (b) the interface region between recrystallized layer and substrate of the annealed sample at 600°C for 24h, (c) the sub-surface region of the annealed sample at 600°C for 24h

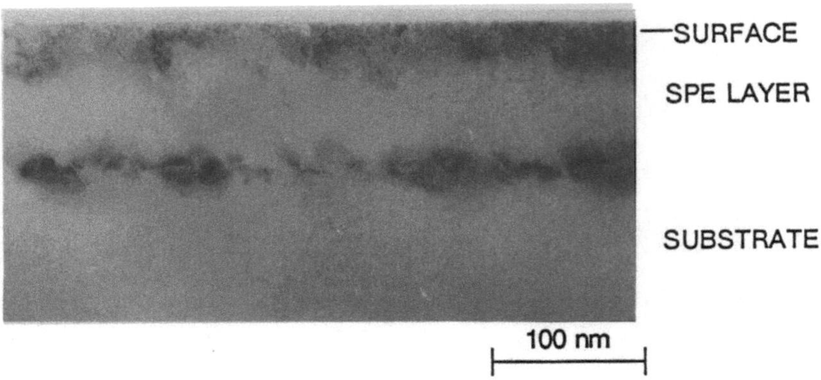

Fig. 6. XTEM micrograph of SiGe/Si structure annealed at 600°C for 24h followed by RTA at 1000°C for 10s

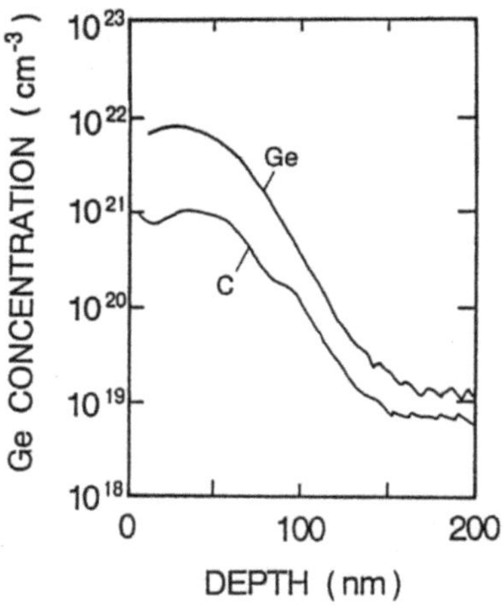

Fig. 7. SIMS profiles of Ge and C annealed at 600°C for 24h followed by RTA at 1000°C for 10s

It has been proposed that the doping of smaller atoms such as boron improves the crystalline quality of SiGe, due to the compensation effect upon the lattice mismatch [4]. As carbon has a diameter close to boron and is not a donor or acceptor impurity in silicon, carbon doping is preferable to the SiGe system. The profiles of Ge and C concentration after furnace-annealing at 600°C for 24 h and an additional RTA at 1000°C for 10 s are shown in Fig. 7. As expected, the depth for the peak concentration of Ge and C are nearly equal, and the concentration of carbon is about one eighth that of germanium. Figure 8 is a XTEM micrograph for a sample co-implanted with Ge and C ions. There is no change in the defects of the interface region. However, the defects near the surface are reduced by adding carbon ions.

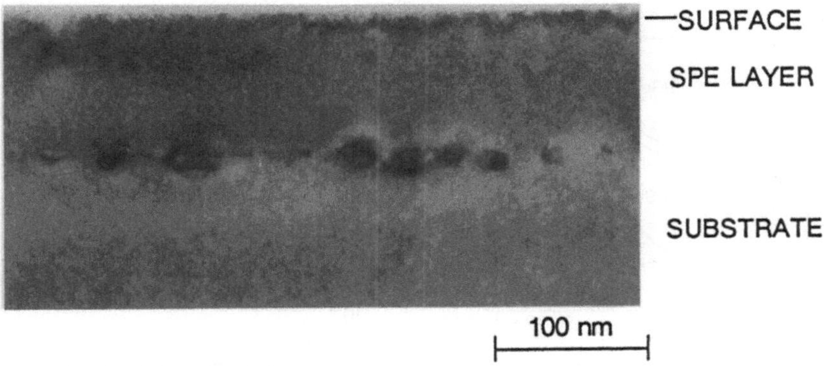

Fig. 8. XTEM micrograph of C-added SiGe/Si structure annealed at 600°C for 24h followed by RTA at 1000°C for 10s

3.2 Electrical characteristics of the p-n junction diodes

Figure 9 shows the current-voltage characteristics for three types of diodes. The diode with the SiGe base (a) does not have as good a reverse characteristics as the diodes with the carbon-added SiGe (b) and pure Si base (c). Figure 10 shows the forward V-I characteristics. The vertical axis shows the current density in the n^{++}-region. The SiGe and the carbon-added SiGe diodes have larger current at lower supply voltages than the Si diode. The ideality factor of both diodes is about 2. This larger current is attributed to recombination-generation caused by the defects at the surface. The current decreases as carbon ions are added. However, the slope does not change. Figure 11 displays the reverse V-I characteristics. The leakage current of the SiGe diode is about 3 orders of magnitude larger than that of a pure Si diode. This is also attributed to large generation current associated with defects in the SiGe layer. The leakage current decreases as carbon ions are added. Figure 12 is the dependence of the breakdown voltage on the carbon ion dose. As a reference, the values for the SiGe diode and the pure Si diode are

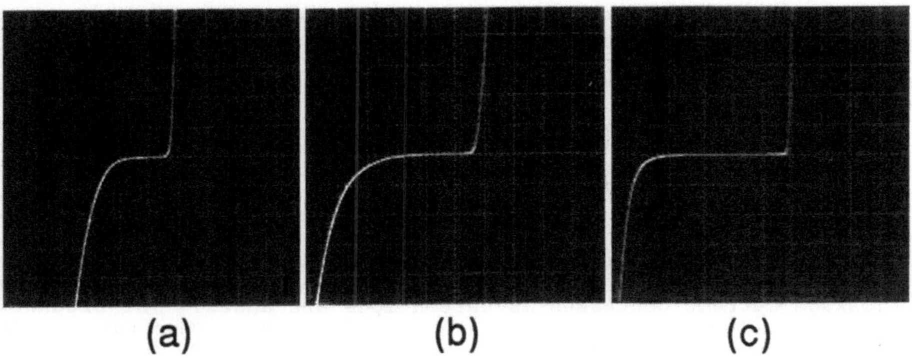

Fig. 9. Electrical characteristics for n^{++}-p^{+} junction diodes :
(a) SiGe, (b) C-added SiGe, (c) pure Si
(Horizontal and vertical scales are 1V/div. and 50μA/div., respectively)

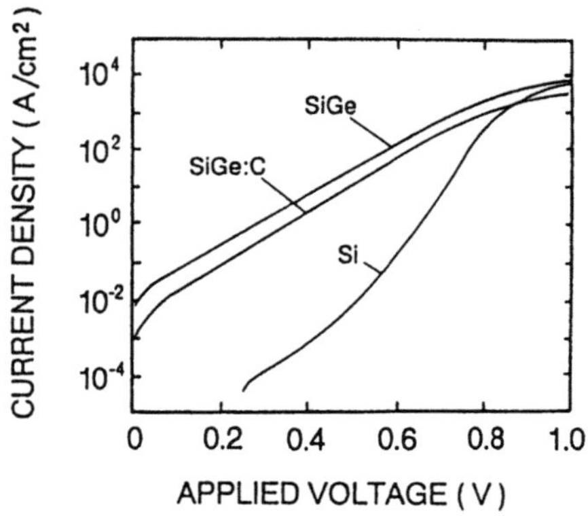

Fig. 10. Forward V-I characteristics for n^{++}-p^{+} junction diodes with SiGe, C-added SiGe and pure Si base

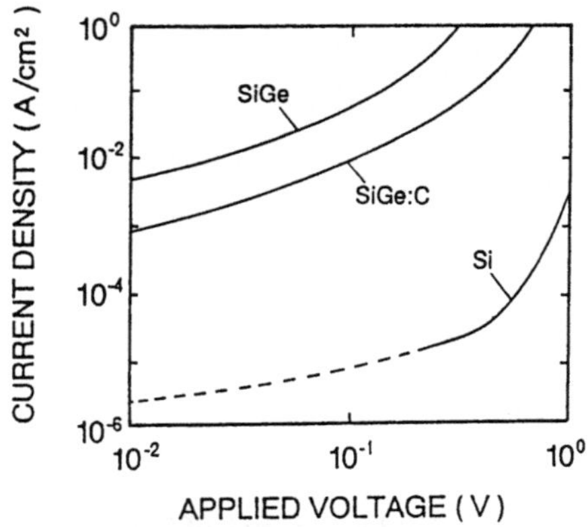

Fig. 11. Reverse V-I characteristics for n^{++}-p^{+} junction diodes with SiGe, C-added SiGe and pure Si base

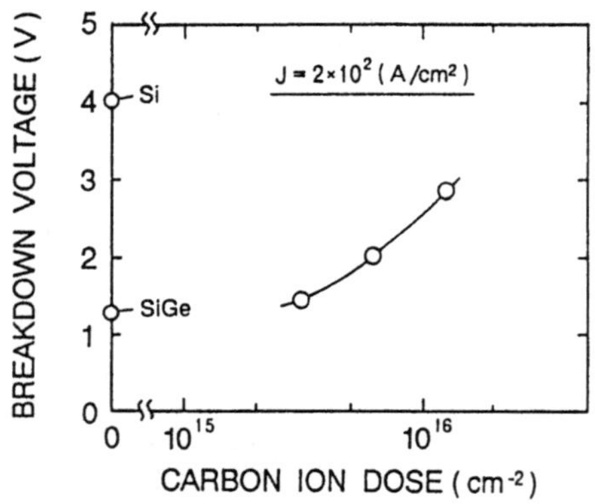

Fig. 12. Dependence of the breakdown voltage on the C ion dose

also plotted. The SiGe diode has the lowest breakdown voltage. However, the breakdown voltages increase with an increase in carbon ion dose.

4. SUMMARY

A SiGe heterostructure was formed by germanium ion implantation followed by thermal annealing. There are two kinds of defects in the recrystallized SiGe layer. One is a misfit dislocation due to the high content of germanium, which makes an angle of 60° with the surface. The other is the residual dislocation caused by the ion implantation.

The p-n junction diodes were fabricated in the SiGe layer. There is a larger recombination current than a pure Si diode. The reverse leakage current decreases as carbon ions are added. The breakdown voltage increases with an increase in carbon ions. Therefore, carbon doping improves the SiGe crystalline quality.

ACKNOWLEDGEMENTS

The authors would like to thank Dr. C. Y. Yang at Santa Clara University and Dr. T. Tokuyama at University of Tsukuba for their stimulating discussions. Thanks are also due to Dr. N. Momma for his encouragement throughout this work.

REFERENCES

1. H. -U. Schreiber, and B. G. Bosch : IEDM Tech. Dig., pp.643-646 (1989)
2. T. I. Kamins, K. Nauka, L. H. Camnitz, J. B. Kruger, J. E. Turner, S. J. Rosner, M. P. Scott, J. L. Hoyt, C. A. King, D. B. Noble, and J. F. Gibbons : IEDM Tech. Dig., pp.647-650 (1989)
3. S. E. Fischer, R. K. Cook, R. W. Knepper, R. C. Range, K. Nummy, D. C. Ahlgren, M. Revitz, and B. S. Meyerson, : IEDM Tech. Dig., pp.890-892 (1989)
4. K. Ohta, J. Sakano, and S. Furukawa : Ext. Abs. 21st Conf. SSDM, pp.555-556 (1989)
5. Y. Fukuda, Y. Kohama, M. Seki, and Y. Ohmachi : Jpn. J. Appl. Phys. 28, pp.L19-L20 (1989)

Low Temperature Selective Deposition of Diamond Thin Films

R. Ramesham, C. Ellis, and T. Roppel

Electrical Engineering Department, Alabama Microelectronics Science and Technology Center, Auburn University, AL 36849, USA

Selective growth of polycrystalline diamond thin films has been achieved at a substrate temperature of 600° C by high pressure microwave plasma-enhanced chemical vapor deposition in a hydrogen/methane plasma. A novel process sequence is delineated to pattern the scratch damaged silicon area to grow diamond selectively. Low temperature growth is performed by cooling the substrate holder with nitrogen gas. As-grown films are analyzed by Raman spectroscopy and scanning electron microscopy.

1. Introduction

Diamond is an important material for innumerable applications because of its remarkable chemical and physical properties such as high thermal conductivity, extreme hardness, high Young's modulus, high electrical resistivity, etc. These features are technologically significant especially in microelectronics and optics, which require minimum mechanical stress in thin diamond films [1,2], good adhesion strength to the substrate, and very high smoothness of the film. For high pressure microwave plasma assisted chemical vapor deposition (HPMACVD), substrate temperature for diamond thin film growth on a variety of substrates is typically greater than 900° C. Mechanical stress in these diamond thin films may be minimized by reducing the diamond growth temperature and this could substantially increase its applicability in microelectronics and optics.

Low temperature diamond deposition by the HPMACVD technique has been tried on a variety of substrates by lowering microwave power and gas pressure [3] and it was reported that

Springer Proceedings in Physics, Volume 56 **Amorphous and Crystalline Silicon Carbide III**
Editors: G.L. Harris, M.G. Spencer, and C.Y. Yang

diamond could be deposited at less than 500° C. In addition, there has been a report on diamond growth achieved at low substrate temperatures by cooling the substrate holder with nitrogen gas [4]. In this paper we report the low temperature selective deposition of diamond by cooling the substrate at 600° C using nitrogen gas. The analysis of as-grown films by Raman spectroscopy and scanning electron microscopy are also reported.

2. Experimental Details

An ASTeX HPMACVD system was used in our experiments to grow polycrystalline diamond thin films at high temperature. Cooling coils were fabricated inside the substrate holder to allow the nitrogen gas to flow through in order to grow the diamond at low temperatures by dissipating the heat during the growth. The schematic diagram of the diamond deposition system, details on the growth procedure and selective deposition process are described in detail elsewhere [2,5].

The steps by which the silicon substrate was prepared for selective diamond deposition are shown in Figure (1). Substrates

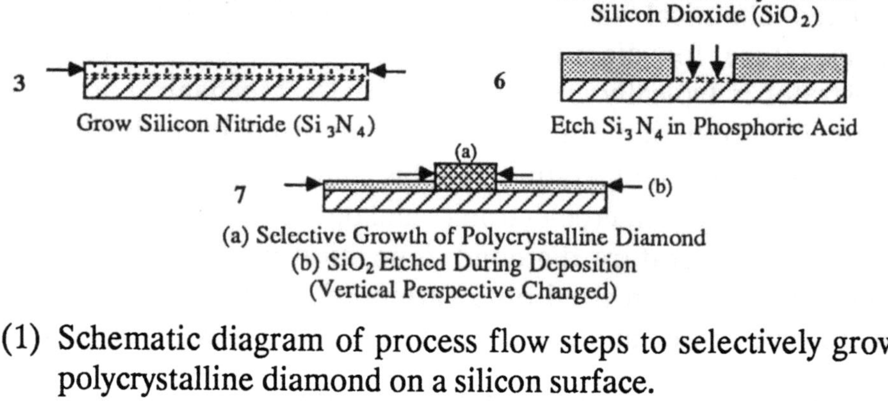

(1) Schematic diagram of process flow steps to selectively grow polycrystalline diamond on a silicon surface.

were one side polished single crystal silicon wafers. The polished side of the wafer was manually scratched using diamond paste containing 0.25 μm mean particle size to enhance the diamond nucleation density substantially. The wafers were then thoroughly washed and dried. Silicon nitride (Si_3N_4) was grown on the scratched side of the wafer at 800° C using a gas mixture of NH_3 and dichlorosilane by low pressure chemical vapor deposition. The Si_3N_4 was patterned and plasma etched in a gas mixture of 90% CF_4 + 10% O_2, and the substrate was cleaned. The silicon wafer with the nitride pattern was oxidized using pyrogenic steam at 1050° C for 10 hrs to form a SiO_2 thickness of 1.3 μm. The scratch density in the oxidized regions was minimized to a large extent by the formation of the oxide and the silicon surface also was smoothed during plasma etching of Si_3N_4 (or scratch edges are rounded). Therefore, the oxidized region was not susceptible to diamond growth over the duration of diamond film deposition. The Si_3N_4 pattern was used as a mask to prevent oxidation, to retain scratching, and eventually to grow diamond in that region selectively. A thin layer of oxide was formed on the Si_3N_4 during the oxidation process and it was etched in a solution of NH_4F + HF (6:1) for an appropriate length of time ($\sim$2 min.). Hot H_3PO_4 ($\sim$180° C) was used to chemically etch the Si_3N_4 completely. The substrate was cooled to room temperature, washed thoroughly and dried. Typical deposition parameters used to grow polycrystalline diamond selectively using microwave plasma system were as follows: Deposition pressure = 59.5 Torr; CH_4 flow rate = 20 SCCM; H_2 flow rate = 500 SCCM; Forward power = 1375 watts; Reflected power = 50 watts; Temperature of the substrate = 600° C (as measured by an IR thermometer); and percentage of CH_4 in H_2 = 4.

3. Results and Discussion

Figure (2) is a scanning electron micrograph showing the morphology of a polycrystalline diamond thin film selectively deposited on a silicon surface at a temperature of 600° C. Figure

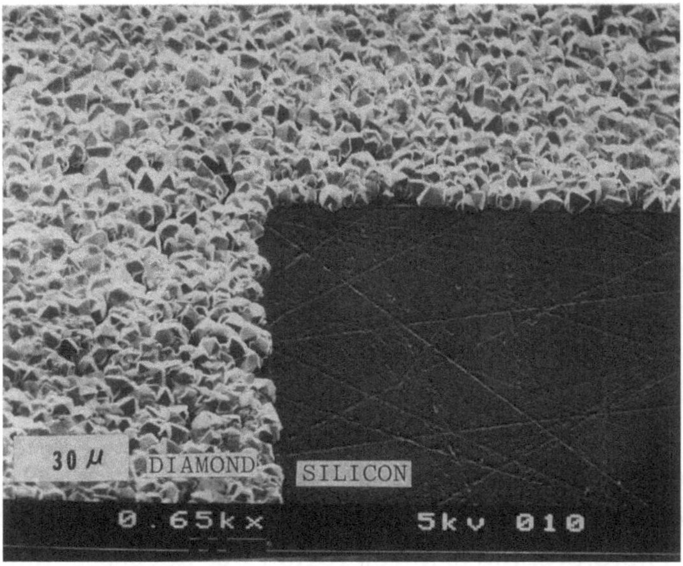

(2) Scanning electron micrograph of selectively deposited, well-faceted morphology of polycrystalline diamond thin films on a silicon substrate. Deposition temperature: 600° C.

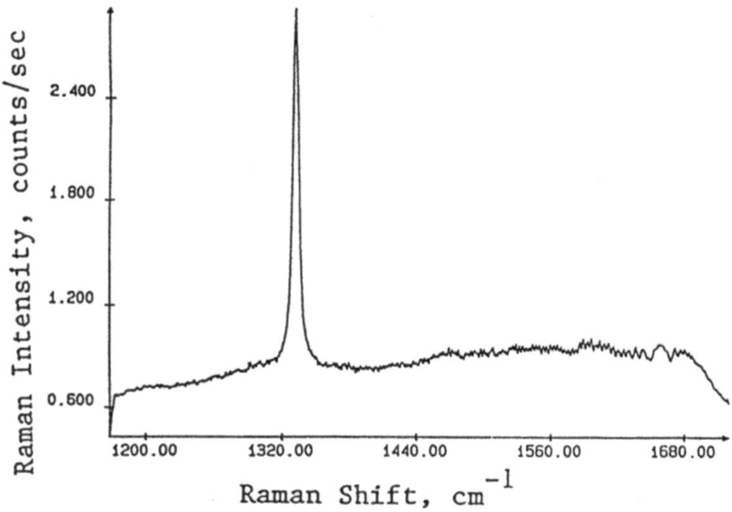

(3) Typical Raman spectrum of polycrystalline diamond thin films grown on a silicon substrate at 600° C.

(3) shows the Raman spectrum. The peak at 1332-1333 cm^{-1} is characteristic of diamond. It is quite interesting to note the absence of a graphite peak in the Raman spectrum, since we generally observed such a peak in the high temperature case [2,5]. This may be explained as follows. Deposition of diamond and graphite compete during growth. Graphite has been etched during growth by atomic hydrogen solvent in the microwave plasma. The diamond deposition rate is reduced as substrate temperature is lowered. There is sufficient time to etch away graphite during growth. Therefore, it may be concluded that the etching rate of graphite is faster than the growth rate of graphite at a substrate temperature of 600° C.

The nucleation of polycrystalline diamond occurred preferentially only in the non-oxidized regions as anticipated. Selectivity is absolutely retained even at low temperature. The edge delineation follows the defined pattern very closely. Figure 4(a, b, and c) clearly demonstrates the selectivity of diamond thin films deposited on a silicon surface at a substrate temperature of 600° C. The etching of SiO_2 during diamond growth may be seen in Figure 4 (a and c) in the regions near the diamond pattern. Diamond growth has not occurred in the completely oxide etched regions within the experiment duration. The oxide etch might have healed the scratch density causing the silicon surface to become passive for diamond growth. Furthermore, we have also observed a decrease in the diamond particle size and growth rate, and an increase in surface smoothness with decreasing substrate temperature during the growth of thin films [2].

In summary, a novel process developed earlier [5] has been used successfully to grow polycrystalline diamond thin films on a silicon surface selectively in a variety of geometries associated with microelectronics applications at a substrate temperature of 600° C. As-grown films are identified as polycrystalline diamond by Raman spectroscopy. Furthermore, in combination with chemical etching of silicon, fabrication of low mechanical stress diamond microstructures is in progress.

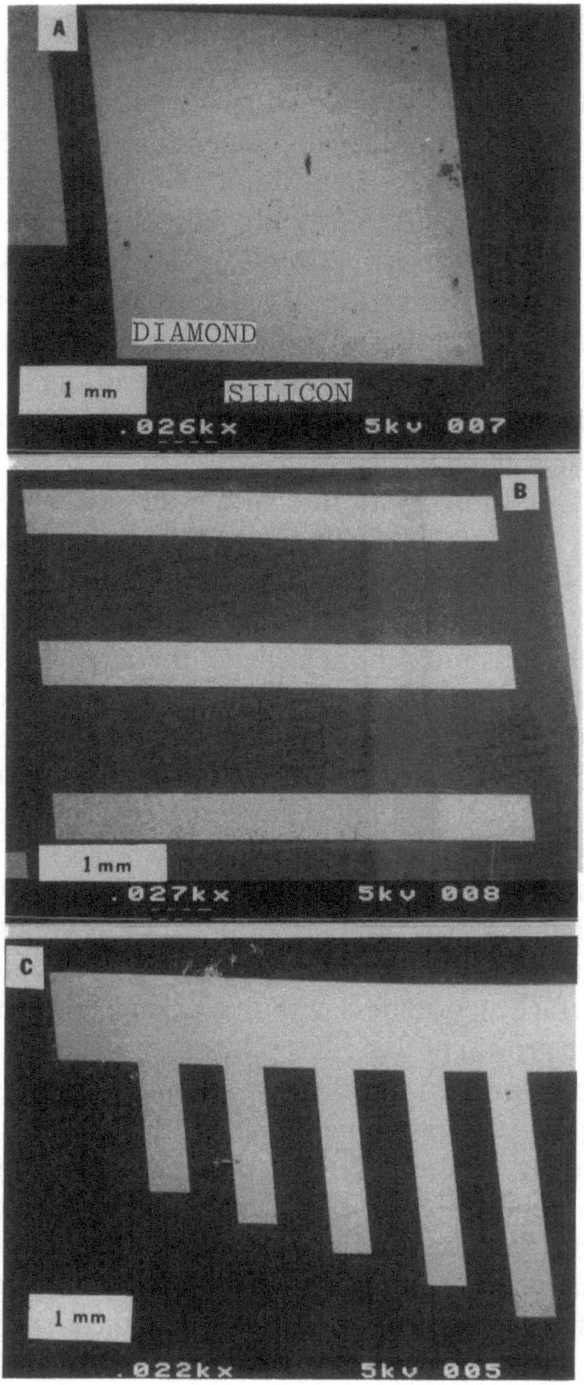
A
DIAMOND
1 mm
SILICON
.026kx 5kv 007
B
1 mm
.027kx 5kv 008
C
1 mm
.022kx 5kv 005

4. Acknowledgements

This work was supported in part by the Strategic Defense Initiative Organization's Office of Innovative Science and Technology (SDI/TNI) through Navy Contract #N60921-86-C-A226 with Naval Surface Warfare Center and in part by the Alabama Microelectronics Science and technology Center. Thanks are due to Dr. D.A. Jaworske and Mr. F. Randy deCelle for their critical comments and assistance in the art work, respectively.

References

1 C.E. Johnson, W.A. Weimer, and D.C. Harris, *Mat. Res. Bull.*, **24**, 1127 (1989).
2 R. Ramesham, C. Ellis, T. Roppel, D.A. Jaworske, and W. Baugh, Spring MRS Meeting, San Francisco, CA, Abstract #M6.11 (1990).
3 Y. Liou, A. Inspektor, R. Weimer, and R. Messier, *Appl. Phys. Lett.*, **55**, 631 (1989).
4 W.L. Hsu, E.A. Fuchs, and K.F. McCarty, Fourth Annual SDIO/IST-ONR, Diamond Technology Initiative Symposium, Crystal City, VA, Paper THP12a, July (1989).
5 J.L. Davidson, C. Ellis, and R. Ramesham, *J. Electron. Mater.*, **18**, 711 (1989).

(4) Scanning electron micrographs (a, b, and c) of selectively deposited polycrystalline diamond thin films in various designs deposited at a substrate temperature of 600° C.

Two-Dimensional Computational Modeling of Silicon Carbide Deposition in a Vertical CVD Reactor

M.E. Aluko and W. Givens, Jr.

Department of Chemical Engineering & Materials Science Research Center of Excellence, Howard University, Washington, DC 20059, USA

We present the results of extensive two-dimensional finite-element computations of the governing mass (overall and species), momentum, and energy conservation equations used to model growth conditions for the chemical vapor deposition of silicon carbide (SiC) in a specially-designed shroud-flow experimental vertical reactor. In particular, the effect of shroud flow on computed flow patterns is explored as a function of reactor pressure, concentric neck flow rate and substrate temperature. A parametric "design" diagram is obtained to guide the choice of operating conditions. Computed growth rates are in reasonable agreement with experiment.

1. INTRODUCTION

Several reports have been published in the literature involving comprehensive models used to obtain reliable predictions of the operating conditions for uniform and reproducible crystal growth in various reactors [1-10]. In this paper, our focus is on a study of chemical vapor deposition of silicon carbide (SiC) in a vertical reactor. We have carried out extensive two-dimensional numerical analyses of the governing mass (overall and species), momentum, and energy conservation equations used to model SiC growth conditions in a specially-designed shroud-flow experimental reactor currently in operation in our Materials Research Center. Following various parametric changes to be outlined below, their effects on gas flow pattern, temperature distribution and deposition rate uniformity were evaluated.

2. MODELING EQUATIONS AND ASSUMPTIONS

A sketch of the vertical CVD reactor model geometry used in the simulations is shown in Figure 1. The reacting components silane (SiH_4) and propane (C_3H_8) flow in hydrogen (as carrier gas) through the neck inlet towards the heated substrate where deposition occurs. The overall reaction which describes the formation of silicon carbide from silane and propane is given by

$$3\,SiH_4 + C_3H_8 \;\text{---->}\; 3\,SiC + 10\,H_2$$

Considerations for obtaining uniform flow field in the reactor may dictate that additional (hydrogen) gas flow through the specially constructed shroud section. Further details of the reactor are given elsewhere [11].

With respect to the fluid mechanics modeling, the general describing equations for fluid motion and heat transfer are given by Bird, Stewart and Lightfoot [12]. In vector form, the equations of continuity, momentum energy and species A balance are as follows:

$$\nabla.(\rho v) = 0$$

$$\rho(v.\nabla v) = -\nabla P + \nabla.\tau + \rho g$$

$$\rho C_p(v.\nabla T) = \nabla.(k\nabla T)$$

$$\nabla.cvx_A = \nabla.\left[cD_{A\,B}\nabla x_A\right]$$

where, for a Newtonian fluid, and assuming ideal gas behavior:

$$\tau = \mu\left[\nabla v + (\nabla v)^T\right] - \frac{2}{3}\mu I(\nabla.v) = 0$$

$$\rho = \frac{PM_W}{RT}$$

D_{AB} is the binary mass diffusivity of species A (silane or propane) in B (hydrogen), while x_A is the mole fraction of species A (c - total gas concentration).

For the geometry shown in Figure 1, the boundary conditions are as follows:

(1) $v_z = v_n$ at neck inlet, $v_z = v_{shrd}$ at the shroud inlet

(2) $\delta v / \delta r = 0$ @ $r = 0$

(3) $T = Tw$ @ $r = R_i$, $0 < z < h$

(4) $T = Tsub$ @ $z = 0$, $0 < r < r_i$

(5) $v_r = v_z = 0$ @ the walls (no slip condition)

(6) $\delta x_A / \delta r = 0$ @ $r = 0$

(7) $\delta x_A / \delta z = 0$ @ $z = 0$

(8) $x_A = 0$ @ $z = 0$, $0 < r < r_i$

In practice we specify a boundary condition at the surface for which a first-order reaction rate R_A is of the form:

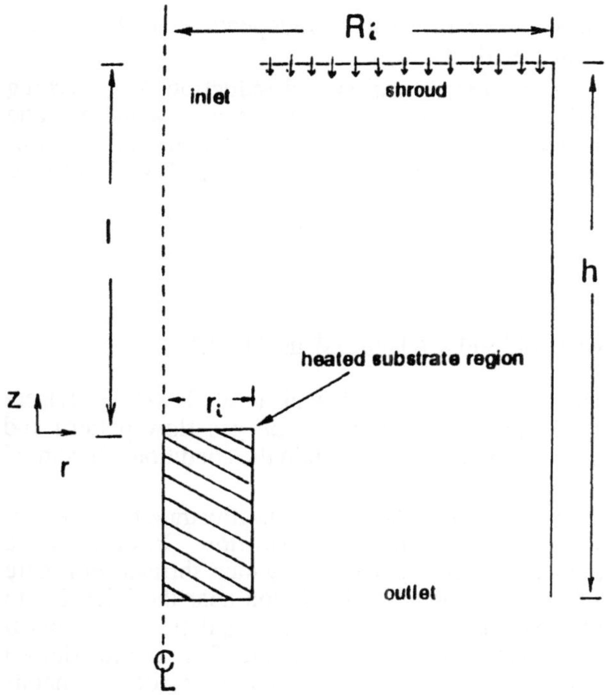

Fig. 1. Sketch of vertical CVD reactor model geometry used in simulations. ($R_i = 0.105$ m; $r_i = 0.016$ m; h = 0.160 m; l = 0.140 m)

$$R_A = -K x_A$$

where the rate constant K is chosen to be very large such that the disappearance of the reactant, SiH_4 or C_3H_8, at the substrate is assumed instantaneous. In essence, the model stipulates that the deposition of silicon carbide is solely dependent on the rate at which the limiting reactant (SiH_4 or C_3H_8) is transported through the carrier gas and to the surface of the substrate.

3. METHOD OF SOLUTION

The above equations are solved by a numerical simulation package, FIDAP [13], which utilizes the standard Galerkin finite element method. The dependent variables are the axial and radial velocities v_z and v_r respectively, the temperature T and the concentration of the "limiting" reactant x_A as a function of the spatial coordinates z and r. Property-dependent fluid properties [μ (viscosity), k (thermal conductivity), and β (volume expansion coefficient)] were assumed following expressions presented by Bird, Stewart and Lightfoot [12] for gases at low density. A variable-size finite element mesh was used in the calculations, with refined mesh at important boundary regions. Due to the assumption of dilute species, the momentum and energy equations were first solved independent of the species equations then imposed on the species field.

The result of the finite elements discretization and reduction of governing equations is a set of algebraic equations describing the gas flow pattern and temperature distribution in the reactor. The average time of solution for the cases simulated is approximately 45 minutes of CPU time on an IBM 3090-I80J mainframe computer.

4. RESULTS

4.1 Effects of process parameters of fluid mechanics & heat transfer

In our study, we have investigated the effects of neck inlet flow rate, system pressure, shroud flow rate and substrate temperature on gas flow pattern and temperature distribution. Our main aim has been to obtain the conditions for shroud flow that lead to a uniform flow pattern.

Figure 2 shows typical streamlines and isotherm plots to illustrate the effect of changing pressure at a fixed inlet neck flow for no shroud flow. In Figure 2a, a huge buoyancy-induced recirculation occurs at 76 torr, which shrinks (in Figure 2b) at 20 torr, but now also features a forced-recirculation near the inlet due to sudden expansion from the inlet to the reactor chamber. Recirculation is somewhat eliminated at 7.6 torr (Figure 2c). However, we note that all of the figures show a wide spread of temperature isotherms, suggesting the possibilty of extraneous thermal reaction processes far removed from the substrate surface. Figure 3 however shows that at a given pressure, increasing the inlet neck flow rate (no shroud flow) would however lead to a more favorable temperature profile, even though recirculation is not eliminated.

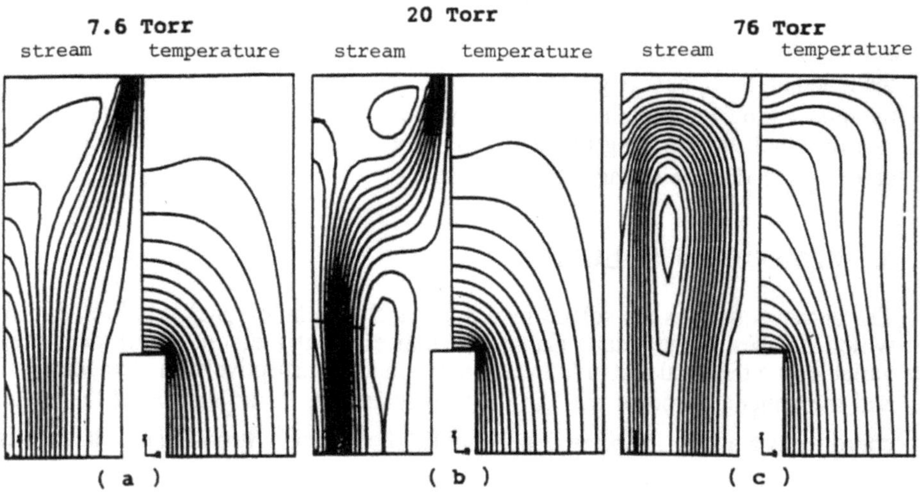

Fig. 2. Effect of pressure on flow pattern for operating at Vn = 2 slm, Vshd = 0 slm, Tsub = 1473 K and pressure (a) 7.6 torr (b) 20 torr and (c) 76 torr.

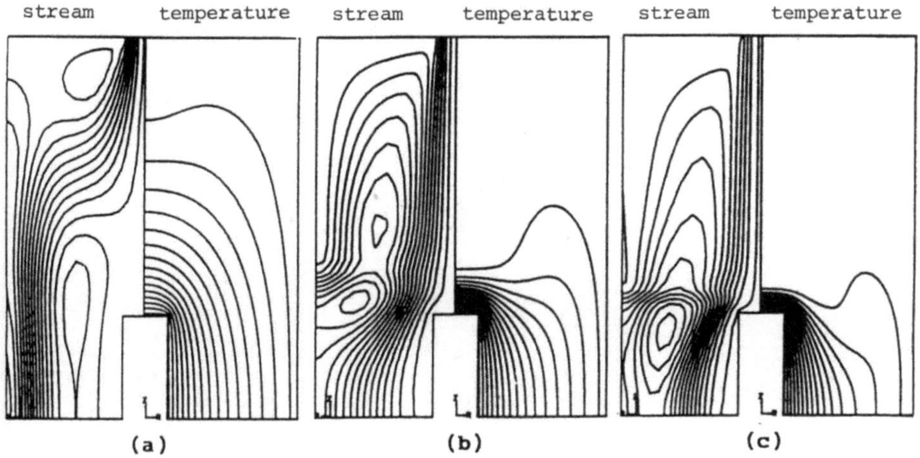

Fig. 3. Effect of neck inlet flow rate on flow pattern for Vshd = 0, P = 20 torr, Tsub = 1473 K and Vn (a) 2 slm (b) 8 slm and (c) 15 slm.

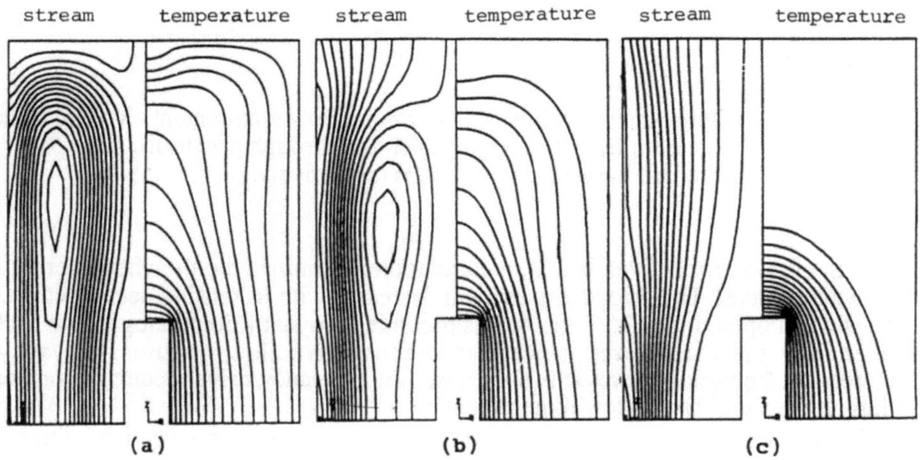

Fig. 4. Effect of shroud flow on flow pattern for Vn = 2 slm, P = 76 torr, Tsub = 1473 K and shroud flow of (a) 0 slm (b) 30 slm and (c) 55 slm.

Figure 4 shows the effect of increasing shroud flow for a given neck inlet flow. The huge recirculation with no shroud flow (Figure 4a) is reduced with some flow (Figure 4b) and eventually disappears with sufficient shroud flow size (Figure 4c).

Following several computations similar to the ones described above, we have obtained, and present in Figure 5, a summary "design" plot of shroud flow rate versus neck inlet flow rate at a set of three fixed pressures for a constant substrate temperature of 1473 K.

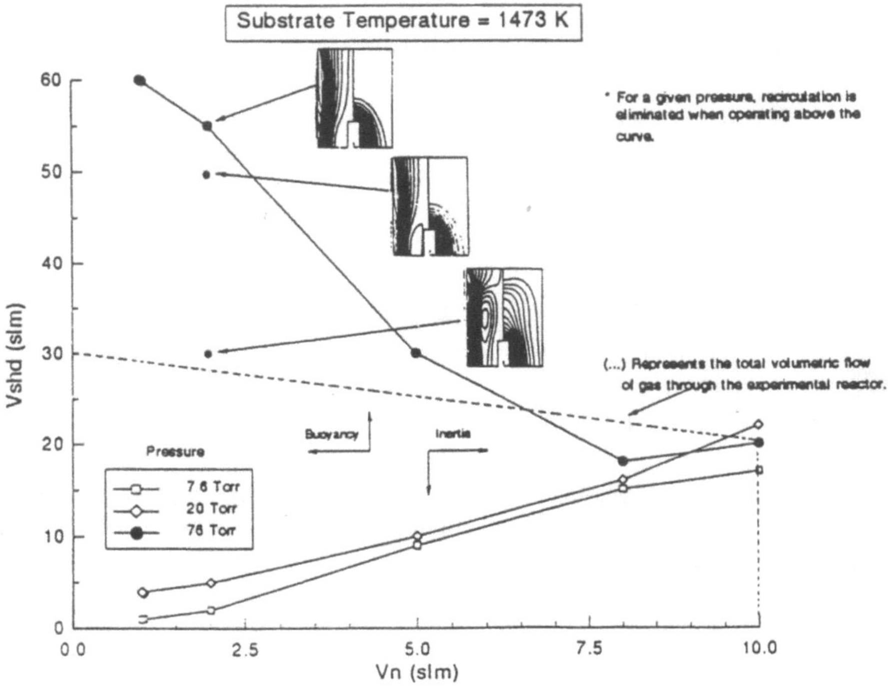

Fig. 5. Optimization with respect to shroud flow requirement. Vn and Vshd represent flow through the neck and through the shroud (in standard liters per minute, slm) respectively.

In the figure, for a given isobar, recirculation is eliminated only when operating above the curve. The shroud requirement decreases with increasing neck inlet flow rates when operating at 76 Torr. At this pressure, buoyancy dominates at low neck inlet flow rates. However, since forced-convective flow opposes buoyancy, increasing the neck inlet velocity suppresses the buoyancy-driven recirculation and thus reduces the shroud flow requirement.

As neck inlet velocity v_n approaches higher values, forced convection begins to dominate and the effect of pressure on the flow pattern diminishes. At this point, the intensity of the forced convective recirculation increases with higher v_n values. Higher shroud flow rates are now required as Vn increases and the slope of the isobar (76 torr) becomes positive.

At further reduced operating pressures lower than 20 torr, buoyancy is minimized and even at low neck inlet flow rates, forced convection is significant. This explains the positive slope of the lower isobars which show that the forced-convective recirculation increases in intensity as neck inlet flow increases, thereby requiring higher shroud flow velocities.

The arrows labelled inertia and buoyancy represent directions of increased predominance of the respective phenomena. The region enclosed by the dotted line represents the present constraint on total gas flow (a maximum of approximately 30 slm through neck and shroud) through the experimental reactor.

4.2 Effect of Process Parameters on Deposition Rate and Uniformities

We focus here on discussing the conditions for which experimental data are available - varying substrate temperatures for a neck flow of 2 standard liters per minute (slm), shroud flow of 10 slm and temperature of 1473 K. Additional information is available elsewhere [11].

Figure 6 shows a comparison between predicted and experimental growth rates as a function of substrate temperature. It is gratifying that despite several simplifying assumptions, the magnitudes are comparable and show similar increasing trends with increasing temperature. However, the closeness of the rate plots for SiH_4 and C_3H_8 does not enable us to discriminate which of these two species is the rate-limiting diffusing species.

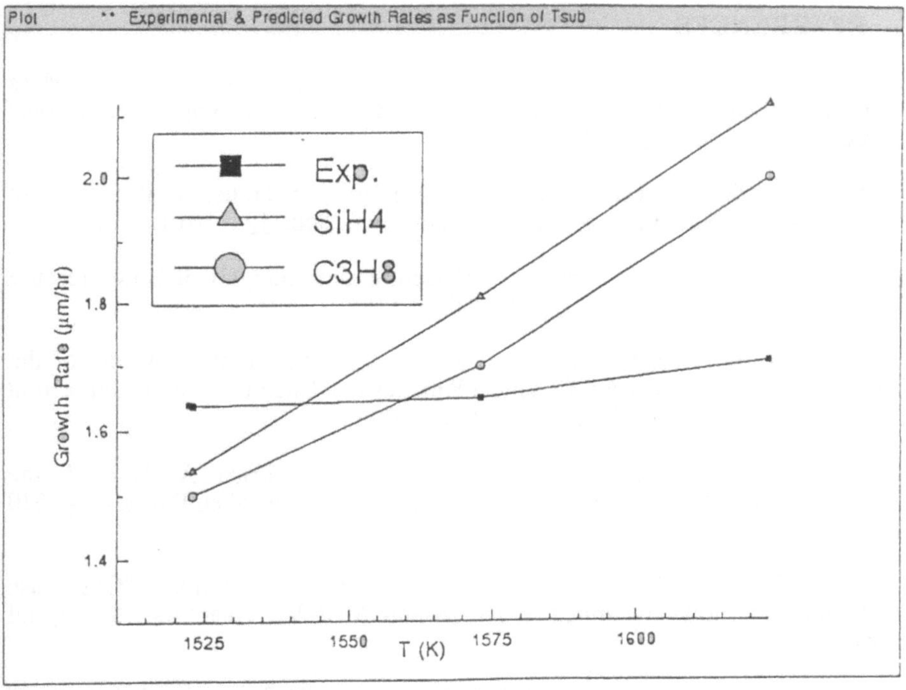

Fig. 6. Predicted and experimental growth rates as a function of substrate temperature.

5. CONCLUSIONS AND FUTURE WORK

A major observation arising from the work so far is that the use of shroud flow is essential for obtaining uniform gas flow, and hence deposition, in the vertical reactor. Figure 5, our "design" plot, is thus very important in determining how much shroud flow should be employed for a given neck flow and reactor pressure (at substrate temperature 1473 K) if uniform gas flow is to be ensured. The plot shows that for high pressure, a surprisingly large and undesirable amount of shroud flow is required to achieve this aim.

Further modeling work is planned on this reactor. This includes the addition of gas-phase kinetics as well as the extension to three-dimensional flows. Further, we continue to work closely with the experimental team in the Center in guiding its choices of experimental conditions for growth.

6. ACKNOWLEDGEMENTS

The authors wish to acknowledge funding support from NSF, NASA and the Department of Chemical Engineering, Howard University. Discussions with Michael Spencer, MSRCE, were all very helpful.

7. REFERENCES

1. Takahashi, R., Y. Koga, and K. Sugawara. " Gas Flow Pattern and Mass Transfer Analysis in a Horizontal Flow Reactor for Chemical Vapor Deposition," J. Electrochem. Soc. 119 1406 (1972).

2. Manke, C.W. and L.F. Donaghey. " Analysis of Transport Processes in Vertical Cylinder Epitaxy Reactors," J. Electrochem. Soc. 124 561 (1977).

3. Jensen, K.F. and D. B. Graves., " Modeling and Analysis of Low Pressure Reactors," J. Electrochem. Soc., 130 1950 (1983).

4. Coltrin, M.E., Kee, R.J. and J. A. Miller. " A Mathematical Model of the Coupled Fluid Mechanics and Chemical Kinetics in a Chemical Vapor Deposition Reactor," J. Electrochem. Soc., 131 425 (1984).

5. Kusumoto, Y., T. Hayashi, and S. Komiya. " Numerical Analysis of the Transport Phenomena in MOCVD Process," Japanese J. Applied Physics, 24 620 (1985).

6. Van De Ven, G., M.J. Rutten, M.J. Raaijmakers and L.J. Giling. "Gas Phase Depletion and Flow Dynamics in Horizontal MOCVD Reactors." J. Crystal Growth, 352 (1986).

7. Houtman, C., D.B. Graves, and K. F. Jensen. " CVD Stagnation Point Flow." J. Electrochem. Soc. 961 (1986).

8. Rhee, S., J. Szekely, and O.J. Ilegbusi. " On Three-Dimensional Transport Phenomena in CVD Processes," J. Electrochem. Soc. 134 2553 (1987).

9. Patnaik, S. and R.A. Brown. " Hydrodynamic Dispersion in Rotating-Disk OMVPE Reactors: Numerical Simulation and Experimental Measurements," J. Crystal Growth, 96 153 (1989).

10. Dilawari, A.H. and J. Szekely. " Computed Results for the Deposition Rates and Transport Phenomena for an MOCVD System with Conical Rotating Substrate," J. Crystal Growth, 97 777 (1989).

11. Givens, W.. "Numerical Modeling of Growth Conditions for Thin-Film Silicon Carbide in an Experimental Vertical Chemical Vapor Deposition Reactor," M.S. Thesis, Department of Chemical Engineering, Howard University (1991).

12 . Bird, R.B, W.E. Stewart , and E.N. Lightfoot. "Transport Phenomena", New York. John Wiley & Sons (1960).

13. FIDAP 5.0 Users Manual, Fluid Dynamics International, Evanston, IL (1990).

Epitaxial Growth of 6H-SiC by Chemical Vapor Deposition

S. Nishino, K. Takahashi, Y. Kojima, and J. Saraie

Department of Electronics and Information Science, Faculty of Engineering and Design, Kyoto Institute of Technology, Matsugasaki, Sakyo-ku, Kyoto 606, Japan

Crystals of 6H-SiC were grown on 6H-SiC with various orientations at 1150°C-1500°C using Si_2H_6-C_2H_2-H_2-TMA CVD system. 6H-SiC was obtained at 1350°-1500 °C on 6H-SiC(0001), off-axis, tilted toward <11$\bar{2}$0>. 6H-SiC was also epitaxially obtained at 1200°C on the 6H-SiC(11$\bar{2}$0) plane. This plane is favorable for low temperature epitaxy, because the stacking sequence of SiC polytypes along the c-axis appears in two dimensions on the same plane. Aluminum doping from TMA was effective for low temperature epitaxy.

1. Introduction

Recently, 6H-SiC was epitaxially grown on the 6H-SiC(0001) plane with off-orientation toward <11$\bar{2}$0> by step-controlled epitaxy. The low temperature epitaxy at 1400 °C was explained by the step-controlled surface, which was made by beveled lapping.[1,2] Powell et al.[3] reported epitaxial growth at low temperatures of 1320°-1390°C onto the a-planes (planes perpendicular to the a-axis). They pointed out that the epitaxy on the a-plane is favorable to minimize the occurrence of stacking faults. Growth on the substrates with off-orientation and a-planes seems interesting for low temperature epitaxy. It is worthwhile surveying the growth of 6H-SiC on various orientations of 6H-SiC for low temperature epitaxy. The lack of large crystals has prevented the use of (11$\bar{2}$0) and (1$\bar{1}$00) planes, which are planes perpendicular to the a-axis. We grew a 6H-SiC ingot by a modified sublimation method.[4] At present, substrates with a large area (6x6 mm^2) of (11$\bar{2}$0) and (1$\bar{1}$00) planes are available. It is interesting to know the favorable growth direction of 6H-SiC by chemical vapor deposition.

In this paper, we report that 6H-SiC can be homoepitaxially grown on 6H-SiC(11$\bar{2}$0) even at 1200°C with a growth rate of 2 μm/h. In this low temperature growth, trimethylaluminum (TMA) doping was necessary. This temperature is much lower than the standard temperature of around 1800°C for the growth of 6H-SiC(0001).[5,6] The growth mechanism for low temperature epitaxy of 6H-SiC is described.

2. Experiment

Growth of 6H-SiC was carried out using the Si_2H_6-C_2H_2-H_2 system under atmospheric pressure in a horizontal reaction

Editors: G.L. Harris, M.G. Spencer, and C.Y. Yang

tube. The reaction tube was made of fused quartz, 30 mm ID and 450 mm long, with a water cooled jacket. The susceptor was made of SiC coated graphite. Substrates were heated by rf induction at 300 kHz. The growth temperature was measured by an optical pyrometer. Crystal growth has two steps; etching and CVD growth. The etching step was accomplished at 1500°C using HCl (20 sccm) addition in H_2 flow for 10 min. After the etching step, the growth of 6H-SiC was carried out at 1150°-1500°C with standard gas composition: Si_2H_6 0.2 sccm, C_2H_2 0.1 sccm, TMA 0.1 sccm , H_2 3 slm. Aluminum doping was always employed using TMA to confirm the epitaxial growth by photoluminescence (PL) measurement. The linear velocity (defined as the ratio of the carrier gas flow rate to the cross section of free space above the substrate) of the carrier gas was about 14 cm/sec.

Hexagonal plates of 6H-SiC with (0001)Si and $(000\bar{1})$C planes grown by the Acheson method were used as substrates. Two off-axis substrates, tilted a few degrees toward the $\langle 11\bar{2}0\rangle$ and $\langle 1\bar{1}00\rangle$ directions, were prepared by bevel-lapping of basal plane of 6H-SiC single crystals. Hereafter we call these substrates off-axis substrates. Hexagonal plates of 6H-SiC with natural flat surfaces of (0001)Si and $(000\bar{1})$C were fixed on a beveled jig and polished with diamond paste. By using a 6H-SiC ingot made by the sublimation method, $(11\bar{2}0)$ and $(1\bar{1}00)$ planes were prepared; these planes contain stacking sequences along the c-axis in the plane. Mesa etched patterns were also made on the SiC substrates by reactive ion etching using the CF_4-O_2 system.

Polytypes of the grown layers were identified by the shape of etch pits, the reflection high energy electron diffraction (RHEED) pattern, the electron channeling pattern (ECP) and PL measurements. The surface morphology of the grown layer was observed by optical microscope (Nomarski interference contrast).

3. Results and Discussion

The temperature dependence of the crystallinity of the grown layer was studied at substrate temperatures of 1150°C to 1500°C. Single crystals of 6H-SiC were grown on the off-axis 6H-SiC(0001) substrate, tilted 6° toward $\langle 11\bar{2}0\rangle$, at 1350°-1500°C. However, 3C-SiC a with twinning defect was grown at lower temperatures than 1350°C. 6H-SiC was grown on the 6H-SiC(1120) substrate at the temperature range of 1200°-1500°C. When the temperature was decreased, the surface became rougher for the layers grown on (0001)Si, $(000\bar{1})$C and off-axis substrates, however, the epilayer on $(11\bar{2}0)$ showed a smooth surface and good crystallinity even grown at 1200°C. Crystallinity and polytypes of the grown layers were confirmed by double diffraction of the RHEED pattern.

A typical mosaic pattern was observed in the layer grown on a natural surface, where domains were separated by step-like or groove-like boundaries as shown in Fig.1(a). The layers with such mosaic patterns were thought to be twin-crystalline (double positioning) 3C-SiC by observation of their triangular shape as shown in Fig.1(a), where arrows show different orientations of the triangle. This layer was

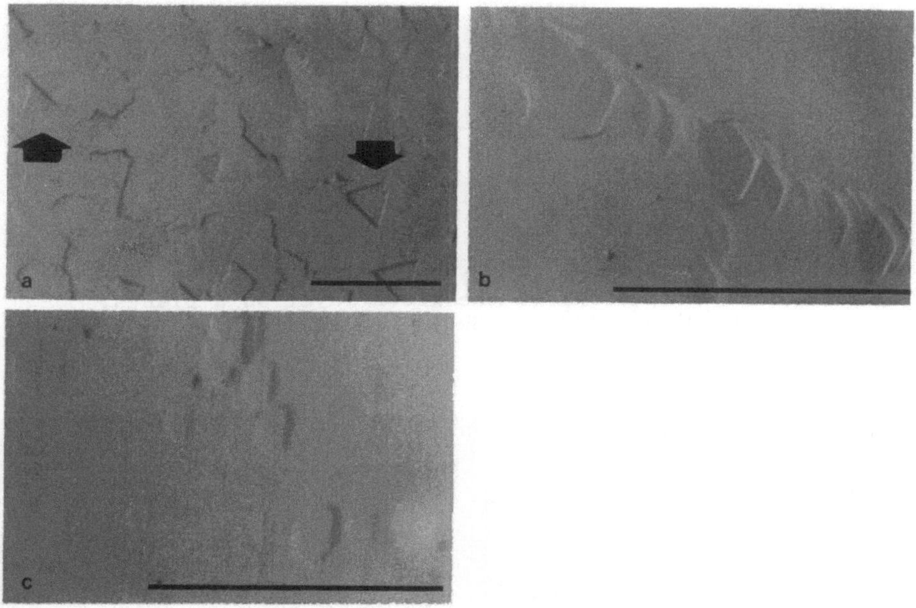

Fig.1 Surface morphology of the grown layer on (0001)Si plane:(a) on-axis, (b) off-axis toward$\langle 11\bar{2}0\rangle$, (c) off-axis toward $\langle 1\bar{1}00\rangle$. Scale bar is 100 μm.

identified as twin (double positioning) crystals of 3C-SiC by RHEED.

On the off-axis substrates of a few degrees, epitaxial layers of very smooth surface without any boundaries were obtained. The grown layers were identified as single crystalline 6H-SiC by analysis of the RHEED patterns. Fig.1(b) shows a layer grown on the off-axis (0001) plane tilted 6° toward $\langle 11\bar{2}0\rangle$. A part of a hexagon is observed, where a side of the hexagon aligns with $\langle 11\bar{2}0\rangle$. Fig.1(c) shows the layer grown on the off-axis (0001) plane tilted 6° toward the $\langle 1\bar{1}00\rangle$ direction. A deformed hexagon is observed, where the edges of the hexagon align perpendicular to the $\langle 1\bar{1}00\rangle$ direction. In general a plane of the crystal is limited by the growth rate of the plane; a flat plane is related to the slowest growth rate of the crystal.

It is interesting to study which is the easy growth direction, so the crystal growth was carried out on a substrate with a mesa-shaped island. After the formation of circular mesa-hillocks, CVD growth was carried out. The surface morphology of the layer grown on $(1\bar{1}00)$ was rather smooth, however, a parallel fringe pattern was observed as shown in Fig.2. The parallel fringes align with the $\langle 1120\rangle$ direction, which is the fastest growth direction of 6H-SiC. The surface morphology of the layer grown on $(11\bar{2}0)$ was very smooth and mirror-like as shown in Fig.2(a). This plane consists of two crystal directions; one is $\langle 1\bar{1}00\rangle$ and the other is $\langle 0001\rangle$.

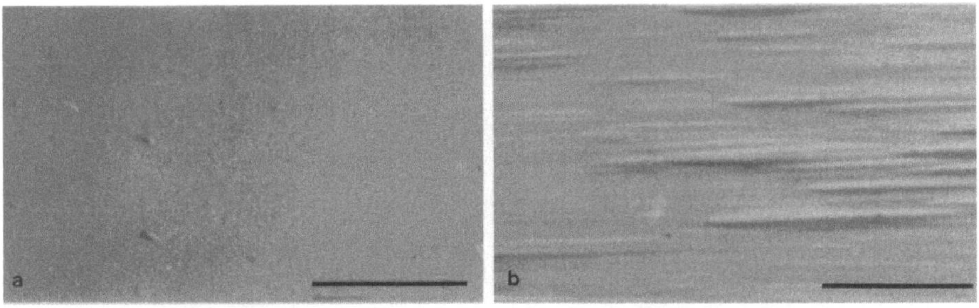

Fig.2. Surface morphology of the grown layer on a-planes: (a) (11$\bar{2}$0) and (1$\bar{1}$00). Scale bar is 100 μm.

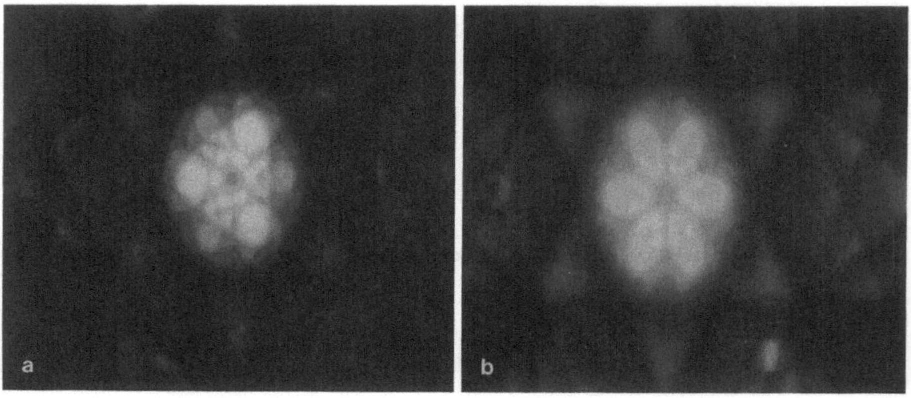

Fig.3. Electron channeling patterns of the grown layers: (a) 3C-SiC on-axis and (b) 6H-SiC off-axis.

Crystallographic information on the bulk specimens was obtained from electron channeling patterns (ECP) taken in the SEM at an acceleration voltage of 30 kV (ABT alpha-30A). The information that can be obtained by this technique includes orientation, lattice parameters and qualitative evaluation of the crystalline perfection of the sample. The ECP's resemble the geometry of Kikuchi patterns and can be indexed by analogous methods. This technique was applied to determine the polytypes of SiC grown on the substrate both on- and off-axis. The qualitative evaluation of the surface perfection is possible because the ECP's are sensitive to the perfection in the top 100 nm of the specimen and any lattice distortion degrades the pattern density. Examples of ECP's are shown in Fig.3. In these patterns, threefold and sixfold symmetry are seen for 3C-SiC on the on-axis substrate and 6H-SiC on the off-axis substrate, respectively. The RHEED pattern of the layer grown on the (11$\bar{2}$0) plane is shown in Fig.4. This pattern is indexed by the reciprocal lattice plane (11$\bar{2}$0), and double diffraction spots are observed along the <0001> direction and identified to be 6H-SiC.

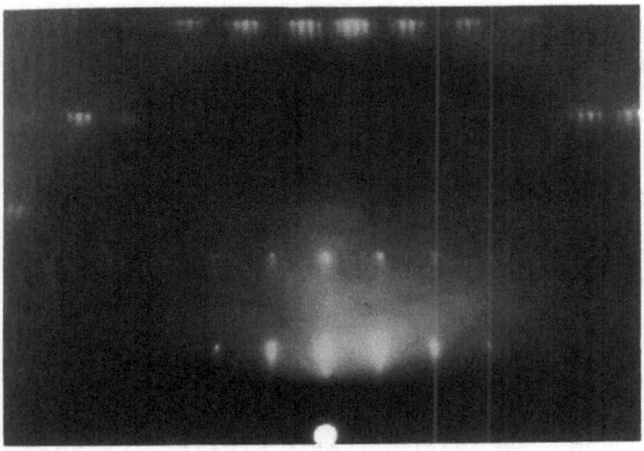

Fig.4. RHEED pattern of the layer grown on the $(11\bar{2}0)$ plane.

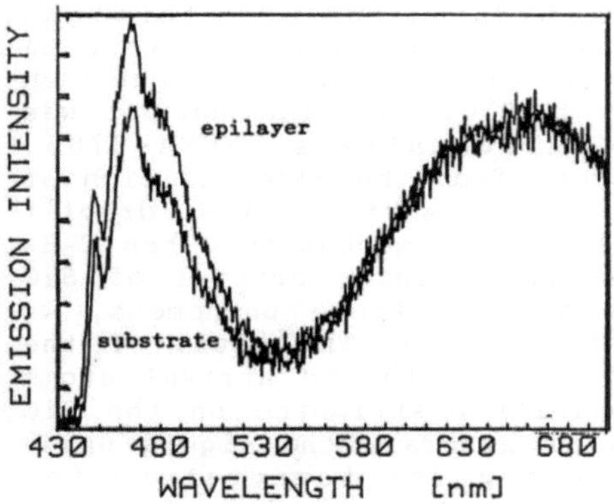

Fig.5. PL spectra of the epilayer on the $(11\bar{2}0)$ plane.

Fig.5 shows the spectra of the substrate and the Al-doped grown layer. In this spectra, the peaks at wavelengths of around 480 nm are mainly due to D-A pair emission, however, the broad band at around 660 nm is also the same emission profile of the substrate. From this comparison, we can confirm that the emission from the epitaxial layer is D-A pair emission from the 6H-SiC and that the 6H-SiC was epitaxially grown on the 6H-SiC$(11\bar{2}0)$ plane.

We found that low temperature epitaxy was much influenced by impurity (TMA) doping. In our case, TMA was always added during the crystal growth to clarify the D-A pair emission from the epilayer by PL measurement. 6H-SiC was epitaxially grown on the 6H-SiC$(11\bar{2}0)$ plane at 1200°C. We speculate that this low temperature epitaxy is related to two effects: one is the substrate orientation and the other is TMA doping. We consider the $(11\bar{2}0)$ plane first. This plane can be obtained by tilting the (0001) plane 90° toward $<11\bar{2}0>$, and this plane can be seen as the extreme case of the off axis (0001) substrate. This plane contains Si atoms and C atoms in the same plane and the arrangement of these atoms reflects the stacking order of the SiC polytypes. So, active Si and C species from the vapor phase can be used to identify the kind of SiC polytypes. In the case of the on- and off-axis substrates, arrival atoms from the vapor must look for a favored site, like a kink or stepped edge, and migrate on the surface, so higher temperatures may be necessary.[1]

Secondly, the effect of TMA doping is related to low temperature epitaxy. Recently we reported low temperature epitaxy of 3C-SiC on a Si substrate at 1150°C using the Si_2H_6-C_2H_2-H_2-TMA CVD system without introduction of buffer layer formation.[7] Although polycrystalline SiC was grown on a Si substrate without TMA addition, crystallinity was strongly enhanced by adding TMA and, in particular, single crystals were grown when the grown layer became p-type. In the case of homoepitaxial growth of 6H-SiC, the flow rate of TMA is almost same amount as that of the reaction gases (Si_2H_6, C_2H_2). One may think that SiC formation is possible using Si_2H_6-TMA system, because TMA is also a hydrocarbon. In this gas combination, SiC did not grow on the substrate, however, Si droplets with an Al dot on top of the droplet were formed on the surface for a 40 min run at 1200°C. The aluminum dot was confirmed by electron probe microanalysis (EPMA-8705, Shimazu). These Al dots were formed from the decomposition of TMA, forming a liquid layer on the SiC surface. A Si droplet was formed by a vapor-liquid-solid (VLS) mechanism. When C_2H_2 was added to the Si_2H_6-TMA system, a single crystal of SiC grew on the SiC substrate. Considering these phenomena, we speculate that liquid Al was attached to kink sites of the surface at the initial stage of the growth and arrival atoms of Si and C species were epitaxially distributed on the SiC substrate at a low temperature. In general, the liquid phase seems to be effective in reducing the temperature for migration of adsorbed atoms on the surface of the substrate, because in the liquid phase atoms with rather low energy can exchange sites and migrate even at a low temperature. Although in our case 6H-SiC was grown by chemical vapor deposition, from the viewpoint of the growth mechanism, it is reasonable to consider that a small amount of liquid Al was formed during SiC deposition. This low temperature growth may be called impurity assisted epitaxy.

4.Conclusion

Epitaxial growth of 6H-SiC was carried out using various orientations of the 6H-SiC substrate. 6H-SiC was epitaxially grown on the 6H-SiC$(11\bar{2}0)$ plane at the low temperature of 1200°C. This $(11\bar{2}0)$ plane is favorable for low temperature epitaxy. Low temperature epitaxy was aided considering by Al doping from TMA.

References

1) N.Kuroda, K.Shibahara, W.Yoo, S.Nishino and H.Matsunami: Extended Abst. 19th Conf. Solid State Devices and Materials (Business Center for Academic Societies Japan, Tokyo,1987)p.227.
2) H.Matsunami, K.Shibahara,N.Kuroda,W.Yoo and S.Nishno, Spring Meeting in Physics **34**,34(1989).
3) J.A.Powell and H.A.Will:J.Appl.Phys.**44**,5177(1973).
4) S.Nishino,K.Kojima and J.Saraie: "Growth and Morphology of 6H-SiC by Sublimation Method", in these Proceedings.
5) S.Nishino, A.Ibaraki, H.Matsunami and T.Tanaka: Jpn.J.Appl.Phys.**19**,L353(1980).
6) S.Nishino, H.Matsunami and T.Tanaka:J.Crystal Growth, **45**,144(1978).
7) K.Takahashi, S.Nishino and J.Saraie: Extended Abst.37th Spring Meeting of Japan Society of Applied Physics and Related Societies,Tokyo,March,31a-T-7,p.314(1990).

Index of Contributors

MIX
Papier aus verantwortungsvollen Quellen
Paper from responsible sources
FSC® C105338

If you have any concerns about our products,
you can contact us on
ProductSafety@springernature.com

In case Publisher is established outside the EU,
the EU authorized representative is:
Springer Nature Customer Service Center GmbH
Europaplatz 3, 69115 Heidelberg, Germany

Printed by Libri Plureos GmbH
in Hamburg, Germany